Springer Series in Operations Research and Financial Engineering

Series Editors:
Thomas V. Mikosch
Sidney I. Resnick
Stephen M. Robinson

T0207215

For further volumes:
http://www.springer.com/series/3182

Henrik Hult • Filip Lindskog • Ola Hammarlid
Carl Johan Rehn

Risk and Portfolio Analysis

Principles and Methods

 Springer

Henrik Hult
Department of Mathematics
Royal Institute of Technology
Stockholm, Sweden

Filip Lindskog
Department of Mathematics
Royal Institute of Technology
Stockholm, Sweden

Ola Hammarlid
Swedbank AB (publ)
SE-105 34 Stockholm
Sweden

Carl Johan Rehn
E. Öhman J:or
Fondkommission AB
Stockholm, Sweden

ISSN 1431-8598
ISBN 978-1-4939-0031-2   ISBN 978-1-4614-4103-8 (eBook)
DOI 10.1007/978-1-4614-4103-8
Springer New York Heidelberg Dordrecht London

Mathematics Subject Classification (2010): 62P05, 91G10, 91G20, 91G70

Printed on acid-free paper

Springer is part of Springer Science+Business Media (www.springer.com)

To our families

Preface

This book presents sound principles and useful methods for making investment and risk management decisions in the presence of hedgeable and nonhedgeable risks.

In everyday life we are often forced to make decisions involving risks and perceived opportunities. The consequences of our decisions are affected by the outcomes of random variables that are to various degrees beyond our control. Such decision problems arise, for instance, in financial and insurance markets. What kind of insurance should you buy? What is an appropriate way to invest money for later stages in life or for building a capital buffer to guard against unforeseen events? While private individuals may choose not to take a quantitative approach to investment and risk management decisions, financial institutions and insurance companies are required to quantify and report their risks. Financial institutions and insurance companies have assets and liabilities, and their investment actions involve both speculation and hedging. In fact, every time a liability is not hedged perfectly, the hedging decision is a speculative decision on the outcome of the hedging error. Although hedging and investment problems are often presented separately in the literature, they are indeed two intimately connected aspects of portfolio risk management. A major objective of this book is to take a coherent and pragmatic approach to investment and risk management integrated in a portfolio analysis framework.

The mathematical fields of probability, statistics, and optimization form a natural basis for quantitatively analyzing the consequences of different investment and risk management decisions. However, advanced mathematics is not a necessity per se for dealing with the problems in this area. On the contrary, a large amount of highly sophisticated mathematics in a book on this topic may lead the reader to draw the wrong conclusions about what is essential (and possible) and what is not. We assume that the reader of this book has a mathematical/statistical knowledge corresponding to undergraduate-level courses in linear algebra, analysis, statistics, and probability. Some knowledge of basic optimization theory will also be useful. The book presents material precisely using basic undergraduate-level mathematics and is self-contained.

There are two fundamental difficulties to finding solutions to the problems in investment and risk management. The first is that the decisions strongly depend on subjective probabilities of the future values of financial instruments and other quantities. Financial data are the consequences of human actions and sentiments as well as random events. It is impossible to know the extent to which historical data explain the future that one is trying to model. This is in sharp contrast to card games or roulette where the probability of future outcomes can be considered as known. Statistics may assist the user in motivating the choice of a particular model or to fit models to historical data, but the probabilities of future events will nevertheless be affected by subjective judgment. As a consequence, it is practically impossible to assess the accuracy of the subjective probabilities that go into the mathematical procedures. Misspecifications of the input to a quantitative procedure for decision making will always be reflected in the output, and critical judgment cannot be replaced by mathematical sophistication.

The second fundamental difficulty is that even when there is a consensus on the probabilities of future events, a decision that is optimal for one decision maker may be far from optimal for another one with a different attitude toward risk. Mathematics can assist in translating a probability distribution and an attitude toward risk and reward into a portfolio choice in a consistent way. However, it is difficult to even partially specify a criterion for a desired trade-off between risk and potential reward in an investment situation. Simple and transparent criteria for financial decision making may be more suitable than more advanced alternatives because they enable the user to fully understand the effects of variations in parameter values and probability distributions. Although designing a quantitative and principle-based approach to financial decision making is by no means easy, the alternatives are often ad hoc and lack transparency.

At this point we emphasize the difference between uncertainty and randomness. Even if we do not know the outcome when throwing a fair six-sided die, we can be rather certain that the probability of each possible outcome is one sixth. However, if we do not know the marking of the die, whether it is symmetric, or the number of sides it has, then we have no clue about the probability distribution generating the outcomes. In particular, uncertainty is closely related to lack of information. Saying that we are unsure about the probability distribution of the future value of an asset does not correspond to assigning a probability distribution with a large variance. Knowing the probability distribution is potentially very valuable since it provides a good basis for taking financial positions that are likely to turn out successful. Conversely, if we are very uncertain about the probability distribution of future values, then we should not take any position at all: we should not play a game that we do not understand. Of course, there is a certain degree of uncertainty in all decision making. If one feels more comfortable with, say, assigning a probability distribution to the difference between two future asset prices rather than to the prices themselves, then clearly it is wiser to take a position on the outcome of the difference of the prices. Intelligent use of statistics, together with a good understanding of whether the data are likely to be representative for future events, may reduce the degree of uncertainty. Techniques from probability theory are useful for quantifying the probability of future events. Techniques from optimization enable one to find optimal decisions and allocations under the assumption that the input to the optimizing procedure is reliable.

Investment and risk management problems are fundamental problems that cannot be ignored. Since it is difficult or impossible to accurately specify the probability distributions that describe the problems we need to solve, we believe that it is essential to focus on the simplest possible principles, methods, and models that still capture the essential features of the problems. Many of the more technically advanced approaches suffer from spurious sophistication when confronted with the real-world problems they are supposed to handle. We have avoided material that is attractive from a mathematical point of view but does not have a clear methodological purpose and practical utility. Our aim has been to produce a text founded in rigorous mathematics that presents practically relevant principles and methods. The material is accessible to students at the advanced undergraduate or Master's level as well as industry professionals with a quantitative background.

The story we want to tell is not primarily told by the theory we present but rather by the examples. The many examples, covering a diverse set of topics, illustrate how principles, methods, and models can be combined to approach concrete problems and to draw useful conclusions. Many of the examples build upon examples presented earlier in the book and form series of examples on a common theme. We want the more extensive examples to be used together with implementations of the methods to address hedging and investment problems with real data. The source code, in the statistical programming language R, that was used to generate the examples and illustrations in the book is publicly available at the authors' Web pages. We have also included exercises that, on the one hand, train the reader in mastering certain techniques and, on the other hand, convey essential ideas. In addition, we have included more demanding projects that assist the reader in obtaining a deeper understanding of the subject matter.

This book is the result of the joint efforts of two academics, Hult and Lindskog, who teamed up with two industry professionals, Hammarlid and Rehn. The material of this book is based on several versions of lecture notes written by Hult and Lindskog for use in courses at KTH. The idea to turn these lecture notes into a book came from Hammarlid and Rehn, and we all underestimated the amount of work required to turn this idea into reality. Essentially all the material from the lecture notes we started off with was either thrown away or rewritten completely. The book was written by Hult and Lindskog but has benefited very much from years of discussions with and valuable feedback from Hammarlid and Rehn. The ordering of the authors reflects the fact that they can be divided into two groups that have contributed differently toward the final result. Within the two groups the authors are simply listed in alphabetic order, and the order there does not have any relevance besides the alphabetical order.

Several people have played an important part in the development of this book. We thank Thomas Mikosch and Sid Resnick for their encouragement and for their valuable feedback on the book. Moreover, their own excellent books have inspired us and provided a goal to aim for. We thank our colleagues Boualem Djehiche and

Harald Lang for supporting our work and for many stimulating discussions. We would also like to thank the students in our courses at KTH for many years of feedback on earlier versions of the material in this book. Vaishali Damle at Springer has played a key role in guiding us toward the completion of this book. Finally, special thanks go to our families for their endless support throughout this long process.

Stockholm, Sweden   Henrik Hult, Filip Lindskog,
                                                               Ola Hammarlid, Carl Johan Rehn

Contents

Part I
Principles

Chapter 1
Interest Rates and Financial Derivatives

In this chapter we present the basic theory of interest rate instruments and the pricing of financial derivatives. The material we have chosen to present here is interesting and relevant in its own right but particularly so as the basis for the principles and methods considered in subsequent chapters.

The chapter consists of two sections. Section 1.1 presents the basic theory of interest rate instruments and focuses on the no-arbitrage valuation of cash flows. Section 1.2 presents the no-arbitrage principle for valuation of financial derivative contracts, contracts whose payoffs are functions of the value of another asset at a specified time in the future, and exemplifies the use of this principle. In a well-functioning market of derivative contracts, the derivative prices can be represented in terms of expected values of the payoffs, where the expectation is computed with respect to a probability distribution for the underlying asset value on which the contracts are written. If many derivative contracts are traded in the market, then we can say rather much about this probability distribution, and individual investors may compare it to their own subjective assessments of the underlying asset value and use the result of the comparison to make wise investment and risk management decisions.

1.1 Interest Rates and Deterministic Cash Flows

Consider a bank account that pays interest at the rate r per year. If yearly compounding is used, then one unit of currency on the bank account today has grown to $(1 + r)^n$ units after n years. Similarly, if monthly compounding is used, then one unit in the bank account today has grown to $(1 + r/12)^{12n}$ units after n years. Compounding can be done at any frequency. If a year is divided into m equally long time periods and if the interest rate r/m is paid at the end of each period, then one unit on the bank account today has grown to $(1 + r/m)^m$ units after 1 year. We say that the annual rate r is compounded at the frequency m. Note that $(1 + r/m)^m$ is increasing in m. In particular, a monthly rate r is better than a yearly

H. Hult et al., *Risk and Portfolio Analysis: Principles and Methods*, Springer Series in Operations Research and Financial Engineering, DOI 10.1007/978-1-4614-4103-8_1, © Springer Science+Business Media New York 2012

rate r for the holder of a savings account. Continuous compounding means that we let m tend to infinity. Recall that $(1 + 1/m)^m \to e$ as $m \to \infty$, which implies that $(1 + r/m)^m \to e^r$ as $m \to \infty$. Unless stated otherwise, interest rates in this book always refer to continuous compounding. That is, one unit deposited in a savings account with a 5% interest rate per year has grown to $e^{0.05t}$ units after t years. Note that the interest rate is just a means of expressing the rate of growth of cash. An investor cares about the rate of growth but not about which type of compounding is used to express this rate of growth.

In reality, things are certainly a bit more involved. The rate of interest on money deposited in a bank account differs from that for money borrowed from the bank. Moreover, the length of the time period also affects the interest rate. In most cases, the lender cannot ignore the risk that the borrower might be unable to live up to the borrower's obligations, and therefore the lender requires compensation in terms of a higher interest rate for accepting the risk of losing money.

1.1.1 Deterministic Cash Flows

Consider a set of times $0 = t_0 < t_1 < \cdots < t_n$, with $t_0 = 0$ being the present time. A deterministic cash flow is a set $\{(c_k, t_k); k = 0, 1, \ldots, n\}$ of pairs (c_k, t_k), where c_k and t_k are known numbers and where c_k represents the amount of cash received at time t_k by the owner of the cash flow. A negative value of c_k means that the owner of the cash flow must pay money at time t_k. Here we consider financial instruments that can be identified with deterministic cash flows. Any two parties can enter an agreement to exchange cash flows, but the contracted cash flow is not deterministic if there is a possibility that one party will fail to deliver the contracted cash flow.

An important instrument corresponding to a deterministic cash flow is the risk-free bond. The bonds issued by governments are typically good proxies. A risk-free bond issued at the present time corresponds to the cash flow

$$\{(-P_0, 0), (c, \Delta t), \ldots, (c, (n-1)\Delta t), (c + F, n\Delta t)\}, \tag{1.1}$$

where $P_0 > 0$ is the present bond price, $c \geq 0$ the periodic coupon amount paid to the bondholder, $F > 0$ the face value or principal of the bond, $\Delta t > 0$ the time between coupon payments, and $T = n\Delta t$ the time to maturity of the bond. Time is typically measured in years with $\Delta t = 0.5$ or $\Delta t = 1$. If $\Delta t = 0.5$, then the bond pays coupons semiannually and $2c$ is the annual coupon amount. If $c = 0$, then the bond is called a zero-coupon bond. Zero-coupon bonds often have less than 1 year to maturity. Buying a bond of the type given by (1.1) at time 0 that was issued at time $-u$, with $u \in (0, \Delta t)$, implies the cash flow

$$\{(-P_0, 0), (c, \Delta t - u), \ldots, (c, (n-1)\Delta t - u), (c + F, n\Delta t - u)\},$$

where P_0 is the price of the bond at time 0. Typically, $P_0 > P_{-u}$ since a buyer who purchases the bond at $-u$ would have to wait longer before receiving money.

Consider a market with an interest rate r per year that applies to all types of investment, loan and deposit (think of an ideal bank account without fees and restrictions on transactions). Then an amount A today is worth $e^{rt}A$ after t years. Similarly, an amount A received in t years from today is worth $e^{-rt}A$ today. We say that $e^{-rt}A$ is the present value of A at time t, and e^{-rt} is the discount factor for cash received at time t. The present value of a cash flow $\{(c_k, t_k); k = 0, \ldots, n\}$ on this market is

$$P_0(r) = \sum_{k=0}^{n} c_k e^{-rt_k}.$$

The internal rate of return is the number r for which $P_0(r) = 0$. Note that the equation $P_0(r) = 0$ does not necessarily determine the internal rate of return uniquely for arbitrary deterministic cash flows. However, if $c_0 < 0$ and $c_k \geq 0$ for $k \geq 1$ with $c_k > 0$ for some k (e.g., the cash flow of a bond), then it is not difficult to verify that the internal rate of return is uniquely determined. For a bond, the internal rate of return is called the yield to maturity of the bond.

Consider a zero-coupon bond with current price $P_0 > 0$ that pays the amount $F > 0$ at t years from now, i.e., the cash flow $\{(-P_0, 0), (F, t)\}$. Clearly, there is a number r_t such that the relation $P_0 = e^{-r_t t} F$ holds. The number r_t is the t-year zero rate (or the t-year zero-coupon bond rate or spot rate), and the number $e^{-r_t t}$ is the discount factor for money received t years from now. Note that the discount factor $e^{-r_t t}$ is the current price for one unit received at time t. The graph of r_t viewed as a function of t is called the zero rate curve (or spot rate curve or yield curve). Market prices show that the zero rate curve is typically increasing and concave (the value of the second-order derivative with respect to t is negative). In particular, the assumption of a flat zero rate curve ($r_t = r$ for all t) is not consistent with market data.

The risk-free bonds discussed above are risk free in the sense that the buyer of such a bond will for sure receive the promised cash flow. However, a risk-free bond is risky if the holder sells the bond prior to maturity since the income from selling the bond is uncertain and depends on the market participants' demand for and valuation of the remaining cash flow. Moreover, the risk-free bond is risk free if held to maturity only in nominal terms. If, for instance, inflation is high, then the cash received at maturity may be worth little in the sense that you cannot buy much for the received amount. A bond is not risk free if it is possible that the issuer of the bond does not manage to pay the bondholder according to the specified cash flow of the bond. Such a bond is called risky or defaultable.

1.1.2   Arbitrage-Free Cash Flows

How are zero rates determined from prices of traded bonds or other cash flows? The simplest way would be to look up prices of zero-coupon bonds with the relevant maturity times. The problem with this approach is that such zero-coupon bond prices are typically not available. The cash flows priced by the market are typically more complicated cash flows such as coupon bonds. Moreover, the total number of cash flow times are often larger than the number of cash flows. Before addressing the question of how to determine zero rates from traded instruments, one must determine whether there exist any zero rates at all that are consistent with the observed prices.

Fix a set of times $0 = t_0 < \cdots < t_n$ and consider a market consisting of m cash flows:

$$\{(c_{k,0}, t_0), (c_{k,1}, t_1), \ldots, (c_{k,n}, t_n)\}, \quad k = 1, \ldots, m.$$

Since the times are held fixed, we represent the cash flows more compactly as m elements $c_1, \ldots, c_m$ in $\mathbb{R}^{n+1}$ (vectors with $n+1$ real-valued components). It is assumed (although this is not entirely realistic) that you can buy and short-sell unlimited amounts of these contracts/cash flows. Short-selling a financial instrument should be interpreted as borrowing the instrument from a lender, then selling it at the current market price and at a later time purchasing an identical instrument at the prevailing market price and returning it to the lender. Here we ignore borrowing fees associated with short-selling. It is also assumed here (again not entirely realistically) that the market prices for buying and selling an instrument coincide and that there are no fees charged for buying and selling.

Under the imposed assumptions one can form linear portfolios of the original cash flows and thereby create new cash flows of the form $c = \sum_{k=1}^{m} h_k c_k$. The h_ks are any real numbers, and negative values correspond to short sales. The market therefore consists of arbitrary linear combinations of the original cash flows and can be represented as a linear subspace $\mathbb{C}$ of $\mathbb{R}^{n+1}$, spanned by the cash flows $c_1, \ldots, c_m$. We say that there exists an arbitrage opportunity if there exists a $c \in \mathbb{C}$ such that $c \neq 0$ ($c_k \neq 0$ for some k) and $c \geq 0$ ($c_k \geq 0$ for all k). Such an element c corresponds to a contract that does not imply any initial or later costs and gives the buyer a positive amount of money. Such a contract cannot exist on a well-functioning market, at least not for long. If it did exist, some market participants would spot it and take advantage of it. Their actions would, in turn, drive the prices to the point where the arbitrage opportunity disappeared. The absence of arbitrage opportunities is equivalent to the existence of discount factors for the maturity times under consideration. This fact is a consequence of the following result from linear algebra.

Theorem 1.1. *Let $\mathbb{C}$ be a linear subspace of $\mathbb{R}^{n+1}$. Then the following statements are equivalent:*

(i) There exists no element $c \in \mathbb{C}$ satisfying $c \neq 0$ and $c \geq 0$.
(ii) There exists an element $d \in \mathbb{R}^{n+1}$ with $d > 0$ satisfying $c^T d = 0$ for all $c \in \mathbb{C}$.

Proof. The implication (ii) $\Rightarrow$ (i) in Theorem 1.1 is easily shown: if $d > 0$ and $c^T d = 0$ for all $c \in \mathbb{C}$, then each nonzero $c \in \mathbb{C}$ must have both a positive component and a negative component. The implication (i) $\Rightarrow$ (ii) is more difficult to show. Assume that (i) holds and let

$$K = \{\mathbf{k} = (k_0, \ldots, k_n)^{\mathrm{T}} \in \mathbb{R}^{n+1} \text{ such that } k_0 + \cdots + k_n = 1 \text{ and } k_i \geq 0 \text{ for all } i\}.$$

From (i) it follows that K and $\mathbb{C}$ have no common element. Let $\mathbf{d}$ be a vector in $\mathbb{R}^{n+1}$ of shortest length among all vectors in $\mathbb{R}^{n+1}$ of the form $\mathbf{k} - \mathbf{c}$ for $\mathbf{k} \in K$ and $\mathbf{c} \in \mathbb{C}$. The proof of the fact that such a vector $\mathbf{d}$ exists is postponed to Lemma 1.1 right after this proof. Take a representation $\mathbf{d} = \mathbf{k}^* - \mathbf{c}^*$, where $\mathbf{k}^* \in K$ and $\mathbf{c}^* \in \mathbb{C}$. For any $\lambda \in [0, 1]$, $\mathbf{k} \in K$, and $\mathbf{c} \in \mathbb{C}$ we notice that $\lambda \mathbf{k}^* + (1 - \lambda)\mathbf{k} \in K$ and $\lambda \mathbf{c}^* + (1 - \lambda)\mathbf{c} \in \mathbb{C}$. By the definition of $\mathbf{k}^*$ and $\mathbf{c}^*$, the function f defined on $[0, 1]$, given by

$$f(\lambda) = \left((\lambda \mathbf{k}^* + (1 - \lambda)\mathbf{k}) - (\lambda \mathbf{c}^* + (1 - \lambda)\mathbf{c})\right)^2$$

has a minimum at $\lambda = 1$. We may write

$$f(\lambda) = (\lambda \mathbf{d} + (1 - \lambda)(\mathbf{k} - \mathbf{c}))^{\mathrm{T}}(\lambda \mathbf{d} + (1 - \lambda)(\mathbf{k} - \mathbf{c}))$$
$$= \lambda^2 \mathbf{d}^{\mathrm{T}}\mathbf{d} + 2\lambda(1 - \lambda)\mathbf{d}^{\mathrm{T}}(\mathbf{k} - \mathbf{c}) + (1 - \lambda)^2(\mathbf{k} - \mathbf{c})^{\mathrm{T}}(\mathbf{k} - \mathbf{c}).$$

The fact that f has a minimum at $\lambda = 1$ implies that

$$f'(1) = 2\left(\mathbf{d}^{\mathrm{T}}\mathbf{d} - \mathbf{d}^{\mathrm{T}}(\mathbf{k} - \mathbf{c})\right) \leq 0.$$

Equivalently, $\mathbf{d}^{\mathrm{T}}\mathbf{k} - \mathbf{d}^{\mathrm{T}}\mathbf{d} \geq \mathbf{d}^{\mathrm{T}}\mathbf{c}$ for any $\mathbf{k} \in K$ and $\mathbf{c} \in \mathbb{C}$. If $\mathbf{d}^{\mathrm{T}}\mathbf{c} \neq 0$ for some $\mathbf{c} \in \mathbb{C}$, then $\mathbf{d}^{\mathrm{T}}(t\mathbf{c}) \neq 0$ for $|t|$ arbitrarily large, which implies that $\mathbf{d}^{\mathrm{T}}\mathbf{k}$ is larger than any positive number for all $\mathbf{k} \in K$. This is clearly false, and we conclude that $\mathbf{d}^{\mathrm{T}}\mathbf{c} = 0$ for all $\mathbf{c} \in \mathbb{C}$, which implies that $\mathbf{d}^{\mathrm{T}}\mathbf{k} - \mathbf{d}^{\mathrm{T}}\mathbf{d} \geq 0$ for all $\mathbf{k} \in K$. It remains to show that the components of $\mathbf{d}$ are strictly positive. With $\mathbf{k} = (1, 0, \ldots, 0)^{\mathrm{T}}$ we get $d_0 \geq \mathbf{d}^{\mathrm{T}}\mathbf{d} > 0$, and similarly for the other components of $\mathbf{d}$ by choosing $\mathbf{k}$ among the standard basis vectors of $\mathbb{R}^{n+1}$. We conclude that the implication (i) $\Rightarrow$ (ii) holds.   □

The following result from analysis is used in the proof of Theorem 1.1.

Lemma 1.1. *There exists a vector $\mathbf{d}$ of shortest length between K and $\mathbb{C}$.*

Proof. For $\mathbf{k}$ in K, let $\mathbf{v}$ be the corresponding vector of shortest length between $\mathbf{k}$ and $\mathbb{C}$. If $\mathbf{c}$ is the orthogonal projection of $\mathbf{k}$ onto $\mathbb{C}$, then $\mathbf{v} = \mathbf{k} - \mathbf{c}$. We will first show that the function f, given by $f(\mathbf{k}) = \mathbf{v}$, is continuous. For any $\mathbf{k}_1, \mathbf{k}_2$ in K, by orthogonality, the corresponding vectors $\mathbf{v}_1, \mathbf{v}_2$ and $\mathbf{c}_1, \mathbf{c}_2$ satisfy

$$|\mathbf{k}_2 - \mathbf{k}_1|^2 = |\mathbf{v}_2 - \mathbf{c}_2 - \mathbf{v}_1 + \mathbf{c}_1|^2 = |\mathbf{v}_2 - \mathbf{v}_1|^2 + |\mathbf{c}_2 - \mathbf{c}_1|^2.$$

In particular,

$$|f(\mathbf{k}_2) - f(\mathbf{k}_1)| = |\mathbf{v}_2 - \mathbf{v}_1| \leq |\mathbf{k}_2 - \mathbf{k}_1|,$$

which proves the continuity of f. Since K is compact and f is continuous, $V = f(K)$ is compact, too. Vector $\mathbf{d}$ is a vector in V of minimal norm. Such a vector exists because it is a minimizer of a continuous function, the norm, over the compact set V.   □

Consider statement (ii) of Theorem 1.1. Clearly the statement holds for some $\mathbf{d}$ if and only if it holds for $\mathbf{d}$ replaced by $t\mathbf{d}$ for any $t > 0$, in particular, for the choice $t = 1/d_0 > 0$. Therefore, Theorem 1.1 says that the market $\mathbb{C}$ has no arbitrage opportunities if and only if there exists a vector $\mathbf{d} = (1, d_1, \ldots, d_n)^{\mathrm{T}}$, $d_k > 0$ for all k, such that $\mathbf{c}^{\mathrm{T}}\mathbf{d} = 0$ for all $\mathbf{c} \in \mathbb{C}$. The components of such a vector $\mathbf{d}$ are the discount factors for the times $t_0, \ldots, t_n$. In particular, an arbitrage-free price of an instrument paying c_k at time t_k, for $k \geq 1$, is

$$P_0 = \sum_{k=1}^{n} c_k d_k. \tag{1.2}$$

Equivalently, $(-P_0, c_1, \ldots, c_n)^{\mathrm{T}}$ belongs to $\mathbb{C}$. There may exist a range of arbitrage-free prices p with each p satisfying (1.2) for some vector $\mathbf{d}$ with the property $\mathbf{c}^{\mathrm{T}}\mathbf{d} = 0$ for all $\mathbf{c} \in \mathbb{C}$. Note that the discount factors d_k, $k = 0, \ldots, n$, may be written $d_k = e^{-r_k t_k}$, where r_k is the zero rate corresponding to payment time t_k.

If there exists precisely one vector $\mathbf{d}$ of discount factors, then $\mathbb{C} = \{\mathbf{c}; \mathbf{c}^{\mathrm{T}}\mathbf{d} = 0\}$, and $\mathbb{C}$ is said to be complete. If $\mathbb{C}$ is complete, then any new cash flow (or contract) $\mathbf{c}$ introduced is either redundant (a linear combination of $\mathbf{c}_1, \ldots, \mathbf{c}_m$) or creates an arbitrage opportunity. Real-world markets are typically not complete: a new contract is not identical to a linear combination of existing contracts.

Suppose that the cash flow corresponds to bonds, i.e., for each $\mathbf{c}_k$ we have that $-c_{k,0}$ is the bond price today, $c_{k,n}$ is the face value plus a coupon, and the other $c_{k,j}$s $(j = 1, \ldots, n - 1)$ are coupons. Under the assumption that this bond market is complete and without arbitrage opportunities, the bond price $-c_{k,0}$ is given by

$$-c_{k,0} = \sum_{j=1}^{n} c_{k,j} e^{-t_j r_j},$$

where r_j are the (unique) zero rates.

Given a market consisting of the cash flows $\mathbf{c}_1, \ldots, \mathbf{c}_m$, it is not difficult to check if the market is arbitrage free and, if so, whether the market is complete

Table 1.1 Specifications of three bonds

Bond	A	B	C
Bond price	99.65	113.43	121.30
Maturity (days)	190	$32 + 2 \cdot 365$	$241 + 3 \cdot 365$
Annual coupon	0	5.5	6.75
Face value	100	100	100

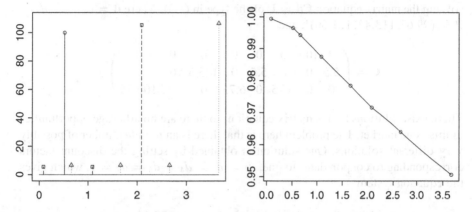

Fig. 1.1 *Left plot*: graphical illustration of cash flows for the three bonds; *right plot*: discount factors in Table 1.2. In the *left plot*, time is on the x-axis and the payment amounts on the y-axis. In the *right plot*, the time to maturity is on the x-axis and the value of the discount factors is on the y-axis

or not. An arbitrage-free (and complete) market is equivalent to the existence (and uniqueness) of a solution $\mathbf{d} = (d_1, \ldots, d_n)^{\mathrm{T}}$ to the matrix equation

$$
\begin{pmatrix} -c_{1,0} \\ \vdots \\ -c_{m,0} \end{pmatrix} = \begin{pmatrix} c_{1,1} & \cdots & c_{1,n} \\ \vdots & \cdots & \vdots \\ c_{m,1} & \cdots & c_{m,n} \end{pmatrix} \begin{pmatrix} d_1 \\ \vdots \\ d_n \end{pmatrix}, \tag{1.3}
$$

where $(c_{k,0}, \ldots, c_{k,n}) = \mathbf{c}_k^{\mathrm{T}}$. The analysis of solutions to matrix equation (1.3) is a standard problem in linear algebra.

Example 1.1 (Bootstrapping zero rates).   Consider a market consisting of the bonds in Table 1.1. From Table 1.1 and Fig. 1.1 we see that there are in total eight nonzero cash flow times

$$(t_1, \ldots, t_8) \approx (0.09, 0.52, 0.66, 1.09, 1.66, 2.09, 2.66, 3.66),$$

where t_1 corresponds to 32 days from now and therefore $32/365 \approx 0.09$ years from now, etc. Therefore, there are also eight undetermined discount factors $d_1, \ldots, d_8$

Table 1.2 Cash flow times (years), discount factors, and zero rates (%) (discount factors obtained as in Example 1.1 by linear interpolation between discount factors)

Time	0.088	0.521	0.660	1.088	1.660	2.088	2.660	3.660
Discount factors	0.999	0.997	0.994	0.987	0.978	0.972	0.964	0.951
Zero rates	0.673	0.674	0.869	1.158	1.317	1.381	1.380	1.384

solving the matrix equation $\mathbf{Cd} = \mathbf{P}$ of the type in (1.3), where $\mathbf{d} = (d_1, \ldots, d_8)^{\mathrm{T}}$, $\mathbf{P} = (99.65, 113.43, 121.30)^{\mathrm{T}}$, and

$$\mathbf{C} = \begin{pmatrix} 0 & 100 & 0 & 0 & 0 & 0 & 0 & 0 \\ 5.5 & 0 & 0 & 5.5 & 0 & 105.5 & 0 & 0 \\ 0 & 0 & 6.75 & 0 & 6.75 & 0 & 6.75 & 106.75 \end{pmatrix}.$$

There exist solutions to this matrix equation, so there are no arbitrage opportunities in this bond market. The problem here is that there is an infinite number of possibly very different solutions. One solution is obtained by setting the discount factors corresponding to coupon dates to one, $d_1 = d_3 = d_4 = d_5 = d_7 = 1$, which gives the equation system

$$\begin{pmatrix} 100 & 0 & 0 \\ 0 & 105.5 & 0 \\ 0 & 0 & 106.75 \end{pmatrix} \begin{pmatrix} d_2 \\ d_6 \\ d_8 \end{pmatrix} = \begin{pmatrix} 99.65 \\ 113.43 - 2 \cdot 5.5 \\ 121.30 - 3 \cdot 6.75 \end{pmatrix}$$

with solution $(d_2, d_6, d_8) \approx (0.9965, 0.9709, 0.9466)$. The corresponding zero rates are, in percentages, with two decimals, $r_1, \ldots, r_8 \approx 0, 0.67, 0, 0, 0, 1.41, 0, 1.50$. This is clearly a silly solution as it would imply that the price of a zero-coupon bond maturing 2.66 years from now with face value 100 is 100. Who would buy this bond?

Let us now take a step back and consider a better approach, which is often referred to as the bootstrap method (note: there are other methods referred to as bootstrap methods that have nothing to do with interest rates). The discount factor $d_2 = 0.9965$ corresponding to the zero-coupon bond is known. Also, the discount factor corresponding to cash flow today is clearly $d_0 = 1$. Therefore, it seems reasonable to assign a value to d_1 by interpolation between the two neighboring discount factors. Let us for simplicity use linear interpolation, which gives

$$d_1 = d_0 + \frac{d_2 - d_0}{t_2 - t_0}(t_1 - t_0) \approx 0.9994.$$

Now we have assigned values to the first two (nontrivial) discount factors, and we need an approach other than linear interpolation between known discount factors to assign values to the remaining ones. The second bond yields the equation

$$113.43 - 5.5d_1 = 5.5d_4 + 105.5d_6,$$

which is an equation with two unknowns. Assuming temporarily that the value of d_4 is given by linear interpolation between the last (in the sense of the order of the cash flow times) known discount factor d_2 and the unknown d_6 we get the equation

$$113.43 - 5.5d_1 = 5.5\left(d_2 + \frac{d_6 - d_2}{t_6 - t_2}(t_4 - t_2)\right) + 105.5d_6,$$

which can be solved for d_6, yielding $d_6 \approx 0.9716$. Now the discount factors d_3, d_4, d_5 are assigned values by linear interpolation between d_2 and d_6:

$$d_k = d_2 + \frac{d_6 - d_2}{t_6 - t_2}(t_k - t_2) \quad \text{for } k = 3, 4, 5.$$

This gives $(d_3, d_4, d_5) \approx (0.9943, 0.9875, 0.9784)$. The last two discount factors d_7 and d_8 are assigned values by repeating the foregoing procedure. This gives $(d_7, d_8) \approx (0.9639, 0.9506)$. The cash flow times, the discount factors, and the corresponding zero rates are given in Table 1.2.

Yield curves are not only derived from bond prices. The next example shows how a yield curve can be extracted from forward prices. In this example, the notion of present price and forward price of an asset is needed. Consider a contract for delivery of an asset at a future time $t > 0$. The forward price $G_0^{(t)}$ of the contract is the price, agreed upon at the current time 0, which will be paid at maturity, time t, of the contract. The present price $P_0^{(t)}$ of the contract is the price that is agreed upon and paid at the current time 0. In the sequel, when there is no risk of confusion about the maturity time, we will sometimes drop the superscript and write G_0 and P_0. The present price is the discounted forward price: $P_0^{(t)} = d_t G_0^{(t)}$, where d_t is the discount factor between 0 and t.

The present price of a share of a stock that does not pay dividends before time t must be identical to the spot price, S_0, for immediate delivery since there is no cost or benefit from holding the asset between time 0 and time t: the forward price must satisfy $d_t G_0^{(t)} = P_0^{(t)} = S_0$. The present price, for delivery at a future time t_2, of one share of a stock that pays a known dividend amount c at time $t_1 < t_2$ is determined by the relation

$$P_0^{(t_2)} = S_0 - d_{t_1}c.$$

The validity of the relation follows from the ensuing argument. Consider first the strategy of, at time 0, buying the share and short-selling a zero-coupon bond that matures at time t_1 with face value c, and at time t_2 selling the share. The initial cost of this strategy is $S_0 - d_{t_1}c$, and it gives the random payoff S_{t_2} at time t_2. On the other hand, consider a contract that delivers one share of the stock at time t_2. Since the contract and the foregoing strategy have identical future cash flows, their initial cash flows must coincide in order not to introduce arbitrage opportunities.

Table 1.3 Forward prices on April 8 for delivery of one share of H&M at different maturity times

Maturity	April 15	May 20	June 17	September 16	December 16	January 20	March 16
Forward price	218.64	209.52	209.92	211.29	212.85	213.50	214.59

Example 1.2 (Zero rates from forward prices). On April 8, the spot price S_0 for buying one share of H&M on the Nasdaq Nordic OMX exchange was 218.60 Swedish kronor. Table 1.3 shows forward prices on that same day for one share of the stock for delivery at different maturities. The company H&M announced that on May 6 it would pay a dividend of $c = 9.50$ kronor per share. This explains the large difference between the current forward prices for the maturity dates April 15 and May 20.

Consider the cash flow times $t_0, \ldots, t_9$ given by

$$t_0 = 0 \quad \text{(Apr 8)}, \qquad t_1 = 0.019 \,\text{(Apr 15)}, \qquad t_2 = 0.063 \,\text{(May 6)},$$
$$t_3 = 0.115 \,\text{(May 20)}, \qquad t_4 = 0.192 \,\text{(Jun 17)}, \qquad t_5 = 0.441 \,\text{(Sep 16)},$$
$$t_6 = 0.690 \,\text{(Dec 16)}, \qquad t_7 = 0.786 \,\text{(Jan 20)}, \qquad t_8 = 0.940 \,\text{(Mar 16)}.$$

The corresponding discount factors are denoted $d_0, \ldots, d_8$. Since there is no dividend paid before t_1, the discount factor d_1 is derived from the relation $d_1 G_0^{(t_1)} = S_0$, where S_0 is the spot price and, hence, also the present price for delivery of one share of H&M at time t_1. The present price for delivery of one share of H&M at t_3 gives the relation $d_3 G_0^{(t_3)} = S_0 - c d_2$. Similarly, for the remaining maturities we have $d_k G_0^{(t_k)} = S_0 - c d_2$ for $k = 4, \ldots, 8$. In all, we have seven equations and eight unknowns, which gives an underdetermined equation system with solution

$$d_1 = \frac{S_0}{G_0^{(t_1)}} \quad \text{and} \quad d_k = \frac{S_0}{G_0^{(t_k)}} - \frac{d_2 c}{G_0^{(t_k)}} \quad \text{for } k = 3, \ldots, 8, \qquad (1.4)$$

parameterized by d_2. To find a reasonable solution among all possible solutions, the bootstrapping procedure presented above suggests expressing d_2 by linear interpolation between d_1 and d_3. The equation

$$d_2 = d_1 + \frac{d_3 - d_1}{t_3 - t_1}(t_2 - t_1),$$

together with the equations for the maturity times t_1 and t_3, gives

$$d_2 = \left(1 + \frac{c}{G_0^{(t_3)}} \frac{t_2 - t_1}{t_3 - t_1} \right)^{-1} \left(\frac{S_0}{G_0^{(t_1)}} \left(1 - \frac{t_2 - t_1}{t_3 - t_1} \right) + \frac{S_0}{G_0^{(t_3)}} \frac{t_2 - t_1}{t_3 - t_1} \right),$$

from which the values of all discount factors can be computed from (1.4) (Fig. 1.2).

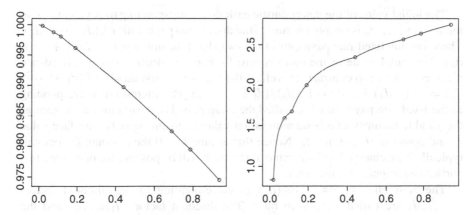

Fig. 1.2 *Left plot*: discount factors in Example 1.2. Time to maturity is on the x-axis; value of discount factors is on the y-axis. The *right plot* shows the zero rates (%) in Example 1.2 corresponding to the linearly interpolated discount factors

Example 1.3 (Interest rate swap). Let $0 = t_0 < t_1 < \cdots < t_n = T$ be a sequence of equally spaced times with $\Delta = t_k - t_{k-1} = T/n$, and let $d_1, \ldots, d_n$ be discount factors giving the value at time 0 of money at times $t_1, \ldots, t_n$.

An interest rate swap is an agreement at time 0 between two parties to exchange floating interest rate payments (a stochastic cash flow) for fixed interest rate payments (a deterministic cash flow) on a notional principal L (US \$100 million, say) until, and including, time t_n with zero initial cost for both parties.

The floating interest rate payments are paid at times $\Delta/m = \delta, 2\delta, \ldots, mn\delta = T$, where typically $m = 2$. The floating-rate payment due at time $k\delta$ is the interest earned between times $(k-1)\delta$ and $k\delta$ on the notional L, i.e., the random amount

$$L\left(\frac{1}{d_{k-1,k}} - 1\right),$$

where $d_{k-1,k}$ denotes the discount factor at time $(k-1)\delta$ between times $(k-1)\delta$ and $k\delta$. To determine the initial value of the floating-rate payments of the swap, we determine the value of a contract that pays the holder a never-ending stream of floating-rate payments at times $k\delta$, for $k = mn+1, mn+2, \ldots$, on principal L. The cash flow of the contract is obtained by investing at time $k\delta$ the amount L in zero-coupon bonds maturing at time $(k+1)\delta$ and at time $(k+1)\delta$, collecting the interest earned, and repeating the procedure with the remaining amount L. The value of this contract is therefore the value $d_n L$ of having the amount L at time $t_n = T$. Similarly, the value of a contract that pays the holder a never-ending stream of floating-rate payments at times $k\delta$, for $k = 1, 2, \ldots$, on principal L is L. Therefore, the initial value of the floating-rate payments of the swap is $L(1 - d_n)$. Notice that the number δ does not show up, so the value of the floating-rate payments does not depend on the frequency of the floating-rate payments.

The initial value of the deterministic cash flow corresponding to payments cL at the times $t_1, \ldots, t_n$ is simply the sum of the discounted payments: $cL(d_1 + \cdots + d_n)$. Therefore, the fixed-rate payments of the swap have the initial value $cL(d_1 + \cdots + d_n)$. The initial value of the swap is zero for both the floating-rate and fixed-rate receiver in the swap contract. Therefore, the number c must satisfy $cL(d_1 + \cdots + d_n) = L(1 - d_n)$, i.e., $c = (1 - d_n)/(d_1 + \cdots + d_n)$. The interest rate corresponding to the fixed-rate payment cL is called the swap rate. The swap rate can be seen as the yield to maturity of a bond with initial value L, maturing at t_n with face value L and coupons cL at times t_k. Notice that at time $t > 0$ the discount factors will typically have changed and the value of the swap will be positive for one of the two parties and negative for the other.

The zero rates $r_k = -\log(d_k)/t_k$ corresponding to the discount factors $d_1, \ldots, d_n$ are called swap zero rates. The discount factors $d_1, \ldots, d_n$ and the corresponding swap zero rates are obtained from a set of swap contracts, with a corresponding set of contracted swap rates, by a bootstrap procedure similar to the one considered in Example 1.1.

There are many versions of interest rate swaps. The most common interest rate swap contract prescribes floating-rate payments every 6 months (3 months) and fixed-rate payments every 12 months (6 months), i.e., at half the frequency of the floating-rate payments. The floating interest rate is an interbank interest rate such as LIBOR (London Interbank Offered Rate) and not defined in terms of government bonds. A practical issue of some importance that we ignored previously is that different day count conventions typically apply to fixed rates and floating rates. When writing $r_k = -\log(d_k)/t_k$ one should specify if t_k equals the actual number of days divided by 360 or 365. Swap data show that two swap contracts with different values of δ, different frequencies of floating-rate payments, that are otherwise identical can have slightly different swap rates. This is at odds with the preceding swap valuation and shows that the credit risk borne by the floating-rate receiver from having to wait longer between the floating-rate payments is taken into account by the market in the valuation of the swap. Here credit risk refers to the risk of a failure to deliver the contracted cash flow.

1.2   Derivatives and No-Arbitrage Pricing

Consider the times 0 and $T > 0$, with 0 being the present time, and let S_T be the spot price of some asset at time T. A contract with payoff $f(S_T)$ at time T for some function f is called a European derivative written on S_T. The derivative price π_f is the amount that is paid now in exchange for the payoff $f(S_T)$ at time T. A European call option on S_T with strike price K is a contract that gives the holder the right, but not the obligation, to purchase the underlying asset at time T for price K. Since this right is only exercised at time T if $S_T > K$, we see that the European call option is a derivative contract with payoff $f(S_T) = \max(S_T - K, 0)$. A European put option on

S_T with strike price K is a derivative contract with payoff $f(S_T) = \max(K-S_T, 0)$. In this case, the holder has the right, but not the obligation, to sell the underlying asset at time T for price K.

We consider a market where m derivative contracts with current prices π_k and payoffs $f_k(S_T)$, for $k = 1, \ldots, m$, and a risk-free zero-coupon bond maturing at time T with face value 1 and current price B_0 can be bought and sold. The bond saves us from difficulties in relating money at time 0 to money at time T. Here we assume that the market participants can buy and short-sell these contracts without paying any fees, and that for each contract the prices for buying and selling the contract coincide.

From the perspective of one of the market participants we want to understand how to assign a price to a new derivative contract in terms of the prices of the m existing derivative contracts and the bond. The market participants can form linear portfolios of the original derivative contracts, and such a portfolio will constitute a new derivative contract with payoff $f(S_T) = \sum_{k=1}^{m} h_k f_k(S_T)$ and price $\pi_f = \sum_{k=1}^{m} h_k \pi_k$. A contract of this type is called an arbitrage opportunity if $\pi_f = 0$, $P(f(S_T) \geq 0) = 1$, and $P(f(S_T) > 0) > 0$. An arbitrage opportunity is a contract that gives the holder a strictly positive probability of making a profit without taking any risk. The probability P is the subjective probability of the market participant under consideration. In particular, the existence of arbitrage opportunities depends on the subjective assessment of which events have probability zero.

Theorem 1.2. *The following statements are equivalent.*

1. *There are no arbitrage opportunities.*
2. *The prices π_f can be expressed as $\pi_f = B_0 E_Q[f(S_T)]$, where the expectation is computed with respect to a probability Q that assigns zero probability to the same events as does the probability P.*

Remark 1.1. (i) The probability Q is called the forward probability. Note that $E_Q[f(S_T)]$ is the forward price of the contract for delivery of $f(S_T)$ at time T.

(ii) There are examples of arbitrage opportunities that do not depend on the subjective probability P. Consider two derivative contracts with prices π_f and π_g and payoffs $f(S_T)$ and $g(S_T)$ satisfying $\pi_f < \pi_g$ and $f(S_T) \geq g(S_T)$ (for example, two European call options such that the one with the higher strike price costs more than the one with the lower strike price). A long position of size one in the cheaper derivative, a short position of size one in the expensive derivative, and a long position with initial value $\pi_g - \pi_f$ in the bond produces a contract with zero initial price and payoff $f(S_T) - g(S_T) + (\pi_g - \pi_f)/B_0 > 0$ at time T.

Proof. We begin by proving the implication (ii) $\Rightarrow$ (i). This implication is the easier one to prove and also probably the most relevant one since it means that as long as one comes up with a model for S_T that produces the observed prices, one can use this model for pricing new contracts without risking the introduction of arbitrage opportunities.

Suppose that (ii) holds, and consider a payoff $f(S_T)$ satisfying $P(f(S_T) \geq 0) = 1$ and $P(f(S_T) > 0) > 0$. We need to show that $\pi_f = B_0 \, E_Q[f(S_T)] \neq 0$. By assumption, it also holds that $Q(f(S_T) \geq 0) = 1$ and $Q(f(S_T) > 0) > 0$. Since $Q(f(S_T) \geq 0) = 1$, we may express $E_Q[f(S_T)]$ as

$$E_Q[f(S_T)] = \int_0^\infty Q(f(S_T) > t)dt,$$

(see Remark 1.2), and since $Q(f(S_T) > 0) > 0$, there exist $\varepsilon > 0$ and $\delta > 0$ such that $Q(f(S_T) > \varepsilon) > \delta$. Therefore,

$$\frac{\pi_f}{B_0} = E_Q[f(S_T)] = \int_0^\infty Q(f(S_T) > t)dt \geq \int_0^\varepsilon Q(f(S_T) > t)dt > \varepsilon \delta > 0,$$

which proves the claim, i.e., the implication (ii) $\Rightarrow$ (i).

Proving the implication (i) $\Rightarrow$ (ii) in a general setting is rather difficult. It becomes much less difficult if we assume that S_T takes values in a finite (but arbitrarily large) set. This is not at all an unrealistic assumption; S_T will take values with finitely many decimals, and it is plausible that $P(S_T > s) = 0$ for all s greater than some sufficiently large number. Let $\{s_1, \ldots, s_n\}$, with $P(S_T = s_k) > 0$ and $P(S_T = s_1) + \cdots + P(S_T = s_n) = 1$, be the set of possible outcomes for S_T. Then every contract can be represented as a vector $\mathbf{x} = (x_0, x_1, \ldots, x_n)^T$ in $\mathbb{R}^{n+1}$. The contract with payoff $f(S_T)$ and price π_f can be represented as the vector $\mathbf{x} = (-\pi_f, f(s_1), \ldots, f(s_n))^T$. Therefore, the set of all contracts constructed from the original m derivative contracts forms a linear subspace of $\mathbb{R}^{n+1}$. Let us denote this linear space by $\mathbb{X}$. We see that $\mathbf{x} \in \mathbb{X}$ is an arbitrage opportunity if $\mathbf{x} \neq \mathbf{0}$ and $\mathbf{x} \geq \mathbf{0}$. Theorem 1.1 says that there are no arbitrage opportunities if and only if there exists a vector $\mathbf{y} \in \mathbb{R}^{n+1}$ with $\mathbf{y} > \mathbf{0}$ such that $\mathbf{x}^T \mathbf{y} = 0$ for all $\mathbf{x} \in \mathbb{X}$. Of course, the same result holds if $\mathbf{y}$ is replaced by $y_0^{-1}\mathbf{y}$. The bond corresponds to the vector $\mathbf{x} = (-B_0, 1, \ldots, 1)^T$. Since $\mathbf{x}^T(y_0^{-1}\mathbf{y}) = 0$, we have $\sum_{k=1}^n y_0^{-1} y_k = B_0$. For $k = 1, \ldots, n$ set $q_k = (B_0 y_0)^{-1} y_k$ and note that $q_k > 0$ and $\sum_{k=1}^n q_k = 1$. In particular, the q_k constitute a probability distribution on the set $\{s_1, \ldots, s_n\}$ of possible outcomes for S_T. With $\mathbf{x} = (-\pi_f, f(s_1), \ldots, f(s_n))^T$ we see that $\mathbf{x}^T(B_0 y_0)^{-1}\mathbf{y} = 0$ is equivalent to $\pi_f = B_0 \sum_{k=1}^n f(s_k)q_k$, which is precisely what Theorem 1.2 says. $\qquad\qquad\qquad\qquad\qquad\qquad\qquad\qquad\qquad\qquad\qquad\square$

Remark 1.2. The representation of the expected value of a nonnegative random variable as an integral of its tail probabilities is not difficult to justify. Consider a random variable $X \geq 0$ with distribution function F, and set $\overline{F} = 1 - F$. If F has a density f, then

$$\int_0^\infty \overline{F}(t)dt = \int_0^\infty \left[\int_t^\infty f(u)du\right]dt = \int_0^\infty f(u)\left[\int_0^u dt\right]du = \int_0^\infty u f(u)du,$$

where we have simply changed the order of integration. The existence of a density f is actually not needed for the result to hold, but it simplifies the presentation.

Theorem 1.2 tells us how to price a new contract with payoff $g(S_T)$ such that no arbitrage opportunity is introduced: simply assign the price $\pi_g = B_0 E_Q[g(S_T)]$ to the derivative contract. The expected value $E_Q[g(S_T)]$ is the expected value of the random variable $g(S_T)$ computed with respect to the probability Q. Theorem 1.2 does not say that this price π_g is the unique arbitrage-free price of the new contract. There are typically many possible representations of the existing prices as discounted expected values, and the different representations are likely to give different prices to new contracts. More precisely: suppose that you assign a probability distribution to S_T with more than m parameters and that there is more than one solution (a set of parameters) to the nonlinear system of equations $\pi_k = B_0 E_Q[f_k(S_T)]$, $k = 1, \ldots, m$, where the left-hand side is the market price of the kth original derivative and the right-hand side is the discounted expected payoff according to your chosen parametric model. Then there are probably several solutions, and the different solutions are likely to give different prices $B_0 E_Q[g(S_T)]$ to a new derivative contract with payoff $g(S_T)$.

Example 1.4 (Rolling dice). Let S_T be the value of a six-sided die. The die is not necessarily fair. Suppose for now that there are two derivative contracts on S_T available on the market, a bet on even numbers (contract A) and a bet on odd numbers (contract B). Both contracts pay 1 if the bet turns up right and 0 otherwise, and the market prices of both contracts are $1/2$. There are no arbitrage opportunities on this market if the subjective probabilities $P(S_T = 1), \ldots, P(S_T = 6)$ are strictly positive. There are infinitely many choices of strictly positive probabilities $Q(S_T = 1), \ldots, Q(S_T = 6)$ such that (ii) of Theorem 1.2 holds. One such choice is given by

$$Q(S_T = 1) = \cdots = Q(S_T = 6) = 1/6.$$

Depending on the subjective view of the probabilities $P(S_T = 1), \ldots, P(S_T = 6)$, there may be opportunities for good deals: portfolios whose expected payoffs are greater than their prices. Consider an agent whose subjective view of the probabilities are such that

$$P(S_T = k) = 0 \quad \text{for } k = 4, 5, 6.$$

To this agent the set of possible outcomes is reduced to $\{1, 2, 3\}$. Note that the observed prices are still consistent with no arbitrage. Suppose a new contract C is introduced paying 1 if the outcome of S_T is 1 or 2, and that the market price of this contract is $1/3$. The original market is still free of arbitrage (the same Q still works). However, on the reduced set of outcomes $\{1, 2, 3\}$ it is not possible to find a probability Q that reproduces the market prices. To the agent who believes in the reduced set of possible outcomes there seems to be an arbitrage opportunity. A portfolio consisting of a long position in C and a short position in A of the same size has a strictly negative price equal to $-1/6$ (you get money now) and has a nonnegative payoff with P-probability 1. The agent now has two choices: try to capitalize on the perceived arbitrage opportunity by going long in C and short in A, or revise the subjective probabilities. This example illustrates that there may be portfolios that are perceived as arbitrage opportunities because the subjective model used to assign probabilities to future events is too simplistic.

Example 1.5 (Calls and digitals). Consider a derivative with payoff $I\{S_T > K\}$ (meaning the value 1 if the event occurs and 0 otherwise) at time T, referred to as a digital or binary option, with current price $D_0(K)$. Consider also two call options with payoffs $\max(S_T - K, 0)$ and $\max(S_T - (K - 1), 0)$ at time T and current prices $C_0(K)$ and $C_0(K - 1)$. Let $x_+ = \max(x, 0)$, and notice that

$$(S_T - K + 1)_+ - (S_T - K)_+ = \begin{cases} 0 & \text{if } S_T < K - 1, \\ S_T - K + 1 & \text{if } S_T \in [K - 1, K], \\ 1 & \text{if } S_T > K. \end{cases}$$

In particular, $(S_T - K + 1)_+ - (S_T - K)_+ \geq I\{S_T > K\}$.

If $C_0(K - 1) - C_0(K) < D_0(K)$, then there are arbitrage opportunities. Buying the call option with strike $K - 1$ and short-selling the call and the digital option with the strike K gives a strictly positive cash flow at time 0, which can be used to buy zero coupon bonds maturing at time T. Moreover, the cash flow from the payoffs of the options at time T is nonnegative. We have thus constructed a contract with zero initial cash flow that gives a strictly positive cash flow at time T. This is an arbitrage opportunity regardless of the probability distribution assigned to S_T.

If $C_0(K - 1) - C_0(K) = D_0(K)$, then there may be arbitrage opportunities. Buying the call option with strike $K - 1$ and short-selling the call and the digital option with the strike K gives zero initial cash flow and a cash flow $(S_T - K + 1)I\{S_T \in [K - 1, K]\} \geq 0$ at time T. If $P(S_T \in [K - 1, K]) > 0$, then this is an arbitrage opportunity.

Example 1.6 (Put–call parity). Suppose there is a risk-free zero-coupon bond maturing at time T with face value 1, a call option with strike price K on the value S_T at time T, and a put option with the same strike price K on S_T. Write B_0, C_0, and P_0 for the current prices of the bond, call option, and put option, respectively. Suppose further that there is a forward contract on S_T with forward price G_0, the amount agreed upon today that is paid at time T in exchange for the random amount S_T.

A position of size $G_0 - K$ in the bond (long or short depending on the sign of $G_0 - K$) and a long position in the forward contract give the price $B_0(G_0 - K)$ for the derivative contract with payoff $S_T - K$. However, the same payoff can be produced by taking positions in the options. A long position in the call option and a short position in the put option correspond to a long position in a derivative contract with price $C_0 - P_0$ and the payoff

$$(S_T - K)_+ - (K - S_T)_+ = S_T - K$$

at time T. In an arbitrage-free market, the prices of two derivative contracts with the same payoffs must coincide. Otherwise a risk-free profit is made by buying the cheaper of the two and short-selling the more expensive one. Therefore,

$$C_0 - P_0 = B_0(G_0 - K).$$

This relation between bond, forward, call option, and put option prices is called the put–call parity.

Example 1.7 (Parametric forward distribution). Suppose you want to use the parametric density function q_θ, whose argument is a real number and whose parameter vector θ is multidimensional, as a model for the forward probability. Suppose further that the nonlinear system of equations in θ

$$\pi_k = B_0 \int f_k(s) q_\theta(s) ds, \quad k = 1, \ldots, m$$

has a solution θ^*. Theorem 1.2 tells us that the market is arbitrage free if for any interval (a, b) it holds that

$$\int_a^b q_{\theta^*}(s) ds = 0 \quad \text{if and only if} \quad \int_a^b p(s) ds = 0,$$

where p is your subjective probability density for the future spot price S_T. In this case you may assign the arbitrage-free price

$$\pi_g = B_0 \int g(s) q_{\theta^*}(s) ds$$

to a derivative contract with payoff $g(S_T)$.

Example 1.8 (Online sports betting). Suppose you are visiting the Web site of an online sports betting agent, the bookmaker, with the intent of betting on a Premier League game, Chelsea vs. Liverpool. The odds offered by the bookmaker are "Chelsea": 2.50, "draw": 3.25, and "Liverpool": 2.70. The corresponding outcome of the game are denoted by 1, X, and 2, and for each of the outcomes it is assumed that you do not assign zero probability to the occurrence of that outcome. This game may be viewed as a market with three digital derivatives with prices $q_1 = 1/2.50$, $q_X = 1/3.25$, and $q_2 = 1/2.70$ and payoffs X_1, X_X, and X_2, where $X_1 = 1$ if the outcome of the game is "Chelsea" and 0 otherwise, and similarly for the other payoffs. Notice that

$$q_1 + q_X + q_2 = \frac{1}{2.50} + \frac{1}{3.25} + \frac{1}{2.70} \approx 1.078.$$

Since the prices do not sum up to one, they cannot be interpreted as probabilities. Equivalently, they cannot be expressed as (discounted) expected payoffs. A natural question, in light of Theorem 1.2, is therefore: is there an arbitrage opportunity? The answer is no. The reason is that you cannot sell the contracts short on this market (the bookmaker is not willing to switch sides with you). To see that there is no arbitrage, one could argue as follows. Consider dividing the initial capital 1 into bets on "Chelsea," "draw," and "Liverpool," where $w_1, w_X, w_2 \geq 0$, with $w_1 + w_X + w_2 = 1$, are the amounts placed on the respective possible outcomes. The portfolio (w_1, w_X, w_2) is an arbitrage opportunity if its post game value

$$\frac{w_1}{q_1} X_1 + \frac{w_X}{q_X} X_X + \frac{w_2}{q_2} X_2$$

is greater than or equal to one for sure and strictly greater than one with a strictly positive probability. Suppose that (w_1, w_X, w_2) is an arbitrage opportunity. For the postgame portfolio value to be greater than or equal to one it is necessary that $w_1/q_1 \geq 1$, $w_X/q_X \geq 1$, and $w_2/q_2 \geq 1$. Therefore,

$$w_1 + w_X + w_2 \geq q_1 + q_X + q_2 > 1,$$

which is a contradiction. We conclude that there are no arbitrage opportunities. The key to arriving at this conclusion is, of course, that the sum of the reciprocal odds is greater than one. The excess $1.078 - 1 = 0.078$ can be interpreted as the margin the bookmaker takes as a profit.

Occasionally, when examining the odds of many different sports betting agents, you may find better odds. If the best available odds happen to be 2.75 on "Chelsea," 3.50 on "Draw," and 2.95 on "Liverpool," then there is an arbitrage opportunity. In the analogy with the digital derivative market, here the sum of the digital derivative prices sum up to a number less than one. Therefore, a portfolio can be formed whose initial value is less than one and whose postgame value is one, from which an arbitrage portfolio can be formed.

1.2.1   The Lognormal Model

Suppose that there exist a risk-free zero-coupon bond with price B_0 that pays the amount 1 at time T and a forward contract on S_T with current forward price G_0. A long position in the bond of size G_0 together with a long position of size one in the forward contract produces a European derivative contract with price $B_0 G_0$ and payoff S_T at time T. Therefore, we are in the setting of Theorem 1.2 [with $m = 1$ and $f_1(s) = s$].

Here we will choose a lognormal distribution for S_T in the representation $B_0 G_0 = B_0 E_Q[S_T]$ and derive arbitrage-free pricing formulas for European derivatives. Note that S_T has a lognormal distribution if $\log S_T$ has a normal distribution. If we choose μT and $\sigma^2 T$ to be the mean and variance of the normal distribution for $\log S_T$, then we may write $\log S_T = \mu T + \sigma\sqrt{T}Z$ for a standard normally distributed random variable Z. Since

$$G_0 = \mathrm{E}_Q[S_T] = \int_{-\infty}^{\infty} e^{\mu T + \sigma\sqrt{T}z}\frac{e^{-z^2/2}}{\sqrt{2\pi}}dz = e^{\mu T + \sigma^2 T/2}\int_{-\infty}^{\infty}\frac{e^{-(z-\sigma\sqrt{T})^2/2}}{\sqrt{2\pi}}dz$$

$$= e^{\mu T + \sigma^2 T/2},$$

we see that $\mu T = \log G_0 - \sigma^2 T/2$ and $\log S_T$ is $\mathrm{N}(\log G_0 - \sigma^2 T/2, \sigma^2 T)$-distributed. In particular, we may write

$$S_T = G_0 e^{\sigma\sqrt{T}Z - \sigma^2 T/2}$$

with Z standard normally distributed, and therefore the price of a derivative on S_T with payoff $g(S_T)$ may be expressed as

$$\pi_g = B_0\,\mathrm{E}_Q[g(S_T)] = B_0\int_{-\infty}^{\infty} g\left(G_0 e^{\sigma\sqrt{T}z - \sigma^2 T/2}\right)\frac{e^{-z^2/2}}{\sqrt{2\pi}}dz. \qquad (1.5)$$

This representation of the derivative price is known as Black's formula (Fisher Black). For call (and put) options, Black's formula turns into a very nice explicit expression. The price C_0 of a call option on S_T with strike price K can be expressed as

$$C_0 = B_0\,\mathrm{E}_Q[\max(S_T - K, 0)]$$

$$= B_0\,\mathrm{E}_Q[(S_T - K)I\{S_T > K\}]$$

$$= B_0\,\mathrm{E}_Q[(G_0 e^{-\sigma^2 T/2 + \sigma\sqrt{T}Z} - K)I\{Z > \gamma\}]$$

$$= B_0 G_0 e^{-\sigma^2 T/2}\,\mathrm{E}_Q[e^{\sigma\sqrt{T}Z}I\{Z > \gamma\}] - K B_0\,\mathrm{E}_Q[I\{Z > \gamma\}],$$

where

$$\gamma = \frac{\log(K/G_0)}{\sigma\sqrt{T}} + \frac{\sigma\sqrt{T}}{2}.$$

Therefore, with Φ denoting the standard normal distribution function,

$$C_0 = B_0 G_0 e^{-\sigma^2 T/2}\int_{\gamma}^{\infty} e^{\sigma\sqrt{T}z}\frac{e^{-z^2/2}}{\sqrt{2\pi}}dz - B_0 K(1 - \Phi(\gamma))$$

$$= B_0 G_0\int_{\gamma}^{\infty}\frac{e^{-(z-\sigma\sqrt{T})^2/2}}{\sqrt{2\pi}}dz - B_0 K\Phi(-\gamma)$$

$$= B_0 G_0 \int_{\gamma-\sigma\sqrt{T}}^{\infty} \frac{e^{-z^2/2}}{\sqrt{2\pi}} dz - B_0 K \Phi(-\gamma)$$

$$= B_0 G_0 \Phi(\sigma\sqrt{T} - \gamma) - B_0 K \Phi(-\gamma).$$

This expression for the call option price, called Black's formula for call options, is typically written as

$$C_0^{\mathrm{B}} = B_0(G_0 \Phi(d_1) - K\Phi(d_2)), \tag{1.6}$$

where

$$d_1 = \frac{\log(G_0/K)}{\sigma\sqrt{T}} + \frac{\sigma\sqrt{T}}{2} \quad \text{and} \quad d_2 = d_1 - \sigma\sqrt{T}.$$

If the underlying asset is a pure investment asset (holding the asset brings neither benefits nor costs), then a buyer of the underlying asset at time 0 does not care whether the asset is delivered at that time or at the later time T. This implies that the spot price S_0 for immediate delivery at time 0 must coincide with the derivative price $B_0 G_0$ for delivery of the asset at time T. If the underlying asset is a pure investment asset, then Black's formula for call option prices is called the Black–Scholes, or the Black–Merton–Scholes formula for call option prices, and reads

$$C_0 = S_0 \Phi(d_1) - B_0 K \Phi(d_2), \tag{1.7}$$

where

$$d_1 = \frac{\log(S_0/(B_0 K))}{\sigma\sqrt{T}} + \frac{\sigma\sqrt{T}}{2} \quad \text{and} \quad d_2 = d_1 - \sigma\sqrt{T}.$$

If the market provides us with the prices C_0 and G_0, or with C_0, S_0, and B_0 if the underlying asset is a pure investment asset, then the model parameter σ is obtained as the solution to a nonlinear equation in one variable [(1.6) or (1.7)] and is called the implied volatility (implied by the market prices). For a given underlying asset and maturity time, an option price is often quoted in volatility rather than in monetary units. The implied volatilities for two call options on S_T with different strike prices typically do not coincide. Therefore, the lognormal model is inconsistent with price data. However, the very simple lognormal model is still surprisingly accurate and is used as a benchmark model with the modification that the volatility parameter σ is viewed as a function of the strike price K (thereby violating the assumption of the lognormal model). The graph of the function $\sigma(K)$ is called the volatility smile or volatility skew.

Table 1.4 Current prices of options maturing in 35 days

Strike	980	990	1,000	1,020	1,040
Call price	63.625	56.625	50.000	37.625	27.250
Strike	1,060	1,080	1,100	1,120	1,140
Call price	18.500	12.000	7.125	3.825	1.875
Strike	980	990	1,000	1,020	1,040
Put price	23.875	26.875	30.375	38.125	47.625

1.2.2 Implied Forward Probabilities

Consider n call option prices $C_0(K_1), \ldots, C_0(K_n)$ on S_T, the forward price G_0 of S_T, and the price B_0 of a zero-coupon bond maturing at time T with face value 1. It is assumed that the set of prices do not give rise to arbitrage opportunities. From Black's formula (1.6) the implied volatilities $\sigma(K_1), \ldots, \sigma(K_n)$ are obtained, and by interpolation and extrapolation among the implied volatilities a volatility smile can be created that can be used together with Black's formula to price any European derivative on S_T. For call options, write $C_0(K) = C_0^B(K, \sigma(K))$, where C_0^B denotes Black's formula and $\sigma(K)$ is the volatility smile evaluated at K. The produced prices are arbitrage free if and only if there is a probability distribution for S_T so that $C_0(K) = B_0 E_Q[\max(S_T - K, 0)]$ for all K. We may write

$$C_0(K) = B_0 E_Q[\max(S_T - K, 0)]$$

$$= B_0 \int_0^\infty Q(\max(S_T - K, 0) > t) dt$$

$$= B_0 \int_K^\infty Q(S_T > t) dt.$$

In particular, the prices are arbitrage free if and only if there exists a distribution function Q, the forward probability distribution function, such that

$$\frac{dC_0}{dK}(k) = -B_0(1 - Q(k)) \quad \text{for all } k \geq 0.$$

Moreover, we see that if $C_0(K)$ is twice differentiable, then the prices are arbitrage free if and only if there exists a density function q such that

$$\frac{d^2 C_0}{dK^2}(k) = B_0 q(k) \quad \text{for all } k \geq 0.$$

Example 1.9 (Implied volatilities). Consider the option prices specified in Table 1.4. The options were the actively traded European call and put options that day on the value of a stock market index 35 trading days later (7 weeks later). For simplicity, the prices in the table are computed as mid prices; the mid price is

Table 1.5 Zero rates derived from put–call parity

Strike	980	990	1,000	1,020	1,040
Zero rate (%)	0.632	0.626	0.529	0.431	0.509

Table 1.6 Implied volatilities using Black's formula

Strike	980	990	1,000	1,020	1,040
Implied vol.	0.274	0.268	0.263	0.250	0.239
Strike	1,060	1,080	1,100	1,120	1,140
Implied vol.	0.227	0.218	0.208	0.198	0.190

the average of the bid price (the highest price at which a buyer is willing to buy) and the ask price (the lowest price at which a seller is willing to sell). The index level at the time, here called the spot, was $S_0 = 1,018.89$.

From the put–call parity in Example 1.6 we see that the put and call prices can be combined to get prices of the derivative that pays one unit of the index at maturity (we ignore commissions and trading costs). The index does not pay dividends, and therefore the spot S_0 equals $B_0 G_0$, where B_0 is the price of a zero-coupon bond that matures at the same time as the options and G_0 is the forward price of the index. Therefore, the put–call parity reads

$$C_0 - P_0 = S_0 - B_0 K.$$

From this relation we can derive B_0 and the zero rate $r = -\log(B_0)/T$, where $T = 35/252$ is the time to maturity (assuming 252 trading days per year). As we have prices on calls and puts for several strikes, each pair will give a possibly different value of r. The extracted zero rates r are presented in Table 1.5. The zero rates are not identical over the range of strikes, but we make a rough approximation and assume the zero rate is equal to 0.5%.

Now we can compute the implied volatilities using Black's formula (1.6). The implied volatilities are presented in Table 1.6. They are also shown in the left-hand plot in Fig. 1.3. The implied volatilities often have a convex looking shape and are therefore often referred to as the volatility smile.

We now turn to the question of how implied volatilities for strikes $K_1, \ldots, K_n$ should be used to price a derivative that is not actively traded on a market. For instance, a digital option with payoff $I\{S_T \geq K\}$, where $K_i < K < K_{i+1}$. The arbitrage-free price of the digital option is given by

$$D_0(K) = B_0 E_Q[I\{S_T \geq K\}] = B_0 Q(S_T \geq K) = B_0(1 - Q(K)),$$

where Q is a choice of pricing probability, satisfying the conditions in Theorem 1.2, and Q is the corresponding distribution function for S_T. If we use the lognormal model, then

$$E_Q[I\{S_T \geq K\}] = Q(S_T \geq K) = \Phi(d_2),$$

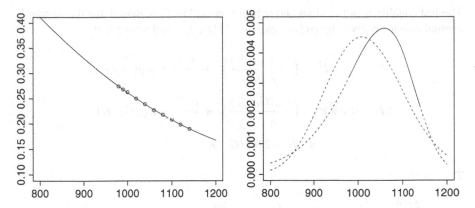

Fig. 1.3 *Left plot*: implied volatilities and graph of fitted second-degree polynomial. The strike price is on the x-axis, and volatility on the y-axis. *Right plot*: graph of implied forward density corresponding to fitted volatility smile, drawn by a *solid curve* within the range of strikes and by a *dashed curve* outside the range of the strikes. The *dashed curve* shows the graph of the density corresponding to the lognormal model with the volatility parameter chosen as the average of the implied volatilities

where

$$d_2 = \frac{\log(G_0/K)}{\sigma\sqrt{T}} - \frac{\sigma\sqrt{T}}{2},$$

but it is far from clear what volatility σ we should use.

A common practice is to use Black's model together with a suitable implied volatility smile $\sigma(k)$ and express the price of a call option with an arbitrary strike price k as $C_0(k) = C_0^B(k, \sigma(k))$. Recall that these prices are arbitrage free if there exists a forward distribution function Q such that

$$Q(k) = 1 + \frac{1}{B_0}\frac{dC_0}{dK}(k). \tag{1.8}$$

If Black's model together, with a suitable implied volatility smile $\sigma(k)$, is used, then $C_0(k) = C_0^B(k, \sigma(k))$ and

$$\frac{dC_0}{dK}(k) = \frac{\partial C_0^B}{\partial K}(k, \sigma(k)) + \frac{\partial C_0^B}{\partial \sigma}(k, \sigma(k))\frac{d\sigma}{dK}(k)$$

$$= -B_0\Phi(d_2) + B_0 G_0\phi(d_1)\sqrt{T}\frac{d\sigma}{dK}(k). \tag{1.9}$$

The last equality is not obvious and requires an explanation. Recall that the standard normal density is given by $\phi(z) = \exp\{-z^2/2\}/\sqrt{2\pi}$, and notice that

$$d_1^2 = \left(\frac{\log(G_0/K)}{\sigma\sqrt{T}}\right)^2 + \frac{\sigma^2 T}{4} + \log(G_0/K),$$

$$(d_1 - \sigma\sqrt{T})^2 = \left(\frac{\log(G_0/K)}{\sigma\sqrt{T}}\right)^2 + \frac{\sigma^2 T}{4} - \log(G_0/K)$$

$$= d_1^2 - 2\log(G_0/K).$$

Therefore,

$$\frac{\partial C_0^{\mathrm{B}}}{\partial K} = -B_0\Phi(d_2) + B_0\left(G_0\phi(d_1)\frac{\partial d_1}{\partial K} - K\phi\left(d_1 - \sigma\sqrt{T}\right)\frac{\partial d_2}{\partial K}\right)$$

$$= -B_0\Phi(d_2)$$

since $\partial d_1/\partial K = \partial d_2/\partial K$, and

$$\frac{\partial C_0^{\mathrm{B}}}{\partial \sigma} = B_0\left(G_0\phi(d_1)\frac{\partial d_1}{\partial \sigma} - K\phi\left(d_1 - \sigma\sqrt{T}\right)\frac{\partial d_2}{\partial \sigma}\right)$$

$$= B_0 G_0 \phi(d_1)\sqrt{T}$$

since $\partial d_1/\partial\sigma = \partial d_2/\partial\sigma + \sqrt{T}$.

With Black's model with a volatility smile $\sigma(k)$ the price of the digital option with payoff $I\{S_T \geq K\}$ follows from (1.8) and (1.9) and is given by

$$D_0(K) = B_0(1 - Q(K)) = B_0\Phi(d_2) - B_0 G_0 \phi(d_1)\sqrt{T}\frac{d\sigma}{dK}(K).$$

Notice that the expression in (1.9) must be nondecreasing in k (recall also that d_1 and d_2 depend on k) and takes values in the interval $[-B_0, 0]$ for the function Q in (1.8) to be a distribution function. In particular, there are conditions that a volatility smile $\sigma(k)$ must satisfy to give rise to arbitrage-free derivative prices.

A natural approach to constructing a volatility smile is to use some interpolation method to interpolate between the implied volatilities. Linear interpolation is one choice, and then $\sigma(k)$, for $k \in [K_i, K_{i+1}]$, is given by

$$\sigma(k) = \sigma(K_i) + \frac{\sigma(K_{i+1}) - \sigma(K_i)}{K_{i+1} - K_i}(k - K_i).$$

However, linear interpolation may lead to a model that admits arbitrage. We now show this claim. For this model to be free of arbitrage it is necessary that the slope of $\sigma(k)$ be nondecreasing, i.e., that the linearly interpolated volatility smile $\sigma(k)$ be a convex function. Suppose, on the contrary, that the slope between K_{i-1} and K_i is larger than the slope between K_i and K_{i+1}. In mathematical terms, suppose that

$$\text{slope}_{i+1} = \frac{\sigma(K_{i+1}) - \sigma(K_i)}{K_{i+1} - K_i} < \frac{\sigma(K_i) - \sigma(K_{i-1})}{K_i - K_{i-1}} = \text{slope}_i.$$

As a consequence,

$$\lim_{k \uparrow K_i} Q(k) = 1 + \frac{1}{B_0} \lim_{k \uparrow K_i} \frac{dC_0}{dK}(k)$$

$$= 1 + \frac{1}{B_0} \lim_{k \uparrow K_i} \left(\frac{\partial C_0^B}{\partial K}(k, \sigma(k)) + \frac{\partial C_0^B}{\partial \sigma}(k, \sigma(k)) \frac{d\sigma}{dk}(k) \right)$$

$$= 1 + \frac{1}{B_0} \left(\frac{\partial C_0^B}{\partial K}(K_i, \sigma(K_i)) + \frac{\partial C_0^B}{\partial \sigma}(K_i, \sigma(K_i)) \text{slope}_i \right)$$

$$> 1 + \frac{1}{B_0} \left(\frac{\partial C_0^B}{\partial K}(K_i, \sigma(K_i)) + \frac{\partial C_0^B}{\partial \sigma}(K_i, \sigma(K_i)) \text{slope}_{i+1} \right)$$

$$= 1 + \frac{1}{B_0} \lim_{k \downarrow K_i} \left(\frac{\partial C_0^B}{\partial K}(k, \sigma(k)) + \frac{\partial C_0^B}{\partial \sigma}(k, \sigma(k)) \frac{d\sigma}{dk}(k) \right)$$

$$= \lim_{k \downarrow K_i} Q(k).$$

We find that the function Q has a negative jump at K_i, and therefore it cannot be a distribution function. We conclude that it is necessary that $\text{slope}_{i+1} \geq \text{slope}_i$, i.e., that

$$\frac{\sigma(K_{i+1}) - \sigma(K_i)}{K_{i+1} - K_i} \geq \frac{\sigma(K_i) - \sigma(K_{i-1})}{K_i - K_{i-1}}$$

for a pricing model with linearly interpolated implied volatilities to be free of arbitrage.

An alternative to linear interpolation between implied volatilities, although still rather ad hoc, is to fit a second-degree polynomial $\sigma(k) = c_0 + c_1 k + c_2 k^2$ to the implied volatilities using least squares. The least-squares-fitted volatilities $\sigma(K_i)$ will not coincide with the original implied volatilities. However, typically the second degree polynomial gives a good enough fit so that the resulting model prices for the call and put options lie between the observed bid and ask prices.

Let us illustrate the procedure on the option data in Example 1.9. The resulting second-degree polynomial and the implied volatilities are shown in the left-hand plot in Fig. 1.3. Here we observe a very good fit to the implied volatilities. Note that in the left-hand plot in Fig. 1.3 the graph of the function $\sigma(k)$ is plotted also outside the range of the strikes. However, to the left of the smallest strike (980) and to the right of the highest strike (1,140) we do not have information on what the volatility smile looks like. Extrapolating outside the range of the data is nothing but a crude guess.

The resulting implied forward distribution Q can now be computed from (1.8), and the corresponding implied density q is given by

$$q(k) = \frac{1}{B_0} \frac{d^2}{dk^2} C_0^B (k, \sigma(k))$$

$$= \frac{1}{B_0} \left(\frac{\partial^2 C_0^B}{\partial K^2}(k, \sigma(k)) + 2 \frac{\partial^2 C_0^B}{\partial K \partial \sigma}(k, \sigma(k)) \frac{d\sigma}{dK}(k) \right.$$

$$\left. + \frac{\partial C_0^B}{\partial \sigma}(k, \sigma(k)) \frac{d^2\sigma}{dK^2}(k) + \frac{\partial^2 C_0^B}{\partial \sigma^2}(k, \sigma(k)) \left(\frac{d\sigma}{dK}(k) \right)^2 \right).$$

Computing the second-order partial derivatives of $C_0^B(K, \sigma)$ from Black's formula is straightforward—but tedious. Therefore, we simply state them and leave it to the reader as an exercise to verify them. The second-order derivatives are given by

$$\frac{\partial^2 C_0^B}{\partial K^2}(K, \sigma) = \frac{B_0}{K\sigma\sqrt{T}} \phi(-d_2),$$

$$\frac{\partial^2 C_0^B}{\partial \sigma \partial K}(K, \sigma) = \frac{B_0 G_0}{K\sigma} d_1 \phi(d_1),$$

$$\frac{\partial^2 C_0^B}{\partial \sigma^2}(K, \sigma) = \frac{B_0 G_0 \sqrt{T}}{\sigma} d_1 d_2 \phi(d_1).$$

The resulting density $q(k)$ is shown in the right-hand plot in Fig. 1.3. The solid part of the curve indicates the range between the smallest and largest strikes (the interval on which we have information from the price data). The dashed part of the curve is the extrapolation outside the range of the strikes of the option data. We compare the density implied by the volatility smile and Black's call option price formula to the density for the lognormal model with the volatility parameter chosen as the average of the implied volatilities. This lognormal density is shown as the dashed curve in the plot to the right in Fig. 1.3. We observe that the effect of the volatility smile, compared to a constant volatility, is that the implied forward density is left-skewed and has more probability mass in the left tail.

1.3   Notes and Comments

To prove the no-arbitrage theorem for deterministic cash flows, Theorem 1.1, we used a proof we learned from Harald Lang. The same idea was also used to prove Theorem 1.2 under the assumption that the spot price S_T takes values in a finite set. Theorem 1.2 is called the First Fundamental Theorem of Asset Pricing and appears here in its simplest form. A more general version of the theorem, without the assumption of a finite set of possible outcomes for S_T and with multiple time periods instead of one, was proved by Robert Dalang, Andrew Morton, and Walter Willinger in [10]. Since then, many alternative proofs of their theorem have appeared.

The material in Sect. 1.2 on no-arbitrage pricing and the lognormal model is a very brief summary of selected topics from the enormous amount of literature written on no-arbitrage pricing of derivatives contracts since the seminal work in [6, 32], and [5] of Fisher Black, Myron Scholes, and Robert Merton in the early 1970s. A natural motivation for the use of lognormal models can be found in the work [42] of Paul Samuelson, which predates that of Fisher Black, Myron Scholes, and Robert Merton.

The reader who seeks more information about financial markets and contracts, including the topics presented here, is recommended to consult the popular textbooks of John Hull, for instance [21].

1.4 Exercises

In the exercises below, it is assumed, wherever applicable, that you can take positions corresponding to fractions of assets.

Exercise 1.1 (Arbitrage in bond prices). (a) Consider a market consisting of the five risk-free bonds shown in Table 1.7. Show that the market is free of arbitrage and determine the zero rates, or construct an arbitrage portfolio.

(b) Consider a market consisting of the three bonds denoted A, D, and E in Table 1.7. Show that the market is free of arbitrage and use the bootstrapping procedure to determine the zero rates, or construct an arbitrage portfolio.

Exercise 1.2 (Put–call parity). (a) Consider a European derivative contract, called a collar, with payoff function f given by

$$f(x) = \begin{cases} K_1 \text{ if } x < K_1, \\ x \quad \text{if } x \in [K_1, K_2], \\ K_2 \text{ if } x > K_2, \end{cases}$$

where $K_1 < K_2$. Express the forward price of a collar in terms of the forward prices of appropriate European call and put options and the forward price of the underlying asset.

Table 1.7 Bond specifications

Bond	A	B	C	D	E
Bond price ($)	98.51	100.71	188.03	111.55	198.96
Maturity (years)	0.5	1	1.5	1.5	2
Annual coupon ($)	0	4	0	12	8
Face value ($)	100	100	200	100	200

Half of the annual coupon is paid every 6 months from today and including the time of maturity; the first coupon payment is in 6 months

Table 1.8 Odds offered by seven bookmakers

Bookmaker	1	2	3	4	5	6	7
Everton	4.30	4.55	4.35	4.30	4.55	4.60	4.70
Draw	3.50	3.55	3.35	3.70	3.30	3.45	3.55
Manchester City	1.85	1.80	1.95	1.80	1.85	1.85	1.75

Table 1.9 Bond specifications

Bond	A	B	C	D
Bond price ($)	98	104	93	98
Maturity (years)	1	2	1	2
Annual coupon ($)	0	5	0	10
Face value ($)	100	100	100	100

The annual coupon is paid every 12 months starting from today and including the time of maturity. The first coupon payment is in 12 months

(b) A risk reversal is a position made up of a long position in an out-of-the-money (worthless if it were to expire today) European call option and a short position of the same size in an out-of-the-money European put option; both options have the same maturity and are written on the same underlying asset. Express the forward price of a risk reversal in terms of the forward prices of the underlying asset and a collar on this asset.

Exercise 1.3 (Sports betting). Consider the odds shown in Table 1.8 of seven bookmakers on the outcome of a football game between Everton and Manchester City. Is it possible to create an arbitrage opportunity by making bets corresponding to long positions?

Exercise 1.4 (Lognormal model). (a) Let Z have a standard normal distribution. For any $a > 0$ and $b \in \mathbb{R}$ compute $E[e^{aZ} I\{Z > b\}]$.

(b) Let R have the lognormal distribution $LN(\mu, \sigma^2)$. For an arbitrary number c, compute

$$E[(R - c)_+], \ \text{Var}((R - c)_+), \ \text{Cov}(R, (R - c)_+), \ \text{Cov}((R - c)_+, (R - d)_+),$$

where $d > c$.

(c) What happens to the preceding quantities if R is replaced by $S = S_0 R$ for a constant $S_0 > 0$?

Exercise 1.5 (Risky bonds). Consider investments in long positions in the four bonds shown in Table 1.9. Bonds A and B are issued by the United States government, and their cash flows are considered risk free. Bonds C and D are issued by a bank in the USA experiencing serious financial difficulties. If the bank were to default, bonds C and D would be worthless.

(a) Determine the current 1- and 2-year US Treasury zero rates and the 1- and 2-year credit spreads for the bonds issued by the bank. Determine the probabilities of default within 1 and within 2 years implied by the bond prices.

(b) An investor is certain that the bank is considered by the government to be too important to the financial system to be allowed to default. However, the investor believes that the market will continue to believe that the bank may default on its bonds. The investor believes that the 1-year Treasury zero rate in 1 year is $N(0.03, 0.01^2)$-distributed and that the 1-year credit spread in 1 year is $N(0.13, 0.03^2)$-distributed. Determine the investment strategy in terms of $10,000 invested in long positions in the bonds and a strategy for how to reinvest any cash flow received in 1 year that maximizes the expected value of the cash flow in 2 years.

(c) Suppose that the investor is wrong and the market's view, corresponding to the current bond prices, is right. Determine the distribution function of the investor's (perceived) optimal cash flow in 2 years in (b).

Project 1 (Implied forward distribution). Find prices of traded European put and call options of the future value of a stock market index. Consider prices of the options that are traded in sufficiently large volumes so that the prices contain relevant information about the future index value.

Use the material in Sect. 1.2.2 to estimate the density function of the future index value implied by the option prices. Make the plots correspond to Fig. 1.3. Make sure that the method used for interpolation and extrapolation of the implied Black's model volatilities does not lead to arbitrage in the pricing model you suggest.

Chapter 2
Convex Optimization

Many of the investment and hedging problems we will encounter can be formulated as a minimization of a function over a set determined by the investor's risk and budget constraints and other restrictions on the type of positions that the investor can take. Such problems become particularly tractable if both the function to be minimized and the set over which the minimization is done are convex. The minimization problem is in this case called a convex optimization problem. This chapter presents basic results for solving convex optimization problems that will be applied in subsequent chapters.

2.1 Basic Convex Optimization

Let $\mathbb{C} \subset \mathbb{R}^n$ be a convex set. That is, for any $\lambda \in [0, 1]$ and $\mathbf{x}, \mathbf{y} \in \mathbb{C}$ it holds that $\lambda \mathbf{x} + (1 - \lambda)\mathbf{y} \in \mathbb{C}$. Let $f : \mathbb{C} \to \mathbb{R}$ and $g_k : \mathbb{C} \to \mathbb{R}$, $k = 1, \dots, m$ be convex functions. That is, for $\lambda \in [0, 1]$ and $\mathbf{x}, \mathbf{y} \in \mathbb{C}$,

$$f(\lambda \mathbf{x} + (1 - \lambda)\mathbf{y}) \leq \lambda f(\mathbf{x}) + (1 - \lambda) f(\mathbf{y}),$$

and similarly for g_k. Consider the optimization problem

$$
\begin{aligned}
&\text{minimize} f(\mathbf{x}) \\
&\text{subject to } g_k(\mathbf{x}) \leq g_{k,0}, \quad k = 1, \dots, m,
\end{aligned}
\tag{2.1}
$$

where $g_{k,0}$ are some constants. The optimization problem (2.1) is a convex optimization problem. To verify this claim, we need to show that the set of points $\mathbf{x}$ satisfying $g_k(\mathbf{x}) \leq g_{k,0}$ for all k is a convex set. Take $\lambda \in [0, 1]$ and two points $\mathbf{x}$ and $\mathbf{y}$ such that $g_k(\mathbf{x}) \leq g_{k,0}$ and $g_k(\mathbf{y}) \leq g_{k,0}$. Then

$$g_k(\lambda \mathbf{x} + (1 - \lambda)\mathbf{y}) \leq \lambda g_k(\mathbf{x}) + (1 - \lambda)g_k(\mathbf{y}) \leq g_{k,0},$$

H. Hult et al., *Risk and Portfolio Analysis: Principles and Methods*, Springer Series in Operations Research and Financial Engineering, DOI 10.1007/978-1-4614-4103-8_2, © Springer Science+Business Media New York 2012

which shows that the set of $\mathbf{x}$ satisfying $g_k(\mathbf{x}) \leq g_{k,0}$ is convex. It remains to show that the intersection $\mathbb{C}_1 \cap \mathbb{C}_2$ of two convex sets $\mathbb{C}_1, \mathbb{C}_2$ is a convex set. Since the empty set is convex, we can without loss of generality consider the case where $\mathbb{C}_1 \cap \mathbb{C}_2$ is nonempty. Take $\lambda \in [0,1]$ and $\mathbf{x}, \mathbf{y} \in \mathbb{C}_1 \cap \mathbb{C}_2$. In particular, $\mathbf{x}, \mathbf{y} \in \mathbb{C}_k$ for $k = 1, 2$, and therefore also $\lambda \mathbf{x} + (1 - \lambda)\mathbf{y} \in \mathbb{C}_k$ for $k = 1, 2$, which means that $\lambda \mathbf{x} + (1 - \lambda)\mathbf{y} \in \mathbb{C}_1 \cap \mathbb{C}_2$. Summing up, we have shown that problem (2.1) is indeed a convex optimization problem.

The following proposition gives sufficient conditions for an optimal solution to (2.1). We write $\nabla f(\mathbf{x})$ for the gradient of f, i.e., the column vector whose kth component is $\partial f(\mathbf{x})/\partial x_k$.

Proposition 2.1. *Suppose that in* (2.1) *f and g_k are convex and differentiable. Suppose further that there exist $\mathbf{x} \in \mathbb{C}$ and $\lambda \in \mathbb{R}^m$ satisfying*

(1) $\nabla f(\mathbf{x}) + \sum_{k=1}^{m} \lambda_k \nabla g_k(\mathbf{x}) = \mathbf{0}$,

(2) $g_k(\mathbf{x}) \leq g_{k,0}$, $k = 1, \ldots, m$,

(3) $\lambda_k \geq 0$, $k = 1, \ldots, m$,

(4) $\lambda_k(g_k(\mathbf{x}) - g_{k,0}) = 0$, $k = 1, \ldots, m$.

Then $\mathbf{x}$ is an optimal solution to (2.1).

Proof. Define the function l, the Lagrangian, from $\mathbb{C}$ to $\mathbb{R}$ by

$$l(\mathbf{x}) = f(\mathbf{x}) + \sum_{k=1}^{m} \lambda_k(g_k(\mathbf{x}) - g_{k,0}).$$

It follows from condition (3) above and the assumptions on f and g_k that l is convex and differentiable. Condition (1) and Proposition 2.2 therefore imply that $\mathbf{x}$ is a global minimum point of l. Condition (2) says that $\mathbf{x}$ is a feasible solution to (2.1), i.e., it does not violate the constraints. We need to show that $f(\mathbf{x}) \leq f(\mathbf{y})$ for all feasible solutions $\mathbf{y}$. By condition (4), we have $f(\mathbf{x}) = l(\mathbf{x})$, and we have already seen that $l(\mathbf{x}) \leq l(\mathbf{y})$. Therefore,

$$f(\mathbf{x}) = l(\mathbf{x}) \leq l(\mathbf{y}) = f(\mathbf{y}) + \sum_{k=1}^{m} \lambda_k(g_k(\mathbf{y}) - g_{k,0}) \leq f(\mathbf{y}),$$

where the last inequality holds because of condition (3) and because $\mathbf{y}$ is a feasible solution to (2.1). $\square$

The following result is used in the proof of Proposition 2.1.

Proposition 2.2. *If $l : \mathbb{C} \to \mathbb{R}$ is a convex and differentiable function and $\mathbf{x}, \mathbf{y} \in \mathbb{C}$, then $l(\mathbf{y}) \geq l(\mathbf{x}) + \nabla l(\mathbf{x})^{\mathrm{T}}(\mathbf{y} - \mathbf{x})$.*

Proof. Take $t \in (0, 1]$ and consider the inequality

$$l(t\mathbf{y} + (1 - t)\mathbf{x}) \le tl(\mathbf{y}) + (1 - t)l(\mathbf{x}).$$

If we subtract $l(\mathbf{x})$ on both sides and divide by t then we arrive at

$$\frac{l(\mathbf{x} + t(\mathbf{y} - \mathbf{x})) - l(\mathbf{x})}{t} \le l(\mathbf{y}) - l(\mathbf{x}). \tag{2.2}$$

Set $\gamma(t) = l(\mathbf{x} + t(\mathbf{y} - \mathbf{x}))$ and note that by the chain rule it holds that $\gamma'(t) = \nabla l(\mathbf{x} + t(\mathbf{y} - \mathbf{x}))^{\mathrm{T}}(\mathbf{y} - \mathbf{x})$. In particular,

$$\nabla l(\mathbf{x})^{\mathrm{T}}(\mathbf{y} - \mathbf{x}) = \gamma'(0) = \lim_{t \to 0} \frac{\gamma(t) - \gamma(0)}{t} = \lim_{t \to 0} \frac{l(\mathbf{x} + t(\mathbf{y} - \mathbf{x})) - l(\mathbf{x})}{t}.$$

However, $l(\mathbf{x} + t(\mathbf{y} - \mathbf{x})) - l(\mathbf{x}) = l(t\mathbf{y} + (1 - t)\mathbf{x}) - l(\mathbf{x})$ and

$$\frac{l(t\mathbf{y} + (1 - t)\mathbf{x}) - l(\mathbf{x})}{t} \le \frac{tl(\mathbf{y}) + (1 - t)l(\mathbf{x}) - l(\mathbf{x})}{t} = l(\mathbf{y}) - l(\mathbf{x}),$$

from which the conclusion follows. $\square$

Finally, consider the problem of maximizing a concave function over a convex set. Let $h : \mathbb{C} \to \mathbb{R}$ be a concave function (i.e., $-h$ is convex), and consider the optimization problem

$$\begin{aligned} &\text{maximize } h(\mathbf{x}) \\ &\text{subject to } g_k(\mathbf{x}) \le g_{k,0}, \quad k = 1, \dots, m, \end{aligned} \tag{2.3}$$

where the functions g_k are convex. With $f = -h$ this is a convex optimization problem identical to (2.1). For further reference we state a minor modification of Proposition 2.1 for this case.

Proposition 2.3. *Suppose that in (2.3) h is concave and g_k are convex and that they are all differentiable. Suppose further that there exist $\mathbf{x} \in \mathbb{C}$ and $\boldsymbol{\lambda} \in \mathbb{R}^m$ satisfying*

(1) $-\nabla h(\mathbf{x}) + \sum_{k=1}^{m} \lambda_k \nabla g_k(\mathbf{x}) = \mathbf{0}$,
(2) $g_k(\mathbf{x}) \le g_{k,0}$, $k = 1, \dots, m$,
(3) $\lambda_k \ge 0$, $k = 1, \dots, m$,
(4) $\lambda_k(g_k(\mathbf{x}) - g_{k,0}) = 0$, $k = 1, \dots, m$.

Then $\mathbf{x}$ is an optimal solution to (2.3).

2.2 More General Convex Optimization

We will encounter more general convex optimization problems in which the functions f and the g_k in (2.1) are replaced by real-valued functions F and G_k that are defined on a set of functions rather than $\mathbb{R}^n$. We will state and prove the analog of Proposition 2.1 in this more abstract setting. Although proving the result in this setting might appear to be substantially harder, it turns out not to be the case.

For $k = 1, \ldots, m$ and $-\infty \leq a < b \leq \infty$ let $\phi, \phi_k : (a, b) \to \mathbb{R}$ be convex and differentiable and let $w, w_k : \mathbb{R}^n \to \mathbb{R}_+$ be nonnegative functions with $\int_{\mathbb{R}^n} w(\mathbf{x}) d\mathbf{x} = \int_{\mathbb{R}^n} w_k(\mathbf{x}) d\mathbf{x} = 1$. For a function $f : \mathbb{R}^n \to (a, b)$ we define

$$F(f) = \int_{\mathbb{R}^n} \phi(f(\mathbf{x})) w(\mathbf{x}) d\mathbf{x},$$

$$G_k(f) = \int_{\mathbb{R}^n} \phi_k(f(\mathbf{x})) w_k(\mathbf{x}) d\mathbf{x} \quad \text{for } k = 1, \ldots, m.$$

Finally, let $\mathbb{F}$ be a set of functions f for which $F(f) < \infty$. Note that F and G_k are convex since ϕ and ϕ_k are convex. Indeed, for $\lambda \in [0, 1]$ and $f, g \in \mathbb{F}$,

$$F(\lambda f + (1 - \lambda) g) = \int_{\mathbb{R}^n} \phi(\lambda f(\mathbf{x}) + (1 - \lambda) g(\mathbf{x})) w(\mathbf{x}) d\mathbf{x}$$

$$\leq \int_{\mathbb{R}^n} [\lambda \phi(f(\mathbf{x})) + (1 - \lambda) \phi(g(\mathbf{x}))] w(\mathbf{x}) d\mathbf{x}$$

$$= \lambda F(f) + (1 - \lambda) F(g),$$

and similarly for G_k. We will consider minimizing F over a convex subset of $\mathbb{F}$ that is determined by a set of inequality constraints just as in (2.1). Consider the optimization problem

$$\begin{aligned} &\text{minimize } F(f) \\ &\text{subject to } G_k(f) \leq G_{k,0}, \quad k = 1, \ldots, m. \end{aligned} \tag{2.4}$$

As in the proof of Proposition 2.1, we introduce the Lagrangian

$$L(f) = F(f) + \sum_{k=1}^{m} \lambda_k (G_k(f) - G_{k,0})$$

$$= \int_{\mathbb{R}^n} \left(\phi(f(\mathbf{x})) w(\mathbf{x}) + \sum_{k=1}^{m} \lambda_k (\phi_k(f(\mathbf{x})) - G_{k,0}) w_k(\mathbf{x}) \right) d\mathbf{x}. \tag{2.5}$$

Before we can state the analog of Proposition 2.1 in the current setting, we need an object that plays the role of $\nabla l(\mathbf{x})^{\mathrm{T}} (\mathbf{y} - \mathbf{x})$ in Proposition 2.2. For $f, g \in \mathbb{F}$ let

$$H(f, g) = \int_{\mathbb{R}^n} \left[\phi'(f(\mathbf{x}))w(\mathbf{x}) + \sum_{k=1}^{m} \lambda_k \phi_k'(f(\mathbf{x}))w_k(\mathbf{x}) \right] (g(\mathbf{x}) - f(\mathbf{x}))d\mathbf{x}.$$

Now we are ready to state the analog of Proposition 2.1. The following proposition gives sufficient conditions for an optimal solution to (2.4).

Proposition 2.4. *Suppose that there exist* $f \in \mathbb{F}$ *and* $\lambda \in \mathbb{R}^m$ *satisfying*

$$
\begin{aligned}
&(1) \ H(f, g) = 0, &&\textit{for all } g \in \mathbb{F}, \\
&(2) \ G_k(f) \leq G_{k,0}, &&k = 1, \dots, m, \\
&(3) \ \lambda_k \geq 0, &&k = 1, \dots, m, \\
&(4) \ \lambda_k(G_k(f) - G_{k,0}) = 0, &&k = 1, \dots, m.
\end{aligned}
$$

Then f *is an optimal solution to* (2.4).

Proof. First note that since F and G_k are convex, (3) implies that also L is convex. By Proposition 2.5, it holds that

$$L(g) \geq L(f) + H(f, g) \quad \text{for all } g \in \mathbb{F}.$$

Therefore, it follows from (1) that f is a global minimum point of L. Condition (2) says that f is a feasible solution to (2.4), i.e., it does not violate the constraints. We need to show that $F(f) \leq F(g)$ for all feasible solutions g. By condition (4), we have $F(f) = L(f)$, and we have already seen that $L(f) \leq L(g)$. Therefore,

$$F(f) = L(f) \leq L(g) = F(g) + \sum_{k=1}^{m} \lambda_k(G_k(g) - G_{k,0}) \leq F(g),$$

where the last inequality holds because of condition (3) and because g is a feasible solution to (2.4). □

Proposition 2.5. *Consider the function L in* (2.5), *where F and G_k are convex and λ_k is nonnegative. Then L has the property that for each $f, g \in \mathbb{F}$ it holds that* $L(g) \geq L(f) + H(f, g)$.

Proof. The assumptions on F, G_k, and λ_k imply that L is convex. The convexity of L implies that, by repeating the argument in the proof of Proposition 2.2,

$$\frac{L(f + t(g - f)) - L(f)}{t} \leq L(g) - L(f)$$

for $t \in (0, 1]$. In particular,

$$L(g) \geq L(f) + \lim_{t \to 0} \frac{L(f + t(g - f)) - L(f)}{t}.$$

It remains to show that the limit on the right-hand side is $H(f, g)$. Indeed,

$$\lim_{t \to 0} \frac{L(f + t(g - f)) - L(f)}{t}$$

$$= \lim_{t \to 0} \frac{F(f + t(g - f)) - F(f)}{t} + \sum_{k=1}^{m} \lambda_k \lim_{t \to 0} \frac{G_k(f + t(g - f)) - G_k(f)}{t}$$

$$= \lim_{t \to 0} \int_{\mathbb{R}^n} \frac{\phi(f(\mathbf{x}) + t(g(\mathbf{x}) - f(\mathbf{x}))) - \phi(f(\mathbf{x}))}{t} w(\mathbf{x}) d\mathbf{x}$$

$$+ \lim_{t \to 0} \sum_{k=1}^{m} \lambda_k \int_{\mathbb{R}^n} \frac{\phi_k(f(\mathbf{x}) + t(g(\mathbf{x}) - f(\mathbf{x}))) - \phi_k(f(\mathbf{x}))}{t} w_k(\mathbf{x}) d\mathbf{x}$$

$$= \int_{\mathbb{R}^n} \lim_{t \to 0} \frac{\phi(f(\mathbf{x}) + t(g(\mathbf{x}) - f(\mathbf{x}))) - \phi(f(\mathbf{x}))}{t} w(\mathbf{x}) d\mathbf{x}$$

$$+ \sum_{k=1}^{m} \lambda_k \int_{\mathbb{R}^n} \lim_{t \to 0} \frac{\phi_k(f(\mathbf{x}) + t(g(\mathbf{x}) - f(\mathbf{x}))) - \phi_k(f(\mathbf{x}))}{t} w_k(\mathbf{x}) d\mathbf{x}$$

$$= \int_{\mathbb{R}^n} \left(\phi'(f(\mathbf{x})) w(\mathbf{x}) + \sum_{k=1}^{m} \lambda_k \phi_k'(f(\mathbf{x})) w_k(\mathbf{x}) \right) (f(\mathbf{x}) - g(\mathbf{x})) d\mathbf{x}$$

$$= H(f, g). \qquad \qquad \square$$

2.3   Notes and Comments

The material presented in Sect. 2.1 can be found in most textbooks on optimization, for instance, [7] by Stephen Boyd and Lieven Vandenberghe and [26] by David Luenberger.

Chapter 3
Quadratic Hedging Principles

Fix a future time T and let L be the value of a liability at that time. One example of L is the portfolio value of derivative instruments issued by a bank. Another example is the value of future claims from insurance products sold by an insurance company. Typically the holder of the liability does not want to speculate on a favorable outcome of this random variable. The ideal approach to managing the risk of an unfavorable outcome of L would be to purchase a portfolio whose value A (A for assets) at the future time T exactly matches that of the liability. In that case, $A = L$, and the risk of an unfavorable outcome of L is removed completely by purchasing the asset portfolio. The problem with this approach is that it is not always possible to find a portfolio of assets whose future value corresponds exactly to that of the liability; one example is when the liability is made up of insurance claims.

The more realistic approach is to look for an asset portfolio whose value A at time T is close to L. For instance, we could search for the portfolio value A that minimizes the expected squared hedging error $\mathrm{E}[(A - L)^2]$. This approach to hedging is called quadratic hedging.

So how do we determine the optimal quadratic hedging portfolio? Let $\mathbf{Z} = (Z_1, \ldots, Z_n)^{\mathrm{T}}$ be a vector whose components are the values at time T of assets that are available on the market. Suppose that we may take A to be any random variable of the form $f(\mathbf{Z})$ for arbitrary functions f. Then we can interpret A as the value at time T of any derivative instrument on $\mathbf{Z}$, and the choice of A that solves the quadratic hedging problem turns out to be the conditional expectation of L given $\mathbf{Z}$. The conditional expectation is presented in Sect. 3.1.

It is quite likely that the derivative instrument corresponding to the optimal quadratic hedge is unavailable. It could be that we simply cannot find a seller of that derivative, that the seller is offering this derivative at a price that we consider too high, or that we fear that the seller may not be able to deliver the derivative payoff. In this case, it is reasonable to restrict the hedging portfolio payoff A to be of the form $h_0 + \mathbf{h}^{\mathrm{T}}\mathbf{Z}$, corresponding to a hedging portfolio made up of a position in a risk-free bond and a combination of positions in assets whose values at time T are the components of $\mathbf{Z}$. In this case, the choice of A that solves the optimal quadratic

H. Hult et al., *Risk and Portfolio Analysis: Principles and Methods*, Springer Series in Operations Research and Financial Engineering, DOI 10.1007/978-1-4614-4103-8_3, © Springer Science+Business Media New York 2012

hedging problem is the value at time T of the portfolio obtained by taking positions corresponding to the regression coefficients of L onto the regressors $Z_1, \ldots, Z_n$. The solution to the quadratic hedging problem then corresponds to solving a linear regression problem. Linear regression is presented in Sect. 3.1.

Equipped with this machinery, we then move on to investigate a number of examples of quadratic hedging. Examples include hedging with futures contracts, hedging of insurance liabilities, hedging of a digital option with call options, and delta hedging. The chapter ends with an extensive section on immunization of cash flows. Immunization is a useful technique to make the value of a portfolio insensitive to changes in the zero-rate curve. It is placed at the end of this chapter, although it is based on first-order Taylor approximations rather than quadratic hedging.

3.1   Conditional Expectations and Linear Regression

Consider a random vector $\mathbf{Z}$ and a random variable L with $E[L^2] < \infty$. The conditional expectation of L given $\mathbf{Z}$, written $E[L \mid \mathbf{Z}]$, is the random variable $g(\mathbf{Z})$, with g a function from $\mathbb{R}^n$ to $\mathbb{R}$ such that $E[g(\mathbf{Z})^2] < \infty$, satisfying the condition

$$E[h(\mathbf{Z})g(\mathbf{Z})] = E[h(\mathbf{Z})L] \qquad (3.1)$$

for all functions h from $\mathbb{R}^n$ to $\mathbb{R}$ such that $E[h(\mathbf{Z})^2] < \infty$. If we take $h(\mathbf{Z}) = 1$ in (3.1), then we find that

$$E[E[L \mid \mathbf{Z}]] = E[L].$$

This relation is called the law of iterated expectations. Furthermore, if we replace L by $I\{L \in B\}$, then we get

$$P(L \in B) = E[I\{L \in B\}] = E[E[I\{L \in B\} \mid \mathbf{Z}]] = E[P(L \in B \mid \mathbf{Z})].$$

We denote by $\mathbb{L}^2(\mathbf{Z})$ the set of random variables $h(\mathbf{Z})$, where h is a function from $\mathbb{R}^n$ to $\mathbb{R}$, satisfying $E[h(\mathbf{Z})^2] < \infty$. Take $Y_1, Y_2 \in \mathbb{L}^2(\mathbf{Z})$. By definition of the conditional expectations $E[L \mid \mathbf{Z}]$ and $E[Y_1 L \mid \mathbf{Z}]$, it holds that

$$E[Y_2 Y_1 E[L \mid \mathbf{Z}]] = E[Y_2 Y_1 L] = E[Y_2 E[Y_1 L \mid \mathbf{Z}]].$$

Since this holds for all $Y_2 \in \mathbb{L}^2(\mathbf{Z})$, we conclude that $E[Y_1 L \mid \mathbf{Z}] = Y_1 E[L \mid \mathbf{Z}]$, i.e., $Y_1 \in \mathbb{L}^2(\mathbf{Z})$ is regarded as a constant when conditioning on $\mathbf{Z}$.

Proposition 3.1. *Consider a random vector $\mathbf{Z}$ and a random variable L with $E[L^2] < \infty$. Suppose that f is any function such that $E[f(\mathbf{Z})^2] < \infty$. Then it holds that*

$$\mathrm{E}\left[(L - f(\mathbf{Z}))^2\right] \geq \mathrm{E}[(L - \mathrm{E}[L \mid \mathbf{Z}])^2],$$

with equality for $f(\mathbf{Z}) = \mathrm{E}[L \mid \mathbf{Z}]$.

The proof is given in Sect. 3.1.2.

We know from Proposition 3.1 that among all the random variables of the form $f(\mathbf{Z})$, the conditional expectation $\mathrm{E}[L \mid \mathbf{Z}]$ is the one that best approximates the random variable L in the sense that the expected value of the squared difference between the two is minimized.

A commonly encountered situation is where we have access only to random variables $f(\mathbf{Z})$ of the form $h_0 + \mathbf{h}^\mathrm{T}\mathbf{Z}$, i.e., translations of linear combinations of the components of $\mathbf{Z}$. In this case we look for constants $h_0, h_1, \ldots, h_n$ that minimize the expected value $\mathrm{E}[(h_0 + \mathbf{h}^\mathrm{T}\mathbf{Z} - L)^2]$. Clearly, the expected value must exist finitely in order for this minimization to make sense, and therefore we assume that the covariance matrix of the vector $\mathbf{Z}$,

$$\Sigma_{\mathbf{Z}} = \mathrm{E}[(\mathbf{Z} - \mathrm{E}[\mathbf{Z}])(\mathbf{Z} - \mathrm{E}[\mathbf{Z}])^\mathrm{T}] = \begin{pmatrix} \mathrm{Cov}(Z_1, Z_1) & \ldots & \mathrm{Cov}(Z_1, Z_n) \\ \vdots & \ddots & \vdots \\ \mathrm{Cov}(Z_n, Z_1) & \ldots & \mathrm{Cov}(Z_n, Z_n) \end{pmatrix},$$

and the vector of covariances between L and $Z_1, \ldots, Z_n$,

$$\Sigma_{L,\mathbf{Z}} = \mathrm{E}[(L - \mathrm{E}[L])(\mathbf{Z} - \mathrm{E}[\mathbf{Z}])] = \begin{pmatrix} \mathrm{Cov}(L, Z_1) \\ \vdots \\ \mathrm{Cov}(L, Z_n) \end{pmatrix},$$

exist finitely. The solution to the minimization problem is simply a standard linear regression of L onto the regressors $Z_1, \ldots, Z_n$.

Proposition 3.2. *For a random vector $\mathbf{Z}$ with an invertible covariance matrix $\Sigma_{\mathbf{Z}}$ and a random variable L with finite variance, the expected value $\mathrm{E}[(h_0 + \mathbf{h}^\mathrm{T}\mathbf{Z} - L)^2]$ attains its unique minimum for $(h_0, \mathbf{h})$ given by*

$$\mathbf{h} = \Sigma_{\mathbf{Z}}^{-1}\Sigma_{L,\mathbf{Z}},$$

$$h_0 = \mathrm{E}[L] - \mathbf{h}^\mathrm{T}\mathrm{E}[\mathbf{Z}].$$

For this choice of $\mathbf{h}$ it holds that $\mathrm{Cov}(h_0 + \mathbf{h}^\mathrm{T}\mathbf{Z} - L, Z_k) = 0$ *for all k.*

The proof is given in Sect. 3.1.2.

Note that Proposition 3.2 says that the minimizer $\hat{A}$ of $\mathrm{E}[(A - L)^2]$ among all random variables of the form $A = h_0 + \mathbf{h}^\mathrm{T}\mathbf{Z}$ is given by

$$\hat{A} = \mathrm{E}[L] + \Sigma_{L,\mathbf{Z}}^\mathrm{T}\Sigma_{\mathbf{Z}}^{-1}(\mathbf{Z} - \mathrm{E}[\mathbf{Z}]).$$

We see that $E[\hat{A}] = E[L]$, and therefore the minimal expected squared hedging error is

$$E[(\hat{A} - L)^2] = \text{Var}(\hat{A} - L)$$
$$= \text{Var}(\hat{A}) + \text{Var}(L) - 2\,\text{Cov}(\hat{A}, L)$$
$$= (\Sigma_{L,\mathbf{Z}}^T \Sigma_{\mathbf{Z}}^{-1}) \Sigma_{\mathbf{Z}} (\Sigma_{L,\mathbf{Z}}^T \Sigma_{\mathbf{Z}}^{-1})^T + \text{Var}(L) - 2\Sigma_{L,\mathbf{Z}}^T \Sigma_{\mathbf{Z}}^{-1} \Sigma_{L,\mathbf{Z}}$$
$$= \text{Var}(L) - \Sigma_{L,\mathbf{Z}}^T \Sigma_{\mathbf{Z}}^{-1} \Sigma_{L,\mathbf{Z}}.$$

Moreover, the last claim in Proposition 3.2 says that the optimal hedging error $\hat{A} - L$ is uncorrelated with the the hedging instruments.

Not surprisingly, the worst possible situation is where $\text{Cov}(L, Z_k) = 0$ for all k. That is, the hedging instruments are uncorrelated with the liability. In that case, there is no opportunity for quadratic hedging. The optimal quadratic hedge is $(h_0, \mathbf{h}) = (E[L], \mathbf{0})$, with $E[(\hat{A} - L)^2] = \text{Var}(L)$.

The next proposition provides the solution to the similar problem of finding the minimizer $\hat{A}$ of $\text{Var}(A - L)$ among all random variables of the form $A = \mathbf{h}^T \mathbf{Z}$.

Proposition 3.3. *For a random vector $\mathbf{Z}$ with an invertible covariance matrix $\Sigma_{\mathbf{Z}}$ and a random variable L with finite variance, the variance $\text{Var}(\mathbf{h}^T \mathbf{Z} - L)$ attains its unique minimum for $\mathbf{h} = \Sigma_{\mathbf{Z}}^{-1} \Sigma_{L,\mathbf{Z}}$.*

The proof is given in Sect. 3.1.2.

The difference between Propositions 3.2 and 3.3 is that in the latter there is no constant term h_0, and we minimize the variance instead of the second-order moment of the hedging error. The optimal positions $h_1, \ldots, h_n$ in the hedging instruments $Z_1, \ldots, Z_n$ are identical in the two situations. The effect of the constant term, h_0, in Proposition 3.2 is simply that it centers the distribution of the hedging error at zero to eliminate a systematic hedging error.

Proposition 3.3 says that the minimizer $\hat{A}$ of $\text{Var}(A - L)$ among all random variables of the form $A = \mathbf{h}^T \mathbf{Z}$ is given by $\hat{A} = \Sigma_{L,\mathbf{Z}}^T \Sigma_{\mathbf{Z}}^{-1} \mathbf{Z}$. The minimal variance is given by

$$\text{Var}(\hat{A} - L) = \text{Var}(\hat{A}) + \text{Var}(L) - 2\,\text{Cov}(\hat{A}, L)$$
$$= \text{Var}(L) - \Sigma_{L,\mathbf{Z}}^T \Sigma_{\mathbf{Z}}^{-1} \Sigma_{L,\mathbf{Z}}.$$

Finally, a brief remark on the important special case where $n = 1$, corresponding to only one hedging instrument. In this case, the optimal hedge is given by

$$h = \frac{\text{Cov}(L, Z)}{\text{Var}(Z)}$$

and the minimal variance is

$$\text{Var}(\hat{A} - L) = \text{Var}(L) - \frac{\text{Cov}(L, Z)^2}{\text{Var}(Z)}$$

$$= \text{Var}(L)\left(1 - \frac{\text{Cov}(L, Z)^2}{\text{Var}(L)\,\text{Var}(Z)}\right)$$

$$= \text{Var}(L)(1 - \text{Cor}(L, Z)^2). \tag{3.2}$$

We see that the optimal position h corresponds to the regression coefficient of L onto Z, and the expression for the variance of the hedged position is very explicit and easily interpreted.

3.1.1 Examples

Fix a future time T and let L be the value of a liability at time T. Proposition 3.1 says that the optimal quadratic hedge of liability L is the conditional expectation of L given $\mathbf{Z}$. The vector $\mathbf{Z}$ need not be a vector of future values of traded instruments. However, we typically want $\text{E}[L \mid \mathbf{Z}]$ to be the future value of an instrument or portfolio that we can buy to realize the quadratic hedge. The following example presents an application of Proposition 3.1.

Example 3.1 (Unit linked life insurance). Consider an insurance company that has sold unit-linked life insurance contracts. If the insured dies during the current year, then the contract pays $\max(S_T, K)$ at time T (the end of the year). Here S_T is the value of an index and K is a guaranteed amount. Let the random number N of insured that die during the next year be $\text{Po}(\lambda)$-distributed (Poisson distributed with mean λ). We assume N is independent of S_T. The value of the liability at time T is therefore $L = N \max(S_T, K)$. Suppose that the insurance company can invest in the index and in any derivative on S_T with payoff $f(S_T)$. Then a good hedge of the liability is obtained by taking

$$f(S_T) = \text{E}[N \max(S_T, K) \mid S_T] = \text{E}[N \mid S_T] \max(S_T, K)$$

$$= \text{E}[N] \max(S_T, K) = \lambda \max(S_T, K)$$

$$= \lambda(\max(S_T - K, 0) + K).$$

We conclude that the insurance company should buy λ call options with strike K and take a position in zero-coupon bonds that pay the amount λK at time T. The hedging error is given by $(\lambda - N) \max(S_T, K)$, and its second-order moment is given by

$$\text{E}[(\lambda - N)^2 \max(S_T, K)^2] = \text{E}[(N - \lambda)^2]\,\text{E}[\max(S_T, K)^2]$$

$$= \text{Var}(N)\,\text{E}[\max(S_T, K)^2]$$

$$= \lambda\,\text{E}[\max(S_T, K)^2].$$

This example illustrates the use of Proposition 3.1, here with Z replaced by S_T.

The next example presents an application of Proposition 3.2. We will encounter many more applications further on.

Example 3.2 (Whole life insurance). Consider a life insurance policy that pays 1 at the random time τ (at the end of the year of death of the policyholder). Suppose the current time is 0 and, for simplicity, that τ can only take the values 1 or 2. Suppose also that there are 1- and 2-year risk-free zero-coupon bonds available on the market.

The value at time 1 of the insurance policy is $L = I\{\tau = 1\} + I\{\tau = 2\}e^{-r_{1,2}}$, where $r_{1,2}$ is the (random) 1-year zero-coupon rate at time 1. Notice that $r_{1,2}$ is observed at time 1. If we buy h_0 units of the 1-year zero-coupon bond, and h_1 units of the 2-year zero-coupon bond, then the value of our assets at time 1 is $A = h_0 + h_1 e^{-r_{1,2}}$. We seek the best possible asset portfolio in the sense that the time 1 value of the assets should match the time 1 value of the liability as well as possible.

Write $Z = e^{-r_{1,2}}$ and note that by Proposition 3.2 there exists a unique minimizer $\hat{A}$ of $E[(A-L)^2]$ among all random variables A of the form $h_0 + h_1 Z$ for coefficients $h_0, h_1 \in \mathbb{R}$. The hedging portfolio h_0, h_1 is given by

$$ h_1 = \frac{\text{Cov}(L,Z)}{\text{Var}(Z)} \quad \text{and} \quad h_0 = E[L] - h_1 E[Z]. $$

Write $p = P(\tau = 1)$ and note that

$$ \text{Cov}(L,Z) = E[(I\{\tau = 1\} + I\{\tau = 2\}Z)Z] - E[I\{\tau = 1\} + I\{\tau = 2\}Z]E[Z] $$
$$ = p\,E[Z] + (1-p)\,E[Z^2] - p\,E[Z] - (1-p)\,E[Z]^2 $$
$$ = (1-p)(E[Z^2] - (E[Z])^2) $$
$$ = (1-p)\,\text{Var}(Z). $$

We see that $h_1 = 1 - p$, $h_0 = p$. In other words, we should buy $P(\tau = 1)$ 1-year zero-coupon bonds and $P(\tau = 2)$ 2-year zero-coupon bonds. It is not difficult to verify that if τ can take any integer value from 1 to n and if zero-coupon bonds with these maturities are available, then the solution to the problem is to buy $P(\tau = k)$ number of k-year zero-coupon bonds.

3.1.2 Proofs of Propositions

Proof of Proposition 3.1. Take $W \in \mathbb{L}^2(\mathbf{Z})$. For such W it follows from (3.1) that $E[W\,E[L \mid \mathbf{Z}]] = E[WL]$. The linearity of the expected value now gives

$$ E[W(L - E[L \mid \mathbf{Z}])] = 0 \quad \text{for all } W \in \mathbb{L}^2(\mathbf{Z}). $$

Take $Y \in \mathbb{L}^2(\mathbf{Z})$ and set $W = Y - \mathrm{E}[L \mid \mathbf{Z}]$. Then

$$\mathrm{E}\left[(L - Y)^2\right] = \mathrm{E}\left[(L - \mathrm{E}[L \mid \mathbf{Z}] - W)^2\right]$$

$$= \mathrm{E}\left[(L - \mathrm{E}[L \mid \mathbf{Z}])^2\right] - 2\,\mathrm{E}\left[W\,(L - \mathrm{E}[L \mid \mathbf{Z}])\right] + \mathrm{E}\left[W^2\right]$$

$$= \mathrm{E}\left[(L - \mathrm{E}[L \mid \mathbf{Z}])^2\right] + \mathrm{E}[W^2],$$

and it is clear that $\mathrm{E}[(L - Y)^2]$ is minimized when $W = 0$, i.e., when $Y = \mathrm{E}[L \mid \mathbf{Z}]$. This concludes the proof. $\qquad\square$

Lemma 3.1. *Let $(\mathbf{A}^{\mathrm{T}}, B)^{\mathrm{T}}$ be a random vector in $\mathbb{R}^{m+1}$, where $\mathbf{A}$ is m-dimensional and B is one-dimensional. Let g be a (strictly) convex function from $\mathbb{R}$ to $\mathbb{R}$ such that the function h from $\mathbb{R}^m$ to $\mathbb{R}$ given by $h(\mathbf{x}) = \mathrm{E}[g(\mathbf{A}^{\mathrm{T}}\mathbf{x} + B)]$ exists finitely. Then h is (strictly) convex.*

Proof. For each $\mathbf{a} \in \mathbb{R}^m$ and $b \in \mathbb{R}$ the function $h_{(\mathbf{a},b)}$ given by $h_{(\mathbf{a},b)}(\mathbf{x}) = g(\mathbf{a}^{\mathrm{T}}\mathbf{x}+b)$ is (strictly) convex. Indeed, for $\lambda \in (0, 1)$,

$$h_{(\mathbf{a},b)}\left(\lambda\mathbf{x} + (1 - \lambda)\mathbf{y}\right) = g\left(\lambda\left(\mathbf{a}^{\mathrm{T}}\mathbf{x} + b\right) + (1 - \lambda)\left(\mathbf{a}^{\mathrm{T}}\mathbf{y} + b\right)\right)$$

$$\leq \lambda g\left(\mathbf{a}^{\mathrm{T}}\mathbf{x} + b\right) + (1 - \lambda)g\left(\mathbf{a}^{\mathrm{T}}\mathbf{y} + b\right)$$

$$= \lambda h_{(\mathbf{a},b)}(\mathbf{x}) + (1 - \lambda)h_{(\mathbf{a},b)}(\mathbf{y}),$$

where the inequality is strict if g is strictly convex. Therefore, it follows from the law of iterated expectations that

$$h\left(\lambda x + (1 - \lambda)y\right) = \mathrm{E}\left[\mathrm{E}[h_{(\mathbf{A},B)}(\lambda\mathbf{x} + (1 - \lambda)\mathbf{y}) \mid (\mathbf{A}, B)]\right]$$

$$\leq \mathrm{E}\left[\mathrm{E}[\lambda h_{(\mathbf{A},B)}(\mathbf{x}) + (1 - \lambda)h_{(\mathbf{A},B)}(\mathbf{y}) \mid (\mathbf{A}, B)]\right]$$

$$= \lambda h(\mathbf{x}) + (1 - \lambda)h(\mathbf{y}),$$

where again the inequality is strict if g is strictly convex. $\qquad\square$

We proceed with the proofs of Propositions 3.2 and 3.3. It is somewhat easier to prove Proposition 3.3 first. Before the proof we note that the variance of a weighted sum of the components of $\mathbf{Z}$ can be expressed as a pure quadratic form in the weights and the covariance matrix of $\mathbf{Z}$. More precisely,

$$\mathrm{Var}(\mathbf{h}^{\mathrm{T}}\mathbf{Z}) = \sum_{j=1}^{n}\sum_{k=1}^{n}\mathrm{Cov}(h_j Z_j, h_k Z_k)$$

$$= \sum_{j=1}^{n}\sum_{k=1}^{n}h_j h_k (\Sigma_{\mathbf{Z}})_{j,k} = \sum_{j=1}^{n}h_j \sum_{k=1}^{n}(\Sigma_{\mathbf{Z}})_{j,k}h_k$$

$$= (h_1 \ldots h_n) \begin{pmatrix} (\Sigma_\mathbf{Z})_{1,1} h_1 + \cdots + (\Sigma_\mathbf{Z})_{1,n} h_n \\ \cdots \\ (\Sigma_\mathbf{Z})_{n,1} h_1 + \cdots + (\Sigma_\mathbf{Z})_{n,n} h_n \end{pmatrix}$$

$$= \mathbf{h}^\mathsf{T} \Sigma_\mathbf{Z} \mathbf{h}.$$

Proof of Proposition 3.3. For an arbitrary portfolio $\mathbf{h}$ the variance of the hedging error is

$$\mathrm{Var}(\mathbf{h}^\mathsf{T}\mathbf{Z} - L) = \mathrm{Var}(\mathbf{h}^\mathsf{T}\mathbf{Z}) + \mathrm{Var}(L) - 2\,\mathrm{Cov}(\mathbf{h}^\mathsf{T}\mathbf{Z}, L)$$

$$= \mathbf{h}^\mathsf{T} \Sigma_\mathbf{Z} \mathbf{h} + \mathrm{Var}(L) - 2\mathbf{h}^\mathsf{T} \Sigma_{L,\mathbf{Z}}.$$

From Lemma 3.1, with $g(y) = y^2$, $\mathbf{x} = \mathbf{h}$, $\mathbf{A} = \mathbf{Z}$, and $B = -L$, we know that $\mathrm{Var}(\mathbf{h}^\mathsf{T}\mathbf{Z} - L)$ is a strictly convex function of $\mathbf{h}$. Since it is differentiable, we obtain its unique minimum by computing partial derivatives, setting the linear expressions in $\mathbf{h}$ to zero, and solving the linear equation. The equation $\nabla \mathrm{Var}(\mathbf{h}^\mathsf{T}\mathbf{Z} - L) = \mathbf{0}$, where the differentiation is with respect to $\mathbf{h}$, reads

$$\Sigma_\mathbf{Z} \mathbf{h} - \Sigma_{L,\mathbf{Z}} = \mathbf{0},$$

from which the conclusion follows.   □

Proof of Proposition 3.2. For an arbitrary portfolio $(h_0, \mathbf{h}^\mathsf{T})^\mathsf{T}$ the second moment of the hedging error is given by

$$\mathrm{E}[(h_0 + \mathbf{h}^\mathsf{T}\mathbf{Z} - L)^2] = \mathrm{Var}(h_0 + \mathbf{h}^\mathsf{T}\mathbf{Z} - L) + (\mathrm{E}[h_0 + \mathbf{h}^\mathsf{T}\mathbf{Z} - L])^2$$

$$= \mathrm{Var}(\mathbf{h}^\mathsf{T}\mathbf{Z} - L) + (\mathrm{E}[h_0 + \mathbf{h}^\mathsf{T}\mathbf{Z} - L])^2.$$

By Proposition 3.3 the variance term is minimized by

$$\mathbf{h} = \Sigma_\mathbf{Z}^{-1} \Sigma_{L,\mathbf{Z}}.$$

The second term is always nonnegative and equal to zero for

$$h_0 = \mathrm{E}[L - \mathbf{h}^\mathsf{T}\mathbf{Z}].$$

This completes the proof.   □

3.2   Hedging with Futures

Consider a coffee producer who will sell the random quantity X pounds of coffee beans in 10 months from now, time T. The outcome of X depends on the size of the coffee crop, which in turn depends on the weather during the next 10 months before harvesting. The spot price S_T in cents per pound of coffee beans in 10 months from now at which the producer can sell the coffee beans depends on the world market price of coffee beans at that time but also on the possible difference in demand for the producer's kind of coffee bean compared to the overall demand for coffee beans. Anyway, the producer will receive the random income XS_T if the beans are sold to the buyers on the market 10 months from now. The variability in the spot price of coffee beans over time represents a financial risk that the coffee producer would like to hedge away.

One way to hedge this risk would be to find a counterpart, a buyer of coffee beans, and write a forward contract for delivery of a certain quantity X_0 of coffee beans 10 months from now. This would give the cash flow $X_0 G_0 + (X - X_0) S_T$ for the producer at the delivery date, where G_0 is the agreed-upon forward price and $(X - X_0) S_T$ is the income from selling the remaining part of the coffee crop on the spot market (or the cost of buying coffee beans on the spot market in order to be able to deliver the promised quantity). Note that if $X = X_0$, meaning that the size of the crop is known in advance, then the forward contract transforms the random future income XS_T into the known income $X_0 G_0$.

However, hedging with forward contracts has its disadvantages. There is always the risk that one of the two parties does not fulfill its part of the agreement. Moreover, if the producer after some time would like to change the agreed-upon quantity X_0 or simply cancel the forward contract, then this is typically difficult since there may be no general agreement on the value of the forward contract prior to maturity.

Fortunately for the coffee producer there is a well-functioning market for futures contracts on coffee beans. A futures contract is similar to a forward contract: it is an agreement to buy/sell a certain quantity of some good at a predetermined future date for a specific price to be paid/received on that date. Many futures contracts are liquidly traded on an exchange. Due to changes in supply and demand, the exchange announces a new futures price each day for each futures contract. In order not to be exposed to credit risk, that the losing party walks away from a loss, the exchange requires the long holder of a futures contract to deposit money on a so-called margin account in response to a price drop, and similarly for the short holder. This means that the futures price is paid/received bit by bit through a daily settlement procedure, unlike for the forward contract, where the forward price is paid/received in full at maturity of the contract. A nice aspect of futures positions is that they can be closed at any time by the long and short holders, resulting in a profit or a loss depending on the cash flow from the daily settlement procedure. The transactions between the long and short holders of a future contract are guaranteed by the exchange, and a futures position closed out by one party does not affect the other party.

Let us give a more precise description of a futures contract. We consider a futures contract for delivery at time T of a certain quantity of some good (coffee beans, say) and let S_T denote the spot price at time T per unit of the good to be delivered. Some futures contracts are settled in cash, and most futures contracts are closed out prior to delivery because many futures traders do not want delivery of the specified good. Although it is an important practical issue here, we do not differentiate between holding a futures contract to maturity and closing it out just before maturity. To make the presentation clearer, we consider here a futures contract for delivery of one single unit of the underlying good (one pound of coffee beans, say). At each resettlement day t the futures price is F_t and the long holder receives the amount $F_t - F_{t-1}$. A negative value of $F_t - F_{t-1}$ means that the long holder pays the amount. The futures price F_t is determined so that the price for entering the futures contract at any time t is zero. In particular, $F_T = S_T$. If the futures contract is held from time 0 until T, then all cash flows before maturity can be postponed until maturity, and interest rate effects can be ignored; then the aggregated cash flow for the long holder is

$$(F_1 - F_0) + (F_2 - F_1) + \cdots + (F_T - F_{T-1}) = F_T - F_0 = S_T - F_0,$$

i.e., the long holder receives S_T and pays F_0, so the futures contract is approximated by a forward contract. However, the very existence of well-functioning markets with exchange-traded futures contracts relies on the inability of the losing party in a futures contract to postpone the settling of daily losses.

It is sometimes possible to avoid having to deal with payments at different times by adopting an appropriate futures strategy. The market participant must have access to a money market account and be able to take both long positions (make deposits) and short positions (receive loans) in the money market account. We denote by $r_{t-1,t}$ the interest rate from time $t - 1$ to time t that is contracted at time $t - 1$ and applies to both deposits and loans.

Consider the following strategy, which represents a combination of dynamic strategies in a futures contract and a money market account.

- At time $t = 0$, the amount F_0 is deposited in the money market account and a long position of $e^{r_{0,1}}$ futures contracts is taken.
- At time $t = 1$, the next resettlement time, the portfolio value is

$$e^{r_{0,1}} F_0 + e^{r_{0,1}}(F_1 - F_0) = e^{r_{0,1}} F_1,$$

where the first term is the value of the money market account and the second term is the income (possibly negative) from the position in the futures contracts. At this date, $t = 1$, the position in the money market account is adjusted to $e^{r_{0,1}} F_1$ and the position in the futures contracts is adjusted to $e^{r_{0,1}+r_{1,2}}$ number of contracts.
- At time $t = 2$, the next resettlement time, the portfolio value is

$$e^{r_{0,1}+r_{1,2}} F_1 + e^{r_{0,1}+r_{1,2}}(F_2 - F_1) = e^{r_{0,1}+r_{1,2}} F_2,$$

where the first term is the value of the money market account and the second term is the income (possibly negative) from the position in the futures contracts.

At this date, $t = 2$, the position in the money market account is adjusted to $e^{r_{0,1}+r_{1,2}} F_2$ and the position in the futures contracts is adjusted to $e^{r_{0,1}+r_{1,2}+r_{2,3}}$ number of contracts.

- The procedure is repeated until and including time $T - 1$, and at time T the portfolio value is

$$e^{r_{0,1}+\cdots+r_{T-1,T}} F_{T-1} + e^{r_{0,1}+\cdots+r_{T-1,T}} (F_T - F_{T-1}) = e^{r_{0,1}+\cdots+r_{T-1,T}} F_T.$$

Summing up, we find that the futures strategy has an initial cost F_0 and delivers the random cash flow $e^{r_{0,1}+\cdots+r_{T-1,T}} F_T$ at time T. If we combine this strategy with a short position in the money market account of value F_0 at time 0 that is closed at time T, then a strategy is achieved that has zero initial cost and delivers the random cash flow $e^{r_{0,1}+\cdots+r_{T-1,T}} (F_T - F_0)$ at time T. The opposite strategy, in which all long positions are replaced by short positions and vice versa, has zero initial cost and delivers the cash flow $e^{r_{0,T}+\cdots+r_{T-1,T}} (F_0 - F_T)$ at time T. It will be called the short futures strategy with unit leverage.

We now return to the hedging problem for the coffee producer described above and assume that there is a futures contract for delivery of k pounds of coffee beans at a future time T. Ideally, T should correspond to 10 months from now. However, such a futures contract may not be available, and in this case the producer looks for a contract with a delivery date that is as early as possible but not before 10 months from now since the coffee producer does not want actual delivery of the coffee beans specified in the futures contract. Let F_0 and F_T denote the futures prices now and in 10 months' time, respectively. The producer may hedge against the variability in the future spot price by following the short futures strategy with leverage h futures contracts with no initial payment, as outlined previously. The position is closed out 10 months from now, resulting in the cash flow

$$X S_T - h k e^{r_{0,1}+\cdots+r_{T-1,T}} (F_T - F_0)$$

at time T, which we call the hedging error. If the maturity date of the futures contract coincides with the date on which the producer sells the coffee beans on the spot market, then $F_T = S_T$, and therefore the hedging error equals

$$(X - h k e^{r_{0,1}+\cdots+r_{T-1,T}}) S_T + h k e^{r_{0,1}+\cdots+r_{T-1,T}} F_0.$$

If the interest rates are known at time 0, then $R_0 = e^{r_{0,1}+\cdots+r_{T-1,T}}$ is the return of a zero-coupon bond maturing at time T.

If the futures contract matures in more than 10 months from now or if the kind of coffee bean produced differs from the one stated in the futures contract, then $F_T \neq S_T$. However, the usefulness of hedging with futures relies on $F_T \approx S_T$ being a good approximation. To determine the optimal hedge, i.e., the leverage of the short futures strategy, the coffee producer needs to decide upon a model for the random vector (X, S_T, F_T) and find the number h that minimizes the variance of the hedging error $X S_T - h k R_0 (F_T - F_0)$.

Proposition 3.3 tells us that the more general hedging problem (with n futures contracts, $F^1, \ldots, F^n$, as hedging instruments) of finding the leverage of the short futures strategy in order to minimize the variance of the hedging error $XS_T - h_1 k_1 R_0(F_T^1 - F_0^1) - \cdots - h_n k_n R_0(F_T^n - F_0^n)$ has the solution $\mathbf{h} = \Sigma_{\mathbf{Z}}^{-1} \Sigma_{XS_T, \mathbf{Z}}$, where $\mathbf{Z}^{\mathsf{T}} = R_0(k_1 F_T^1, \ldots, k_n F_T^n)$. In particular, when $n = 1$, the variance of the hedging error is minimized by choosing

$$h = \frac{\text{Cov}(XS_T, kR_0 F_T)}{\text{Var}(kR_0 F_T)} = \frac{\text{Cov}(XS_T, F_T)}{kR_0 \text{Var}(F_T)}.$$

The reduction in the variance of the price at which the producer sells his coffee crop is given by [see (3.2)]

$$\frac{\text{Var}(XS_T - hkR_0(F_T - F_0))}{\text{Var}(XS_T)} = \frac{\text{Var}(XS_T)(1 - \text{Cor}(XS_T, kR_0 F_T)^2)}{\text{Var}(XS_T)}$$

$$= 1 - \text{Cor}(XS_T, F_T)^2.$$

In some cases, it may be reasonable to assume that the quantity X and the spot price S_T are independent and, further, that $F_T = S_T$. This gives $h = E[X]/(kR_0)$, i.e., hedge by following the short futures strategy with leverage corresponding to the expected size of the coffee crop. Moreover, the formula

$$\text{Var}(XS_T) = E[X^2 S_T^2] - E[XS_T]^2$$

$$= E[X^2]E[S_T^2] - E[X]^2 E[S_T]^2$$

$$= \text{Var}(X)\text{Var}(S_T) + E[X^2]E[S_T]^2 + E[X]^2 E[S_T^2] - 2E[X]^2 E[S_T]^2$$

$$= \text{Var}(X)\text{Var}(S_T) + \text{Var}(X)E[S_T]^2 + \text{Var}(S_T)E[X]^2$$

for the variance of a product of independent random variables implies that the variance of the hedging error is given by

$$\text{Var}(XS_T - hkR_0(F_T - F_0))$$

$$= \text{Var}(XS_T)(1 - \text{Cor}(XS_T, F_T)^2)$$

$$= \text{Var}(XS_T)\left(1 - \frac{\text{Cov}(XS_T, F_T)^2}{\text{Var}(XS_T)\text{Var}(S_T)}\right)$$

$$= \text{Var}(XS_T)\left(1 - \frac{(E[XS_T^2] - E[XS_T]E[S_T])^2}{\text{Var}(XS_T)\text{Var}(S_T)}\right)$$

$$= \text{Var}(XS_T)\left(1 - \frac{E[X]^2(E[S_T^2] - E[S_T]^2)^2}{\text{Var}(XS_T)\text{Var}(S_T)}\right)$$

$$= \text{Var}(XS_T)\left(1 - \frac{E[X]^2 \text{Var}(S_T)}{\text{Var}(XS_T)}\right)$$

$$= \text{Var}(XS_T) - \text{E}[X]^2 \, \text{Var}(S_T)$$

$$= \text{Var}(X) \, \text{Var}(S_T) + \text{Var}(X) \, \text{E}[S_T]^2.$$

In particular, we observe that if the variance of X is much bigger than the variance of S_T, then hedging will not reduce the variance by much because

$$\frac{\text{Var}(XS_T - hkR_0(F_T - F_0))}{\text{Var}(XS_T)} = \frac{\text{Var}(X) \, \text{Var}(S_I) + \text{Var}(X) \, \text{E}[S_T]^2}{\text{Var}(X) \, \text{Var}(S_T) + \text{Var}(X) \, \text{E}[S_T]^2 + \text{Var}(S_T) \, \text{E}[X]^2}$$

will be close to one. The opposite is also true: if $\text{Var}(X)$ is small compared to $\text{Var}(S_T)$, then variance is significantly reduced.

Example 3.3 (Hedging failures). Here we consider a stylized example illustrating the importance of understanding the inherent risks associated with hedging with futures contracts.

Consider a coffee producer offering to a major buyer to deliver X_0 pounds of coffee beans at time $T > 0$ at a price of G_0 per pound to be paid at time T. In essence, the company has a short position in a forward contract and receives the cash flow $X_0(G_0 - S_T)$ at time T, where S_T is the spot price at that time.

To hedge the risk of an increase in the coffee price, the producer takes a long position in a futures contract of size k so that $kF_T \approx X_0S_T$. The contracted income X_0G_0 at time T for delivery of the coffee beans is negotiated so that $X_0G_0 - kF_0 > 0$. If the buyer of the coffee beans pays as contracted, the producer manages to keep the balance on the margin account, and $kF_T = X_0S_T$, then the coffee produces makes the net profit $X_0G_0 - kF_0 > 0$ regardless of the future price of coffee. For simplicity we ignore the effect of interest rates.

Case 1. Suppose that at time $t < T$ the futures price has dropped substantially from the initial price F_0 and the producer receives another margin call to deposit more money, for simplicity assumed to be $-k(F_t - F_{t-1}) > 0$, on the margin account. The margin calls following the depreciation in the value of the producer's futures position has led to a shortage of cash for the producer, and at time t the producer is unable to meet the latest margin call. Therefore, the futures position is closed out and leaves the producer with a debt of size $-k(F_t - F_{t-1})$ to his broker. Due to the producer's poor financial condition following the loss on the futures position, raising capital to service the producer's debt to his broker turns out to be costly and results in a loss of size C_T at time T. The net result at time T is therefore

$$X_0G_0 - kF_0 - C_T + kF_{t-1} - X_0S_T.$$

The intended gain $X_0G_0 - kF_0$ is likely to be offset by the loss C_T, and if the coffee price increases between time t and time T, then the loss $-(kF_{t-1} - X_0S_T)$ could be substantial.

Case 2. Suppose that at time $t \leq T$ the intended buyer of coffee beans announces a failure to meet her contracted obligation to pay the amount X_0G_0 at time T, and the coffee producer sees no alternative but to cancel the forward contract and the futures position, resulting in the net outcome $k(F_t - F_0)$, which amounts to an unintended speculation in the futures price. To avoid the negative effects of one party's violating its obligation in a forward contract, it is common for a so-called collateralization agreement to be made that resembles the daily resettlement procedure in futures markets. However, there is no guarantee that the two parties will agree on the value of the forward contract prior to maturity, and the financially stronger party may impose its valuation on the weaker party.

3.3   Hedging of Insurance Liabilities

We will now study hedging of insurance liabilities represented as a general cash flow and investigate how to hedge the future value of this cash flow with a bond portfolio. The results are then illustrated in a series of insurance examples.

Consider a discrete-time model where C_k is the random amount paid out at the end of year k, for $k = 1, \ldots, n$. The value of the liabilities at the end of year 1 is C_1 plus the time 1 value of the remaining cash flow $C_2, \ldots, C_n$. We write $\pi_1(C_k)$ for the time 1 value of C_k. This is random as seen from today but known at time 1. Hence, the value at time 1 of the liability cash flow is

$$L = \sum_{k=1}^{n} \pi_1(C_k). \tag{3.3}$$

We choose $\pi_j(C_k)$, for $j < k$, to be the best estimate of C_k made at time j discounted to money at time j. More precisely, $\pi_j(C_k)$ can be chosen to be the conditional expectation of C_k given the information available at time j, multiplied by the discount factor $e^{-r_{j,k}(k-j)}$; $r_{j,k}$ is the time j zero-rate for a zero-coupon bond maturing at time k with a face value 1. The information at time j may be coded as a vector $\mathbf{I}_j$ of random variables whose outcomes become known between now and time j. As before, we write $\mathrm{E}[C_k \mid \mathbf{I}_j]$ for the conditional expectation. The information at time $j' > j$ can similarly be coded as the random vector $\mathbf{I}_{j'}$ whose components are the components of $\mathbf{I}_j$ but also random variables whose outcomes become known between times j and j'.

We will now determine the quadratic hedge of the liability $L = \sum_{k=1}^{n} \pi_1(C_k)$ in terms of time 1 prices of zero-coupon bonds maturing at times $1, \ldots, n$. We assume that, for each k, the value $\pi_1(C_k)$ is the best estimate at time 1 of C_k, discounted to the money value at time 1. That is,

$$\pi_1(C_k) = \mathrm{E}[C_k \mid \mathbf{I}_1] e^{-r_{1,k}(k-1)} \quad \text{for all } k. \tag{3.4}$$

Suppose further that the time 1 best estimates of C_k are uncorrelated with the time 1 bond prices. That is,

$$\mathrm{Cov}(\mathrm{E}[C_j \mid \mathbf{I}_1], e^{-r_{1,k}(k-1)}) = 0 \quad \text{for all } j, k. \tag{3.5}$$

If we write $\mathbf{Z} = (e^{-r_{1,2}(2-1)}, \ldots, e^{-r_{1,n}(n-1)})^{\mathsf{T}}$, then we know from Proposition 3.2 that the unique portfolio $(h_0, \mathbf{h})$, which minimizes $\mathrm{E}[(A - L)^2]$ among all A of the form $h_0 + \mathbf{h}^{\mathsf{T}}\mathbf{Z}$, is given by

$$\mathbf{h} = \Sigma_{\mathbf{Z}}^{-1}\Sigma_{L,\mathbf{Z}} \quad \text{and} \quad h_0 = \mathrm{E}[L] - \mathbf{h}^{\mathsf{T}}\,\mathrm{E}[\mathbf{Z}],$$

and the corresponding value of the asset portfolio is

$$\hat{A} = \mathrm{E}[L] + \Sigma_{L,\mathbf{Z}}^{\mathsf{T}}\Sigma_{\mathbf{Z}}^{-1}(\mathbf{Z} - \mathrm{E}[\mathbf{Z}]).$$

Let us compute explicitly this quadratic hedge. We have

$$\mathrm{Cov}(L, Z_j) = \sum_{k=1}^{n} \mathrm{Cov}\left(\mathrm{E}\left[C_k \mid \mathbf{I}_1\right]e^{-r_{1,k}(k-1)}, e^{-r_{1,j+1}j}\right)$$

$$= \sum_{k=1}^{n} \mathrm{E}[C_k]\,\mathrm{Cov}\left(e^{-r_{1,k}(k-1)}, e^{-r_{1,j+1}j}\right)$$

$$= \left(\mathbf{c}^{\mathsf{T}}\Sigma_{\mathbf{Z}}\right)_j,$$

where $\mathbf{c}^{\mathsf{T}} = (\mathrm{E}[C_2], \ldots, \mathrm{E}[C_n])$ and $\Sigma_{\mathbf{Z}} = \mathrm{Cov}(\mathbf{Z})$. Similarly,

$$\mathrm{E}[L] = \sum_{k=1}^{n} \mathrm{E}\left[\mathrm{E}[C_k \mid \mathbf{I}_1]e^{-r_{1,k}(k-1)}\right]$$

$$= \sum_{k=1}^{n} \mathrm{E}[\mathrm{E}[C_k \mid \mathbf{I}_1]]\,\mathrm{E}\left[e^{-r_{1,k}(k-1)}\right]$$

$$= \mathrm{E}[C_1] + \sum_{k=2}^{n} \mathrm{E}[C_k]\,\mathrm{E}\left[Z_{k-1}\right]$$

$$= \mathrm{E}[C_1] + \mathbf{c}^{\mathsf{T}}\,\mathrm{E}[\mathbf{Z}].$$

If we set $\Sigma_{L,\mathbf{Z}} = \mathrm{Cov}(L, \mathbf{Z})$, then $\Sigma_{L,\mathbf{Z}} = \mathbf{c}^{\mathsf{T}}\Sigma_{\mathbf{Z}}$, and we see that the quadratic hedge is given by

$$\mathbf{h} = \Sigma_{\mathbf{Z}}^{-1}\Sigma_{L,\mathbf{Z}} = \mathbf{c} \quad \text{and} \quad h_0 = \mathrm{E}[L] - \mathbf{h}^{\mathsf{T}}\,\mathrm{E}[\mathbf{Z}] = \mathrm{E}[C_1].$$

We conclude that for each k we should buy $\mathrm{E}[C_k]$ units of the zero-coupon bond with maturity k. Our findings can be summarized as follows.

Proposition 3.4. *Consider a stochastic cash flow* $(C_1, \ldots, C_n)$, *zero-coupon bonds maturing at times* $1, \ldots, n$ *with common face value* 1, *and the time* 1 *liability* L *given by (3.3) and (3.4). If (3.5) holds, then the time* 1 *value of the bond portfolio* *consisting of* $h_k = \mathrm{E}[C_k]$ *zero-coupon bonds maturing at time* k *is the optimal quadratic hedge of the liability* L *among all portfolios consisting of these bonds. The hedging error, the time* 1 *value of the bond portfolio minus that of the liability, is*

$$\hat{A} - L = \sum_{k=1}^{n}(\mathrm{E}[C_k] - \mathrm{E}[C_k \mid \mathbf{I}_1])e^{-r_{1,k}(k-1)}.$$

Note that this result holds regardless of whatever view one has of the time 1 zero rates $r_{1,2}, \ldots, r_{1,n}$.

Example 3.4 (Multiple business lines). Consider a liability consisting of N payment streams active at time 1. For instance, each payment stream could come from a separate line of business. Think, for instance, of an insurance company issuing motor vehicle insurance, fire insurance, property insurance, etc. If $C_{j,k}$ is the payment at time k of the jth payment stream and π_1 is linear, then the value of the liability at time 1 is

$$L = \sum_{k=1}^{n} \pi_1(C_k) = \sum_{k=1}^{n}\sum_{j=1}^{N} \pi_1(C_{j,k}).$$

If the valuation of $C_{j,k}$ is such that $\pi_1(C_{j,k}) = \mathrm{E}[C_{j,k} \mid \mathbf{I}_1]e^{-r_{1,k}(k-1)}$, then the value of the liability is given by

$$L = \sum_{k=1}^{n} e^{-r_{1,k}(k-1)} \sum_{j=1}^{N}\mathrm{E}[C_{j,k} \mid \mathbf{I}_1].$$

Example 3.5 (Pure endowment). Consider a portfolio of n_x identical T-year pure endowment insurance contracts with sum insured equal to 1. Each such policy pays the sum insured if the insured is alive at time T. All the n_x policyholders are assumed to be of age x at time 0. Consider the current (at time 0) mortality rate, or force of mortality, μ_0 such that $\mu_0(x)$ is the current death intensity of an age x individual. The probability of survival at least t years for an insured of age x at time 0 is

$$p_0(t, x) = \exp\left\{-\int_0^t \mu_0(x + u)du\right\}.$$

The cash flow generated by the portfolio of the n_x pure endowment contracts consists of a single payment $C_T = N_T$ at time T, where N_T is the number of survivors at time T, and $C_k = 0$ for all $k \neq T$. Viewed from time 0, C_T has a binomial distribution with parameters n_x and $p_0(t, x)$. The value $\pi_0(C_T)$ of the portfolio at time 0 is given by

$$\pi_0(C_T) = \mathrm{E}[N_T]e^{-r_{0,T}T} = p_0(T, x)n_x e^{-r_{0,T}T}$$

$$= \exp\left\{-\int_0^T \mu_0(x + u)du\right\} n_x e^{-r_{0,T}T}.$$

Let N_1 be the number of survivors after the first year. If the mortality rate in 1 year is μ_1, then the value of the portfolio in 1 year is

$$\pi_1(C_T) = E[N_T \mid \mathbf{I}_1]e^{-r_{1,T-1}(T-1)} = p_1(T-1, x+1)N_1 e^{-r_{1,T-1}(T-1)}$$

$$= \exp\left\{-\int_1^T \mu_1(x+u)du\right\} N_1 e^{-r_{1,T-1}(T-1)}.$$

We see that $\pi_1(C_T)$ is the product of three random variables. The first is expressed in terms of the mortality rate μ_1, which may be random, the second is the number of survivors N_1, which has a binomial distribution with parameters n_x and $p_0(1,x)$, and the third is the value 1 year from now of an, at that time, $T-1$ year zero-coupon bond with face value 1.

Example 3.6 (Pure endowment, continued). Suppose the mortality rate is deterministic with $\mu_1(t) = \mu_0(t) = \mu(t)$. It is reasonable to assume that the number of survivors N_1 is independent of the zero-coupon bond prices, so (3.5) holds. Then Proposition 3.4 implies that the optimal quadratic hedging portfolio is a long position of d zero-coupon bonds with maturity T and face value 1, where

$$d = \exp\left\{-\int_1^T \mu(x+u)du\right\} E[N_1]$$

$$= \exp\left\{-\int_1^T \mu(x+u)du\right\} \exp\left\{-\int_0^1 \mu(x+u)du\right\} n_x$$

$$= \exp\left\{-\int_0^T \mu(x+u)du\right\} n_x.$$

The hedging error, that is, the value of the liability minus the hedging portfolio, is

$$\pi_1(C_T) - de^{-r_{1,T-1}(T-1)}$$

$$= \exp\left\{-\int_1^T \mu(x+u)du\right\} e^{-r_{1,T-1}(T-1)}\left(N_1 - n_x \exp\left\{-\int_0^1 \mu(x+u)du\right\}\right).$$

Example 3.7 (Whole life insurance). Consider a portfolio of n_x identical whole life insurance contracts that pays 1 at the end of the year of death of the insured. Suppose that the times of deaths of the insured individuals are independent and that the mortality rate is deterministic, with $\mu_1(t) = \mu_0(t) = \mu(t)$. Let N_k be the number of survivors at the end of year k. The cash flow generated by the portfolio is

$$n_x - N_1 \text{ at the end of year 1,}$$

$$N_1 - N_2 \text{ at the end of year 2, etc.}$$

The amount to be paid at the end of year k is $C_k = N_{k-1} - N_k$. At the end of year 1 we observe N_1, and we want to compute $E[C_k \mid N_1]$. At time 1 the age of each individual is $x + 1$, and for any $k \geq 2$ the distribution of $(C_2, \ldots, C_k, N_k)$ conditional on N_1 is multinomial with parameters N_1 and $q_2, \ldots, q_k, \widetilde{q}_k$. That is,

$$P(C_2 = c_2, \ldots, C_k = c_k, N_k = n_k \mid N_1) = \frac{N_1}{c_2! \ldots c_k! n_k!} q_2^{c_2} \cdots q_k^{c_k} \widetilde{q}_k^{n_k}$$

when $c_2 + \ldots c_k + n_k = N_1$, and 0 otherwise. Here

$$q_j = p_1(j - 2, x + 1) - p_1(j - 1, x + 1)$$

$$= \exp\left\{-\int_1^{j-1} \mu(x + u)du\right\} - \exp\left\{-\int_1^{j} \mu(x + u)du\right\},$$

$$\widetilde{q}_k = p_1(k - 1, x + 1) = \exp\left\{-\int_1^{k} \mu(x + u)du\right\}.$$

In particular, $E[C_k \mid N_1] = q_k N_1$, and the liability L is given by

$$L = \sum_{k \geq 1} E[C_k \mid N_1] e^{-r_{1,k-1}(k-1)}$$

$$= (n_x - N_1) + \sum_{k \geq 2} q_k N_1 e^{-r_{1,k-1}(k-1)}$$

$$= (n_x - N_1)$$

$$+ N_1 \sum_{k \geq 2} e^{-r_{1,k-1}(k-1)} \left(\exp\left\{-\int_1^{k-1} \mu(x + u)du\right\}\right.$$

$$\left. - \exp\left\{-\int_1^{k} \mu(x + u)du\right\}\right).$$

Assuming the number of survivors N_1 at time 1 is uncorrelated with the zero-coupon bond prices, Proposition 3.4 gives the optimal quadratic hedge as

$$n_x - E[N_1] = n_x(1 - p_0(1, x)) = n_x \left(1 - \exp\left\{-\int_0^1 \mu(x + u)du\right\}\right)$$

zero-coupon bonds with maturity 1 and, for each $k \geq 2$, invest in

$$E[C_k] = E[N_1] \left(\exp\left\{-\int_1^{k-1} \mu(x + u)du\right\} - \exp\left\{-\int_1^{k} \mu(x + u)du\right\}\right)$$

$$= n_x \left(\exp\left\{-\int_0^{k-1} \mu(x + u)du\right\} - \exp\left\{-\int_0^{k} \mu(x + u)du\right\}\right)$$

zero-coupon bonds with maturity k.

Example 3.8 (Nonlife insurance). Consider an insurance company selling basic nonlife insurance. For simplicity we assume that there are 100,000 insureds over the period 1 January (time 0) to 31 December (time 1). During the year there will be a random number of claims incurred, most of which will be reported during the year but some of which will be reported later. All incurred claims will be treated as a liability of the insurance company to its policyholders. In this example, we ignore the fact that there may be claims from past years that will be settled during the current year. The insurance company invests in a simple bond portfolio to hedge the value of its liabilities at the end of the year.

Let Z_k, for $k \geq 1$, be independent and identically distributed random variables representing claim sizes of the insurance claims. Let N be the total number of claims incurred between time 0 and 1. We simplify the problem somewhat by assuming that N has a Poisson distribution with parameter λ, where λ is approximately the number of insureds times the probability that a randomly chosen insured will experience an event during the year that will lead to a claim. To the kth claim there is a random delay τ_k indicating the time period when the claim is settled. Suppose that the delay times are independent and that for each k, $\tau_k - 1$ has a geometric distribution with parameter θ, i.e.,

$$P(\tau_k - 1 = l) = \theta(1 - \theta)^l, \quad l \geq 0.$$

Then the total number of claims to be paid at time 1 is

$$N_1 = \sum_{k=1}^{N} I\{\tau_k = 1\} = \sum_{k=1}^{N} I\{\tau_k - 1 = 0\}$$

and

$$P(N_1 = j) = \sum_{n=j}^{\infty} P(N_1 = j \mid N = n)\,P(N = n)$$

$$= \sum_{n=j}^{\infty} \binom{n}{j} \theta^j (1 - \theta)^{n-j} \frac{\lambda^n}{n!} \exp\{-\lambda\}$$

$$= \frac{(\lambda\theta)^j}{j!} \exp\{-\lambda\} \sum_{n=0}^{\infty} \frac{(1 - \theta)^n \lambda^n}{n!}$$

$$= \frac{(\lambda\theta)^j}{j!} \exp\{-\lambda\theta\}.$$

That is, N_1 is $Po(\lambda\theta)$-distributed. Similarly, the number of claims to be paid in year 2 conditional on $N_1 = j$ is given by

$$P(N_2 = k \mid N_1 = j)$$

$$= \frac{P(N_1 = j, N_2 = k)}{P(N_1 = j)}$$

$$= \frac{1}{P(N_1 = j)} \sum_{n=j+k}^{\infty} P(N_1 = j, N_2 = k \mid N = n)\, P(N = n)$$

$$= \frac{1}{\frac{(\lambda\theta)^j}{j!} e^{-\lambda\theta}} \sum_{n=j+k}^{\infty} \binom{n}{j}\binom{n-j}{k} \theta^j (1-\theta)^k \theta^k (1-\theta)^{2(n-j-k)} \frac{\lambda^n}{n!} e^{-\lambda}$$

$$= \frac{1}{\frac{(\lambda\theta)^j}{j!} e^{-\lambda\theta}} \frac{(\lambda\theta)^{j+k}(1-\theta)^k}{j!k!} e^{-\lambda} \sum_{n=j+k}^{\infty} \frac{[\lambda(1-\theta)^2]^{n-j-k}}{(n-j-k)!}$$

$$= \frac{[\lambda\theta(1-\theta)]^k}{k!} e^{-\lambda\theta(1-\theta)}.$$

Thus, the conditional distribution of N_2 given $N_1 = j$ is $\mathrm{Po}(\lambda\theta(1-\theta))$, which does not depend on j. In particular, N_1 and N_2 are independent. Proceeding like this, we see that $N_1, N_2, \ldots$, are independent, with N_k having a $\mathrm{Po}(\lambda\theta(1-\theta)^{k-1})$ distribution. As a consequence, the amounts $C_1, C_2, \ldots$ to be paid in periods $1, 2, \ldots$ from claims occurring in period 1 are independent, and each C_k can be represented as

$$C_k \stackrel{\mathrm{d}}{=} \sum_{j=1}^{N_k} Z_j,$$

where $\stackrel{\mathrm{d}}{=}$ means that the random variables on the left- and right-hand sides have the same distribution function (equality in distribution). In particular,

$$E[C_k \mid I_1] = E[C_k] = \lambda\theta(1-\theta)^{k-1} E[Z_1], \quad k \geq 2.$$

Therefore, the value of the liability at time 1 is given by

$$L = \sum_{i=1}^{N_1} Z_i + E[Z_1]\lambda\theta \sum_{k=2}^{\infty} e^{-r_{1,k}}(1-\theta)^{k-1}.$$

Since $E[C_k \mid I_1]$ is deterministic (known already at time 0), (3.5) is satisfied and Proposition 3.4 implies that the quadratic hedge is to buy $E[C_k]$ number of k bonds at time 0. We see that the quadratic hedge completely eliminates the second term of the liability, and the hedging error at time 1 is given by

$$\hat{A} - L = \sum_{k=1}^{n} \left(E[C_k] - E[C_k \mid I_1] \right) e^{-r_{1,k}(k-1)}$$

$$= E[Z_1]\lambda\theta - \sum_{i=1}^{N_1} Z_i.$$

In this example, there is no remaining interest rate risk, only insurance risk coming from the uncertainty in the number and severity of claims during the first year.

As a variation on this example we may assume that each claim can be settled in one of a fixed number m of possible future time periods. Let τ_i denote the period when the ith claim is settled, and let $p_k = P(\tau_i = k)$. Assume further that, for each n, the distribution of the vector $(N_1, \ldots, N_m)$, given the total number of claims $N = n$, is a multinomial distribution,

$$P(N_1 = n_1, \ldots, N_m = n_m \mid N = n) = \frac{n!}{n_1! \ldots n_m!} p_1^{n_1} \cdots p_m^{n_m}.$$

Then it follows that $N_1, \ldots, N_m$ are independent, with N_k having a $\mathrm{Po}(\lambda p_k)$ distribution. Indeed, with $n = n_1 + \cdots + n_m$ and $p_1 + \cdots + p_m = 1$,

$$P(N_1 = n_1, \ldots, N_m = n_m) = P(N_1 = n_1, \ldots, N_m = n_m \mid N = n) P(N = n)$$

$$= \frac{n!}{n_1! \ldots n_m!} p_1^{n_1} \cdots p_m^{n_m} \frac{\lambda^n}{n!} e^{-\lambda}$$

$$= \frac{(p_1 \lambda)^{n_1}}{n_1!} e^{-\lambda p_1} \cdots \frac{(p_m \lambda)^{n_m}}{n_m!} e^{-\lambda p_m},$$

which is the product of Poisson probabilities. Moreover,

$$E[C_k \mid I_1] = E[C_k] = \lambda p_k\, E[Z_1], \quad k = 2, \ldots, m.$$

Again, $E[C_k \mid I_1]$ is deterministic and (3.5) is satisfied. Proposition 3.4 tells us that the quadratic hedge is to buy $E[C_k] = \lambda p_k\, E[Z_1]$ number of k bonds. The resulting hedging error is

$$\hat{A} - L = E[Z_1]\lambda p_1 - \sum_{i=1}^{N_1} Z_i$$

and depends only on the outcome of claims settled during the first year.

3.4 Hedging of a Digital Option with Call Options

Let today be time 0 and let S_1 be the spot price of some asset at time 1. Suppose that there are call options with payoffs $(S_1 - K_l)_+$, for some $0 < K_1 < \cdots < K_n$, at time 1 available on the market and that we may take arbitrary long and short positions in these contracts. Suppose that we issue a digital option with payoff $I\{S_1 > K\}$ at time 1 and that we want to hedge its payoff at time 1 as much as possible by forming a suitable portfolio of call options.

First, note that $I\{S_1 > K\}$ can be approximated by

$$\frac{(S_1 - (K - \delta))_+ - (S_1 - K)_+}{\delta} = \begin{cases} 0 & \text{if } S_1 < K - \delta, \\ (S_1 - K + \delta)/\delta & \text{if } S_1 \in [K - \delta, K], \\ 1 & \text{if } S_1 > K, \end{cases}$$

so we can hedge the digital option arbitrarily well if call options with strikes arbitrarily close to K are available on the market. This is unfortunately not realistic as it would make digital options redundant since they could then be formed as call option portfolios. However, an approximate hedge is obtained by looking for strikes $K_l < K_{l+1}$ close to K and approximating $I\{S_1 > K\}$ by

$$f(S_1, K_l, K_{l+1}) = \frac{(S_1 - K_l)_+ - (S_1 - K_{l+1})_+}{K_{l+1} - K_l}, \tag{3.6}$$

which corresponds to taking a long position in the call option with strike K_l of size $(K_{l+1} - K_l)^{-1}$ and a short position in the call option with strike K_{l+1} of the same size. Note that $f(S_1, K_l, K_{l+1}) \geq I\{S_1 > K\}$ if $K_{l+1} \leq K$, whereas $f(S_1, K_l, K_{l+1}) \leq I\{S_1 > K\}$ if $K_l \geq K$.

The accuracy of this approximate hedge depends not only on the approximation of the payoff function but also, and possibly quite a lot, on the probability distribution we assign to S_1. For instance, if the probability distribution we assign to S_1 is such that the event $\{S_1 > K\}$ has a probability of zero, whereas the event $\{S_1 = K\}$ has a positive probability, then the preceding hedge would be a rather foolish choice.

To determine the quadratic hedge, we must compute the expected values of the liability $L = I\{S_1 > K\}$ and the hedging instruments $Z_l = (S_1 - K_l)_+, l = 1, \ldots, n$ as well as all the covariances. To this end we must compute the expectations

$$E[I\{S_1 > K\}] = P(S_1 > K),$$

$$E[(S_1 - K_l)_+] = E[I\{S_1 > K_l\}S] - K_l P(S_1 > K_l),$$

$$E[I\{S_1 > K\}(S_1 - K_l)_+] = E[I\{S_1 > \max(K, K_l)\}S_1] - K_l P(S_1 > \max(K, K_l)),$$

$$E[(S_1 - K_l)_+(S_1 - K_m)_+] = E[I\{S_1 > K_m\}S_1^2] - (K_l + K_m) E[I\{S_1 > K_m\}S_1]$$

$$+ K_l K_m P(S_1 > K_m).$$

In the last equation, we have assumed $0 < K_l \leq K_m$.

To get any further we need a stochastic model for S_1. Let us assume that S_1 has a lognormal distribution; $\log S_1$ is $N(\mu, \sigma^2)$-distributed. Then, with $\widetilde{K}_l = (\log K_l - \mu)/\sigma$, we may write

$$P(S_1 > K_l) = P(\mu + \sigma Z > \log K_l) = 1 - \Phi((\log K_l - \mu)/\sigma) = \Phi(-\widetilde{K}_l).$$

The remaining terms are not difficult to compute:

$$E[I\{S_1 > K_l\}S_1] = \int_{\widetilde{K}_l}^{\infty} e^{\mu+\sigma z}e^{-z^2/2}\frac{1}{\sqrt{2\pi}}dz$$

$$= e^{\mu+\sigma^2/2}\int_{\widetilde{K}_l}^{\infty} e^{-(z-\sigma)^2/2}\frac{1}{\sqrt{2\pi}}dz$$

$$= e^{\mu+\sigma^2/2}\Phi(\sigma - \widetilde{K}_l)$$

and

$$E[I\{S_1 > K_l\}S_1^2] = e^{2\mu+2\sigma^2}\Phi(2\sigma - \widetilde{K}_l).$$

With $L = I\{S_1 > K\}$ and $Z_l = (S_1 - K_l)_+$ we can determine the optimal time 1 quadratic hedge by Proposition 3.2 as

$$\mathbf{h} = \Sigma_{L,\mathbf{Z}}^{\mathrm{T}}\Sigma_{\mathbf{Z}}^{-1},$$

$$h_0 = E[L] - \mathbf{h}^{\mathrm{T}}E[\mathbf{Z}].$$

From the preceding computations we find that $E[L] = \Phi(-\widetilde{K})$, $E[Z_l] = e^{\mu+\sigma^2/2}\Phi(\sigma - \widetilde{K}_l) - K_l\Phi(-\widetilde{K}_l)$,

$$(\Sigma_{L,\mathbf{Z}})_l = E[I\{S_1 > K_l\}S_1] - E[I\{S_1 > K_l\}]E[S_1]$$

$$= e^{\mu+\sigma^2/2}\Phi(\sigma - \max(\widetilde{K}, \widetilde{K}_l)) - K_l\Phi(-\max(\widetilde{K}, \widetilde{K}_l))$$

$$- \Phi(-\widetilde{K})(e^{\mu+\sigma^2/2}\Phi(\sigma - \widetilde{K}_l) - K_l\Phi(-\widetilde{K}_l)),$$

and

$$(\Sigma_{\mathbf{Z}})_{l,m} = E[(S_1 - K_l)_+(S_1 - K_m)_+] - E[(S_1 - K_l)_+]E[(S_1 - K_m)_+]$$

$$= e^{2\mu+2\sigma^2}\Phi(2\sigma - \widetilde{K}_m) - (K_l + K_m)e^{\mu+\sigma^2/2}\Phi(\sigma - \widetilde{K}_m)$$

$$+ K_lK_m\Phi(-\widetilde{K}_m) - (e^{\mu+\sigma^2/2}\Phi(\sigma - \widetilde{K}_l) - K_l\Phi(-\widetilde{K}_l))$$

$$\cdot (e^{\mu+\sigma^2/2}\Phi(\sigma - \widetilde{K}_m) - K_m\Phi(-\widetilde{K}_m)) \quad \text{for } l \le m.$$

Now we have all we need to compute the quadratic hedge. But what does it look like in a numerical example? Our intuition tells us to expect that a good hedge is similar to (3.6). Next follows a numerical example that investigates if the quadratic hedge is of a similar type.

Example 3.9 (Numerical illustration). Let $\log S_1$ be $N(\mu, \sigma^2)$-distributed with $\mu = 0.05$ and $\sigma = 0.3$, and consider a digital option with payoff $I\{S_1 > K\}$ with

Table 3.1 Robustness of the quadratic hedge

(μ, σ)	h_0	h_1	h_2	h_3	h_4	h_5	Var
(0.05, 0.3)	0.006	−0.926	12.718	−12.975	1.469	−0.293	0.0096
(0.05, 0.5)	0.003	−0.881	12.641	−12.912	1.400	−0.258	0.0058
(0.3, 0.7)	0.003	−0.878	12.633	−12.890	1.375	−0.240	0.0039
(−0.05, 0.1)	0.014	−0.995	12.844	−13.443	2.626	−1.636	0.0177

$K = 1.05$. Suppose that a risk-free zero-coupon bond and five call options on S_1 are available as hedging instruments, and suppose that the strikes are given by $0.9, 1.0, 1.1, 1.2, 1.3$. Let h_0 be the position in the bond, and let $h_1, \ldots, h_5$ be the positions in the call options. The optimal quadratic hedge is given, with $\mathbf{h} = (h_1, \ldots, h_5)^T$, by

$$\mathbf{h}^T = \Sigma^T_{L,\mathbf{Z}} \Sigma^{-1}_{\mathbf{Z}} \quad \text{and} \quad h_0 = \mathrm{E}[L] - \mathbf{h}^T \, \mathrm{E}[\mathbf{Z}].$$

Evaluating $(h_0, h_1, \ldots, h_5)$ numerically, using the expressions above, gives (rounded off to three decimals)

$$(0.006, -0.926, 12.718, -12.975, 1.469, -0.293).$$

The variance of the payoff of the digital option is approximately 0.250, and the variance of the optimally hedged position is approximately 0.0096. The optimal quadratic hedge matches rather well to the natural hedge (3.6) with $K_l = K_2 = 1.0$ and $K_{l+1} = K_3 = 1.1$. This hedge corresponds to the position $(0, 0, 10, -10, 0, 0)$, i.e., a long position of 10 call options with strike 1.0 and a short position of 10 call options with strike 1.1. The variance of the hedged position using this hedge is approximately 0.0105.

So how robust is the optimal quadratic hedge to variations in the probability distribution assigned to S_1? Table 3.1 shows the effect of varying the parameters μ and σ on the positions $h_0, h_1, \ldots, h_5$ of the optimal quadratic hedge and on the variance of the hedged positions.

As we can see from Table 3.1, the quadratic hedge is not sensitive to the choice of the parameters μ and σ. The portfolio weights do not change much as the parameter values are varied.

3.5   Delta Hedging

Let S_0, S_t, and S_T be the spot prices of some pure investment asset at times $0 < t < T$, where 0 is the current time. The asset could be an exchange rate between two currencies or a share of a non-dividend-paying stock. Consider a financial contract with payoff $g(S_T)$ at time T, and suppose that the price of the contract at time $t < T$ can be expressed as a function f of S_t, where the function f is known at time 0 and has a derivative f'. The assumption that the price of the contract at time t can be expressed as $f(S_t)$ with f known at time 0 is not entirely realistic but often a reasonable approximation if t is small.

Example 3.10 (Black's formula). If the price of the financial contract is expressed through Black's formula (1.5) and time is measured in years, then

$$\pi_0 = e^{-r_{0,T}T} \int_{-\infty}^{\infty} g(S_0 e^{r_{0,T}T + \sigma_0 \sqrt{T} z - \sigma_0^2 T/?}) \phi(z) dz,$$

where $r_{0,T}$ is the current zero rate for a zero-coupon bond maturing at time T, σ_0 is the current implied volatility for this derivative, and ϕ is the standard normal density function. The derivative price at time t is given by

$$\pi_t = e^{-r_{t,T}(T-t)} \int_{-\infty}^{\infty} g(S_t e^{r_{t,T}(T-t) + \sigma_t \sqrt{T-t} z - \sigma_t^2 (T-t)/2}) \phi(z) dz$$

$$\approx e^{-r_{0,T}(T-t)} \int_{-\infty}^{\infty} g(S_t e^{r_{0,T}(T-t) + \sigma_0 \sqrt{T-t} z - \sigma_0^2 (T-t)/2}) \phi(z) dz$$

$$= f(S_t),$$

where the approximation $\pi_t \approx f(S_t)$ is likely to be accurate if t is small and changes in the derivative price are most likely due to changes in the spot price of the underlying asset and, to a lesser extent, to changes in the term structure of interest rates and changes in the implied volatility.

Suppose that we want to hedge $f(S_t)$ by taking positions h_1 in the underlying asset and h_0 in a zero-coupon bond maturing at t in order to minimize $E[(h_0 + h_1 S_t - f(S_t))^2]$. We assume that both $f(S_t)$ and S_t have finite and nonzero variances. We know, from Proposition 3.2, that the optimal quadratic hedge (h_0, h_1) is given by

$$h_1 = \frac{\text{Cov}(f(S_t), S_t)}{\text{Var}(S_t)} \quad \text{and} \quad h_0 = E[f(S_t)] - h_1 E[S_t].$$

It is rather common that the optimal position (h_0, h_1) is difficult to compute, both because of the nonlinear function f and because we may have difficulties assigning a probability distribution to S_t. It would greatly simplify things if we could use a first-order Taylor approximation to approximate $f(S_t) \approx f(E[S_t]) + f'(E[S_t])(S_t - E[S_t])$. In this case,

$$h_1 \approx f'(E[S_t]) \quad \text{and} \quad h_0 \approx f(E[S_t]) - f'(E[S_t]) E[S_t].$$

We also observe that in this case, the hedging error is $h_0 + h_1 S_t - f(S_t) = f(E[S_t]) + f'(E[S_t])(S_t - E[S_t]) - f(S_t)$, which is the approximation error when approximating $f(S_t)$ with the first-order approximation of f around the point $E[S_t]$, evaluated at S_t. Moreover, if t is small, then it is reasonable to approximate $E[S_t] \approx S_0$, and therefore

$$h_1 \approx f'(S_0) \quad \text{and} \quad h_0 \approx f(S_0) - f'(S_0) S_0.$$

This hedging approach is called delta hedging. Notice that for this approximative solution to the quadratic hedging problem we do not need to assign any probability distributions to the future spot prices. However, this nice feature of the delta hedging approach comes at a cost: the approximation is only accurate for small t, which calls for frequent adjustments of the hedge, which in turn produces hedging costs.

The delta hedging approach deserves more than the heuristic arguments presented so far.

Proposition 3.5. *Let f be differentiable in a neighborhood of S_0, and suppose that $f(S_t)$ and S_t have finite and nonzero variances for t sufficiently small. Then*

$$\frac{\text{Cov}(f(S_t), S_t)}{\text{Var}(S_t)} = f'(\text{E}[S_t]) + o(\text{Var}(S_t)). \tag{3.7}$$

In particular, if $\lim_{t \to 0} \text{E}[S_t] = S_0$ and $\lim_{t \to 0} \text{Var}(S_t) = 0$, then

$$\lim_{t \to 0} \frac{\text{Cov}(f(S_t), S_t)}{\text{Var}(S_t)} = f'(S_0).$$

In the proposition, the notation $o(\cdot)$ is used. A function $f(x)$ is in $o(g(x))$ if $\lim_{x \to 0} f(x)/g(x) = 0$.

Proof. Write $f(S_t) = f(\text{E}[S_t]) + f'(\text{E}[S_t])(S_t - \text{E}[S_t]) + R$, where by construction the error term is

$$R = \left(\frac{f(S_t) - f(\text{E}[S_t])}{S_t - \text{E}[S_t]} - f'(\text{E}[S_t]) \right) (S_t - \text{E}[S_t]).$$

We get $\text{Cov}(f(S_t), S_t) = f'(\text{E}[S_t]) \text{Var}(S_t) + \text{Cov}(R, S_t)$, where, for any $M > 0$, we can write

$$\text{Cov}(R, S_t) = \text{E}[(R - \text{E}[R])(S_t - \text{E}[S_t]) I\{|S_t - \text{E}[S_t]| \leq M \text{Var}(S_t)\}] \tag{3.8}$$

$$+ \text{E}[(R - \text{E}[R])(S_t - \text{E}[S_t]) I\{|S_t - \text{E}[S_t]| > M \text{Var}(S_t)\}]. \tag{3.9}$$

Since $\text{Cov}(R, S_t)$ is a finite number, for a given small $\varepsilon > 0$ we can take M large enough so that the absolute value of the second term (3.9) is smaller than ε. We now deal with the first term (3.8) and first notice that its absolute value is bounded from above by

$$|\text{E}[R] \text{E}[(S_t - \text{E}[S_t]) I\{|S_t - \text{E}[S_t]| \leq M \text{Var}(S_t)\}]| = o(\text{Var}(S_t))$$

plus the absolute value of

$$\text{E}\left[\left(\frac{f(S_t) - f(\text{E}[S_t])}{S_t - \text{E}[S_t]} - f'(\text{E}[S_t]) \right) (S_t - \text{E}[S_t])^2 \mid |S_t - \text{E}[S_t]| \leq M \text{Var}(S_t) \right],$$

which in turn is bounded from above by

$$\max_{|x-\mathrm{E}[S_t]|\leq M\,\mathrm{Var}(S_t)}\left|\frac{f(x)-f(\mathrm{E}[S_t])}{x-\mathrm{E}[S_t]}-f'(\mathrm{E}[S_t])\right|\,\mathrm{Var}(S_t)=o(\mathrm{Var}(S_t)).$$

Therefore,

$$\left|\frac{\mathrm{Cov}(f(S_t),S_t)}{\mathrm{Var}(S_t)}-f'(\mathrm{E}[S_t])\right|\leq o(\mathrm{Var}(S_t))+\varepsilon.$$

Since $\varepsilon>0$ was arbitrary, we conclude that

$$\left|\frac{\mathrm{Cov}(f(S_t),S_t)}{\mathrm{Var}(S_t)}-f'(\mathrm{E}[S_t])\right|=o(\mathrm{Var}(S_t)). \qquad \square$$

Example 3.11 (Delta hedging of call options). Recall the Black–Scholes formula
(1.7) for call options: $C_0 = S_0\Phi(d_1) - B_0 K\Phi(d_1 - \sigma\sqrt{T})$. The partial derivative
of C_0 with respect to S_0 is given by

$$\frac{\partial C_0}{\partial S_0} = \Phi(d_1) + \left(S_0\phi(d_1) - B_0 K\phi(d_1 - \sigma\sqrt{T})\right)\frac{\partial d_1}{\partial S_0},$$

where $\phi(z) = \exp\{-z^2/2\}/\sqrt{2\pi}$ is the standard normal density function. Moreover,

$$d_1^2 = \left(\frac{\log(S_0/(B_0 K))}{\sigma\sqrt{T}}\right)^2 + \frac{\sigma^2 T}{4} + \log(S_0/(B_0 K)),$$

$$(d_1 - \sigma\sqrt{T})^2 = \left(\frac{\log(S_0/(B_0 K))}{\sigma\sqrt{T}}\right)^2 + \frac{\sigma^2 T}{4} - \log(S_0/(B_0 K))$$

$$= d_1^2 - 2\log(S_0/(B_0 K)).$$

Therefore,

$$\frac{\partial C_0}{\partial S_0} = \Phi(d_1) + \left(S_0\phi(d_1) - B_0 K\phi(d_1 - \sigma\sqrt{T})\right)\frac{\partial d_1}{\partial S_0}$$

$$= \Phi(d_1) + \left(S_0\frac{e^{-d_1^2/2}}{\sqrt{2\pi}} - B_0 K e^{\log(S_0/B_0 K)}\frac{e^{-d_1^2/2}}{\sqrt{2\pi}}\right)\frac{\partial d_1}{\partial S_0}$$

$$= \Phi(d_1).$$

We conclude that delta hedging of call options in the Black–Scholes model amounts
to holding $\Phi(d_1)$ units of the underlying asset.

3.5.1   *Dynamic Hedging of a Call Option*

Suppose we have issued a European call option with strike price K on the value
of a share of a stock at time $T > 0$ that does not pay dividends from now until
maturity of the option. We want to delta hedge the call option at times $k\Delta = kT/n$
for $k = 0, \ldots, n - 1$ and analyze the distribution of the hedging cost.

Suppose that the market prices the call option such that the price C_0 corresponds
to using the Black–Scholes formula with implied volatility σ_I. This means that C_0
is the discounted expected option payoff given that the value of the share at time T
is lognormally distributed, $\mathrm{LN}(\log S_0 + (r - \sigma_I^2/2)T, \sigma_I^2 T)$, where r is the time-T
zero-coupon bond rate and S_0 is the current share price.

Suppose also that we believe that the log-returns $\log(S_{t+\Delta t}/S_t)$ of the spot price
over nonoverlapping periods of length Δt are independent and $\mathrm{N}(\mu\Delta t, \sigma_S^2 \Delta t)$-
distributed so that the spot price at maturity of the option is $\mathrm{LN}(\log S_0 + \mu T, \sigma_S^2 T)$-
distributed. Suppose also that we believe that the implied volatility of the option will
not change from now until maturity.

To make the argument clearer, we set the interest rate to zero. We will analyze
the costs of repeated delta hedging over the lifetime of the option. Therefore, we
cannot ignore transaction costs. To make things simple, we suppose that transaction
costs only come from the bid–ask spread: when modifying the delta hedge position
in the stock we buy for the higher ask price and sell for the lower bid price. We
assume that the bid–ask spread does not vary over time. Moreover, we take the time
t share price S_t to be a middle price (the average of the bid and ask prices) and also
assume that the settlement price on which the option payoff is calculated is a middle
price. We take the time t bid and ask prices to be $S_t - \lambda$ and $S_t + \lambda$, respectively,
where λ is constant.

At time 0 we approximate

$$C_{\Delta t} \approx C_0 + \frac{\partial C_0}{\partial S_0}(S_{\Delta t} - S_0).$$

This means that we should take a position $\partial C_0/\partial S_0$ in the stock and a position
$C_0 - (\partial C_0/\partial S_0)S_0$ in cash. The income C_0 from issuing the option matches precisely
the initial hedging costs if there are no transaction costs. With transaction costs, the
initial hedging cost is $\lambda \partial C_0/\partial S_0$. At time Δt we approximate

$$C_{2\Delta t} \approx C_{\Delta t} + \frac{\partial C_{\Delta t}}{\partial S_{\Delta t}}(S_{2\Delta t} - S_{\Delta t}).$$

This means that we should modify the position in the underlying stock from
$\partial C_0/\partial S_0$ to $\partial C_{\Delta t}/\partial S_{\Delta t}$ and modify the cash position from $C_0 - (\partial C_0/\partial S_0)S_0$ to
$C_{\Delta t} - (\partial C_{\Delta t}/\partial S_{\Delta t})S_{\Delta t}$. The cost (possibly negative) for modifying the hedge is

$$\left(\frac{\partial C_{\Delta t}}{\partial S_{\Delta t}} - \frac{\partial C_0}{\partial S_0}\right)S_{\Delta t} + C_{\Delta t} + \left|\frac{\partial C_{\Delta t}}{\partial S_{\Delta t}} - \frac{\partial C_0}{\partial S_0}\right|\lambda - \frac{\partial C_{\Delta t}}{\partial S_{\Delta t}}S_{\Delta t} - \left(C_0 - \frac{\partial C_0}{\partial S_0}S_0\right)$$

$$= C_{\Delta t} - C_0 - \frac{\partial C_0}{\partial S_0}(S_{\Delta t} - S_0) + \left|\frac{\partial C_{\Delta t}}{\partial S_{\Delta t}} - \frac{\partial C_0}{\partial S_0}\right|\lambda.$$

At time $k\Delta t$ we approximate

$$C_{(k+1)\Delta t} \approx C_{k\Delta t} + \frac{\partial C_{k\Delta t}}{\partial S_{k\Delta t}}\left(S_{(k+1)\Delta t} - S_{k\Delta t}\right).$$

This means that we should modify the position in the stock from $\partial C_{(k-1)\Delta t}/\partial S_{(k-1)\Delta t}$ to $\partial C_{k\Delta t}/\partial S_{k\Delta t}$ and the cash position from $C_{(k-1)\Delta t} - (\partial C_{(k-1)\Delta t}/\partial S_{(k-1)\Delta t})S_{(k-1)\Delta t}$ to $C_{k\Delta t} - (\partial C_{k\Delta t}/\partial S_{k\Delta t})S_{k\Delta t}$. The cost for modifying the hedge is

$$\left(\frac{\partial C_{k\Delta t}}{\partial S_{k\Delta t}} - \frac{\partial C_{(k-1)\Delta t}}{\partial S_{(k-1)\Delta t}}\right)S_{k\Delta t} + \left|\frac{\partial C_{k\Delta t}}{\partial S_{k\Delta t}} - \frac{\partial C_{(k-1)\Delta t}}{\partial S_{(k-1)\Delta t}}\right|\lambda + C_{k\Delta t} - \frac{\partial C_{k\Delta t}}{\partial S_{k\Delta t}}S_{k\Delta t}$$

$$- \left(C_{(k-1)\Delta t} - \frac{\partial C_{(k-1)\Delta t}}{\partial S_{(k-1)\Delta t}}S_{(k-1)\Delta t}\right)$$

$$= C_{k\Delta t} - C_{(k-1)\Delta t} - \frac{\partial C_{(k-1)\Delta t}}{\partial S_{(k-1)\Delta t}}\left(S_{k\Delta t} - S_{(k-1)\Delta t}\right) + \left|\frac{\partial C_{k\Delta t}}{\partial S_{k\Delta t}} - \frac{\partial C_{(k-1)\Delta t}}{\partial S_{(k-1)\Delta t}}\right|\lambda.$$

At time T we pay the option payoff $C_T = \max(S_T - K, 0)$ in cash and receive money from selling off the position in the stock and cash. The cost is therefore

$$\max(S_T - K, 0) - \left(\frac{\partial C_{(n-1)\Delta t}}{\partial S_{(n-1)\Delta t}}(S_T - \lambda) + C_{(n-1)\Delta t} - \frac{\partial C_{(n-1)\Delta t}}{\partial S_{(n-1)\Delta t}}S_{(n-1)\Delta t}\right)$$

$$= C_{n\Delta t} - C_{(n-1)\Delta t} - \frac{\partial C_{(n-1)\Delta t}}{\partial S_{(n-1)\Delta t}}\left(S_{n\Delta t} - S_{(n-1)\Delta t}\right) + \frac{\partial C_{(n-1)\Delta t}}{\partial S_{(n-1)\Delta t}}\lambda.$$

The aggregate hedging cost is therefore

$$\max(S_T - K, 0) - C_0 - \sum_{k=1}^{n}\frac{\partial C_{(k-1)\Delta t}}{\partial S_{(k-1)\Delta t}}\left(S_{k\Delta t} - S_{(k-1)\Delta t}\right)$$

$$+ \lambda\left(\frac{\partial C_0}{\partial S_0} + \frac{\partial C_{(n-1)\Delta t}}{\partial S_{(n-1)\Delta t}} + \sum_{k=1}^{n-1}\left|\frac{\partial C_{k\Delta t}}{\partial S_{k\Delta t}} - \frac{\partial C_{(k-1)\Delta t}}{\partial S_{(k-1)\Delta t}}\right|\right).$$

Let us investigate the performance of the delta hedge numerically. In the first numerical, example the market's implied volatility coincides with ours, $\sigma_I = \sigma_S$. In the second example, we believe the market's implied volatility is too high, $\sigma_I > \sigma_S$, and we could make a profit if our view were correct.

Example 3.12 (Equal volatilities). Suppose that $S_0 = 100$, $K = 110$, $T = 0.5$, $\sigma_I = 0.2$, $r = 0$, $n = 50$, $\Delta t = T/50$, $\mu = -\sigma_I^2/2$, and $\sigma_S = \sigma_I$. This means that the distribution of the log returns coincides with the view of the market in the sense that $C_0 = \mathrm{E}[\max(S_T - K, 0)]$ and

$$S_{k\Delta t} = S_0 \prod_{j=1}^{k} \exp\left\{\mu \Delta t + \sigma_S \sqrt{\Delta t}\, Z_j\right\} = S_0 \prod_{j=1}^{k} \exp\left\{-\frac{\sigma_I^2}{2}\Delta t + \sigma_I \sqrt{\Delta t}\, Z_j\right\},$$

where $Z_1, \ldots, Z_k$ are independent and $N(0, 1)$-distributed. In particular, $\mathrm{E}[S_{k\Delta t}] = S_0$. The upper left plot in Fig. 3.1 shows a histogram of the aggregate hedging result based on 10^4 simulated spot price trajectories when trading costs are not included. The upper right plot in Fig. 3.1 shows the corresponding result when trading costs are included, $\lambda = 0.5$.

Example 3.13 (Unequal volatilities). Suppose that $S_0 = 100$, $K = 110$, $T = 0.5$, $\sigma_I = 0.2$, $r = 0$, $n = 50$, $\Delta t = T/50$, $\mu = 0$, and $\sigma_S = 0.15$. This means that the distribution of the log returns is such that $C_0 > \mathrm{E}[\max(S_T - K, 0)]$, we expect the spot price to fluctuate rather slowly ($\sigma_S < \sigma_I$), and

$$S_{k\Delta t} = S_0 \prod_{j=1}^{k} \exp\{\mu \Delta t + \sigma_S \sqrt{\Delta t}\, Z_j\} = S_0 \prod_{j=1}^{k} \exp\left\{\sigma_S \sqrt{\Delta t}\, Z_j\right\},$$

where $Z_1, \ldots, Z_k$ are independent and $N(0, 1)$-distributed. In particular, $\mathrm{E}[S_{k\Delta t}] > S_0$. The lower left plot in Fig. 3.1 shows a histogram of the aggregate hedging result based on 10^4 simulated spot price trajectories when trading costs are not included. The lower right plot in Fig. 3.1 shows the corresponding result when trading costs are included, $\lambda = 0.5$.

Examples 3.12 and 3.13 show it is only wise to issue or short-sell a call option and delta hedge the option if we consider the difference $\sigma_I - \sigma_S$ between the implied volatility and our subjective assessment of the volatility to be positive and large enough. Moreover, we must take the bid–ask spread of the underlying stock into account in order to accurately assess whether a short position in the option together with dynamic delta hedging is a good deal. If before maturity of the option the implied volatility changes so that $\sigma_I - \sigma_S$ is no longer large enough, then we should close the short position in the option and the position in the stock. Otherwise, we are likely to start accumulating losses.

3.6   Immunization of Cash Flows

Let $\mathbf{r} = (r_1, \ldots, r_n)^{\mathrm{T}}$ be a vector whose components are the current zero rates for the maturity times $0 < t_1 < \cdots < t_n$, and let $\Delta \mathbf{r}$ be a vector whose components are instantaneous changes in the zero rates. We consider a liability whose present

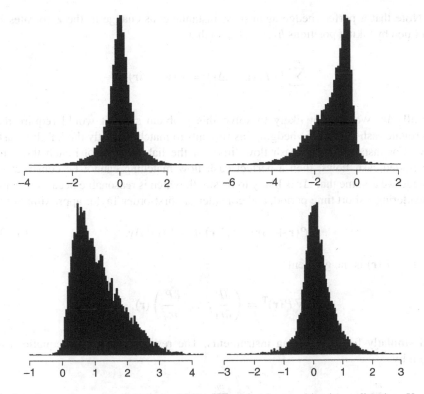

Fig. 3.1 Aggregate results of dynamic delta hedging of a short position in a call option. *Upper plots*: hedging results for $\sigma_I = \sigma_S = 0.2$. *Lower plots*: hedging results for $\sigma_I = 0.2 > 0.15 = \sigma_S$. The plots to the *right* show the hedging results when transaction costs are included

value is given by $P(\mathbf{r})$. If the liability is a deterministic cash flow $\{(c_k, t_k) : k = 1, \ldots, n\}$, then

$$P(\mathbf{r}) = \sum_{i=1}^{n} c_i e^{-r_i t_i}.$$

We face the risk that the value of the liability will increase due to an unfavorable outcome of $\Delta \mathbf{r}$. Therefore, we want to purchase a hedging portfolio that costs $P(\mathbf{r})$ with the property that the current net value of the hedging portfolio and the liability is zero and such that the net value is immune (or at least insensitive) to zero rate changes $\Delta \mathbf{r}$. The hedging portfolio is made up of positions $h_1, \ldots, h_m$ in m hedging instruments whose present values are given by $P_k(\mathbf{r})$ for $k = 1, \ldots, m$. The hedging instruments may be thought of as traded bonds. We will present a widely used approach, called immunization, to the hedging of the liability over a short time period.

Note that a perfect hedge against an instantaneous change in the zero rates is obtained by taking positions $h_1, \ldots, h_m$ so that

$$\sum_{k=1}^{m} h_k P_k(\mathbf{r} + \Delta \mathbf{r}) = P(\mathbf{r} + \Delta \mathbf{r})$$

for all $\Delta \mathbf{r}$. We are not likely to solve this problem since it would require the aggregate cash flow of the hedging instruments to match perfectly the liability cash flow. For instance, if the cash flow times of the liability do not match those of the traded cash flows, then a perfect cash flow matching cannot be constructed. Instead we assume that $\Delta \mathbf{r}$ is likely to be small, which is reasonable because we are considering a short time period, and consider the first-order Taylor approximation:

$$P(\mathbf{r} + \Delta \mathbf{r}) \approx P(\mathbf{r}) + \nabla P(\mathbf{r})^{\mathsf{T}} \Delta \mathbf{r}, \qquad (3.10)$$

where $\nabla P(\mathbf{r})$ is the gradient

$$\nabla P(\mathbf{r})^{\mathsf{T}} = \left(\frac{\partial P}{\partial r_1}, \ldots, \frac{\partial P}{\partial r_n} \right)(\mathbf{r}),$$

and similarly for the hedging instruments. The resulting system of equations is given by

$$\sum_{k=1}^{m} h_k P_k(\mathbf{r}) = P(\mathbf{r}) \quad \text{and} \quad \sum_{k=1}^{m} h_k \nabla P_{k,}(\mathbf{r})^{\mathsf{T}} \Delta \mathbf{r} = \nabla P(\mathbf{r})^{\mathsf{T}} \Delta \mathbf{r}.$$

In vector notation the two equations are written as

$$\mathbf{h}^{\mathsf{T}} \mathbf{P}(\mathbf{r}) = P(\mathbf{r}) \quad \text{and} \quad \mathbf{h}^{\mathsf{T}} \nabla \mathbf{P}(\mathbf{r}) \Delta \mathbf{r} = \nabla P(\mathbf{r})^{\mathsf{T}} \Delta \mathbf{r},$$

where $\mathbf{P}(\mathbf{r}) = (P_1(\mathbf{r}), \ldots, P_m(\mathbf{r}))^{\mathsf{T}}$ and

$$\nabla \mathbf{P}(\mathbf{r}) = \begin{pmatrix} \frac{\partial P_1}{\partial r_1}(\mathbf{r}) & \cdots & \frac{\partial P_1}{\partial r_n}(\mathbf{r}) \\ \vdots & \ddots & \vdots \\ \frac{\partial P_m}{\partial r_1}(\mathbf{r}) & \cdots & \frac{\partial P_m}{\partial r_n}(\mathbf{r}) \end{pmatrix}.$$

Instead of considering the change $\Delta \mathbf{r}$ of the zero rates as a random vector, we select a number of deterministic scenarios denoted $\Delta \mathbf{r}_1, \ldots, \Delta \mathbf{r}_q$ and look for positions $h_1, \ldots, h_m$ that make the portfolio immune to these scenarios. To find such positions $h_1, \ldots, h_m$, we must solve the following system of equations:

$$\sum_{k=1}^{m} h_k P_k(\mathbf{r}) = P(\mathbf{r}), \tag{3.11}$$

$$\sum_{k=1}^{m} h_k \nabla P_k(\mathbf{r})^{\mathrm{T}} \Delta \mathbf{r}_l = \nabla P(\mathbf{r})^{\mathrm{T}} \Delta \mathbf{r}_l \quad \text{for } l = 1, \ldots, q. \tag{3.12}$$

Since both the left- and right-hand sides are linear in $\Delta \mathbf{r}_l$, only the directions $\Delta \mathbf{r}_l / |\Delta \mathbf{r}_l|$ of the scenario vectors are relevant, not their norms $|\Delta \mathbf{r}_l|$.

The changes in the zero rates for different maturity times have a strong positive dependence; typically they all move up or they all move down. Without a detailed analysis it therefore seems wise to first consider the normalized scenario $\Delta \mathbf{r}_1 = -\mathbf{1}$, which corresponds to a downward parallel shift of the zero-rate curve. As protection against a parallel shift, we take positions $h_1, \ldots, h_m$, solving

$$\mathbf{h}^{\mathrm{T}} \mathbf{P} = P \quad \text{and} \quad \mathbf{h}^{\mathrm{T}} \nabla \mathbf{P} \mathbf{1} = \nabla P^{\mathrm{T}} \mathbf{1}.$$

Here we have suppressed the dependence on the current zero rates by writing $P = P(\mathbf{r})$ and $\mathbf{P} = \mathbf{P}(\mathbf{r})$. Since there are only two equations to be satisfied, immunization against small parallel shifts requires only two hedging instruments. It suffices to pick indices j, k and solve

$$\begin{pmatrix} h_j & h_k \end{pmatrix} \begin{pmatrix} P_j \\ P_k \end{pmatrix} = P \quad \text{and} \quad \begin{pmatrix} h_j & h_k \end{pmatrix} \begin{pmatrix} \nabla P_j^{\mathrm{T}} \mathbf{1} \\ \nabla P_k^{\mathrm{T}} \mathbf{1} \end{pmatrix} = \nabla P^{\mathrm{T}} \mathbf{1}.$$

The two equations may be written in matrix notation as

$$\begin{pmatrix} P_j & P_k \\ \sum_{i=1}^{n} \frac{\partial P_j}{\partial r_i} & \sum_{i=1}^{n} \frac{\partial P_k}{\partial r_i} \end{pmatrix} \begin{pmatrix} h_j \\ h_k \end{pmatrix} = \begin{pmatrix} P \\ \sum_{i=1}^{n} \frac{\partial P}{\partial r_i} \end{pmatrix}. \tag{3.13}$$

Finding a portfolio h_j, h_k that is immune to parallel shifts of the zero-rate curve is therefore simply a matter of inverting the 2×2 matrix.

Example 3.14 (Deterministic cash flows). Consider a deterministic cash flow that produces the cash flow c_k at time t_k for $k = 1, \ldots, n$. The current value of the cash flow is

$$P = P(\mathbf{r}) = \sum_{k=1}^{n} c_k e^{-r_k t_k},$$

and the partial derivatives are given by

$$\frac{\partial P}{\partial r_k} = \frac{\partial P}{\partial r_k}(\mathbf{r}) = -t_k c_k e^{-r_k t_k}.$$

We conclude that if the liability and the hedging instruments can be represented as deterministic cash flows, then all the expressions needed to determine the immunization portfolio are easy to compute.

Remark 3.1 (Duration). A normalized measure of sensitivity of an interest rate security to an instantaneous change $\Delta \mathbf{r}$ in the zero rate is

$$D(P, \Delta \mathbf{r}) = -\frac{\nabla P(\mathbf{r})^{\mathrm{T}} \Delta \mathbf{r}}{P(\mathbf{r})}.$$

Clearly, this choice of sensitivity measure makes sense only when $P(\mathbf{r}) \neq 0$. The sensitivities $D = D(P, \mathbf{1})$ and $D_k = D(P_k, \mathbf{1})$ are called the durations of the liability and hedging instruments, respectively. The durations measure the sensitivities of the values of the liability and hedging instrument to a small parallel shift in the zero-rate curve.

If $P(\mathbf{r})$ is the present value of a deterministic cash flow that produces the cash flow c_k at time t_k, for $k = 1, \ldots, n$, then it follows from the computations in Example 3.14 that

$$D = \sum_{k=1}^{n} t_k \frac{c_k e^{-r_k t_k}}{\sum_{j=1}^{n} c_j e^{-r_j t_j}}.$$

Therefore, the duration D is the weighted mean value of the cash flow times, where the kth weight is the fraction of the present value of the entire cash flow that refers to the kth cash flow time. In particular, the duration of a zero-coupon bond is the maturity time of the bond.

The immunization equation (3.13) can be written in terms of durations as follows:

$$\begin{pmatrix} P_j & P_k \\ P_j D_j & P_k D_k \end{pmatrix} \begin{pmatrix} h_j \\ h_k \end{pmatrix} = \begin{pmatrix} P \\ PD \end{pmatrix}.$$

The solution is given by

$$\begin{pmatrix} h_j \\ h_k \end{pmatrix} = \frac{1}{P_j P_k (D_k - D_j)} \begin{pmatrix} P_k P(D_k - D) \\ PP_j(D - D_j) \end{pmatrix}.$$

In particular, choosing the bonds such that $D_j < D < D_k$ ensures that the solution corresponds to long positions in both bonds.

Example 3.15 (Interest rate swap). In this example we illustrate how interest rate swaps can be useful for immunization against the effects of parallel shifts of the zero-rate curve for an insurance company whose liability cash flows stretch far into the future.

Many life insurers, and certain nonlife insurers, face liabilities corresponding to payments 10, 20, and even more than 30 years into the future. If the liabilities are valued using a zero-rate curve corresponding to government bonds, then the assets of the insurer should include government bonds with long maturities in order to make the net value of assets and liabilities insensitive to changes in the zero-rate curve. However, due to the rather limited supply of government bonds with long maturities, it is difficult and costly to set up such a bond portfolio.

Suppose the insurer's liabilities have the current value $P_L(\mathbf{r})$ and that the asset portfolio intended to match the liabilities has current value $P_A(\mathbf{r})$. Similar to Eqs. (3.11) and (3.12), the insurer would be immune to instantaneous parallel shifts of the zero-rate curve if

$$P_A(\mathbf{r}) = P_L(\mathbf{r}) \quad \text{and} \quad \nabla P_A(\mathbf{r})^{\mathrm{T}} \mathbf{1} = \nabla P_L(\mathbf{r})^{\mathrm{T}} \mathbf{1}.$$

Here we assume that the insurer's asset portfolio mainly consists of long positions in bonds with relatively short maturities, whereas it faces liabilities that may require payments to policyholders for a very long time, so that the duration of the assets is shorter than that of the liabilities. In this case,

$$P_A(\mathbf{r}) = P_L(\mathbf{r}) \quad \text{and} \quad 0 < \nabla P_A(\mathbf{r})^{\mathrm{T}}(-\mathbf{1}) < \nabla P_L(\mathbf{r})^{\mathrm{T}}(-\mathbf{1}).$$

The second inequality implies that if interest rates fall (by a parallel shift), then the asset portfolio will be worth less than the liability. Moreover, the sensitivity of the value of the liabilities to an increase or decrease in interest rates is greater than that of the assets.

We will now show how the mismatch between the assets' and liabilities' sensitivities to interest rate changes can be handled by the insurer entering an interest swap agreement as the swap's fixed-rate receiver. Zero rates from government bonds are not the same as zero rates from interest rate swaps in the local currency. However, changes in the two zero-rate curves are highly correlated, and it is a reasonable approximation to assume that, over a short time period, the swap zero rates can be expressed as $\mathbf{r} + \mathbf{s}$, where $\mathbf{s}$ is a constant vector representing the credit spreads for different maturity times (representing the fact that a commercial bank may have a higher probability than a government to fail to deliver a contracted cash flow).

From Example 1.3 (with the same kind of swap and notation) we know that the initial value of the swap for the fixed-rate receiver is

$$L\left(c \sum_{k=1}^{n} e^{-(r_k+s_k)t_k} + e^{-(r_n+s_n)t_n} - 1\right) = 0,$$

where c is the number for which the initial value is zero. Therefore, the value of the asset portfolio after entering the swap remains the same, and $P_A(\mathbf{r}) = P_L(\mathbf{r})$. However, by adding the swap to the asset portfolio, the sensitivity of the asset portfolio value as a function of the zero rates, in the direction $-\mathbf{1}$, changes from $\nabla P_A(\mathbf{r})^{\mathrm{T}}(-\mathbf{1})$ to

$$\nabla P_A(\mathbf{r})^{\mathrm{T}}(-1) + Lc\sum_{k=1}^{n} t_k e^{-(r_k+s_k)t_k} + Lt_n e^{-(r_n+s_n)t_n} > \nabla P_A(\mathbf{r})^{\mathrm{T}}(-1).$$

In particular, by entering into a swap agreement with an appropriate notional principal as the fixed-rate receiver, the sensitivity of the insurer's assets to parallel shifts in the zero-rate curve can be modified to equal that of the liabilities.

3.6.1 Immunization and Principal Component Analysis

It is reasonable to consider scenarios other than a parallel shift. But how do we select good scenarios $\Delta\mathbf{r}_1,\ldots,\Delta\mathbf{r}_q$ in the sense that they correspond to likely zero rate changes? Additionally, there should be as little redundancy as possible among the scenarios. We now present a useful approach to finding suitable scenarios called principal component analysis (PCA).

The symmetric and positive-definite matrix $\mathrm{Cov}(\Delta\mathbf{r})$ may be expressed as the product $\mathrm{Cov}(\Delta\mathbf{r}) = \mathbf{ODO}^{\mathrm{T}}$, where $\mathbf{D}$ is a diagonal matrix with the (strictly positive) eigenvalues $\lambda_1,\ldots,\lambda_n$ of $\mathrm{Cov}(\Delta\mathbf{r})$ as diagonal elements and $\mathbf{O}$ is an orthogonal matrix (meaning that $\mathbf{OO}^{\mathrm{T}} = \mathbf{O}^{\mathrm{T}}\mathbf{O} = \mathbf{I}$ is the identity matrix) whose columns $\mathbf{o}_1,\ldots,\mathbf{o}_n$ are eigenvectors of $\mathrm{Cov}(\Delta\mathbf{r})$, orthogonal and of length one. We may without loss of generality assume that the columns of $\mathbf{D}$ and $\mathbf{O}$ are ordered so that the diagonal elements in $\mathbf{D}$ appear in descending order. Set $\Delta\mathbf{r}^* = \mathbf{O}^{\mathrm{T}}(\Delta\mathbf{r}-\mathrm{E}[\Delta\mathbf{r}])$ and note that

$$\mathrm{Cov}(\Delta\mathbf{r}^*) = \mathrm{E}[\mathbf{O}^{\mathrm{T}}(\Delta\mathbf{r}-\mathrm{E}[\Delta\mathbf{r}])(\Delta\mathbf{r}-\mathrm{E}[\Delta\mathbf{r}])^{\mathrm{T}}\mathbf{O}] = \mathbf{O}^{\mathrm{T}}\mathrm{Cov}(\Delta\mathbf{r})\mathbf{O} = \mathbf{D},$$

i.e., the components of $\Delta\mathbf{r}^*$ are uncorrelated and have variances $\lambda_1 \geq \cdots \geq \lambda_n$, in that order. The transformation of $\Delta\mathbf{r}$ into $\Delta\mathbf{r}^* = \mathbf{O}^{\mathrm{T}}(\Delta\mathbf{r} - \mathrm{E}[\Delta\mathbf{r}])$ has a natural geometric interpretation. Consider a sample of independent copies of $\Delta\mathbf{r}$ and its scatter plot. The point cloud is first centered to have zero mean. Suppose that it has the shape of an ellipsoid. Multiplication by an orthogonal matrix corresponds to rotating the ellipsoidal point cloud. Here, it is rotated until the main axes are parallel to the coordinate axes.

We may write, with $\mathbf{e}_k$ being the kth standard unit vector in $\mathbb{R}^n$,

$$\sum_{k=1}^{n}\Delta r_k \mathbf{e}_k = \Delta\mathbf{r} = \mathbf{OO}^{\mathrm{T}}\Delta\mathbf{r} = \mathrm{E}[\Delta\mathbf{r}] + \mathbf{O}\Delta\mathbf{r}^* = \mathrm{E}[\Delta\mathbf{r}] + \mathbf{O}\sum_{k=1}^{n}\Delta r_k^* \mathbf{e}_k$$

$$= \mathrm{E}[\Delta\mathbf{r}] + \sum_{k=1}^{n}\Delta r_k^* \mathbf{o}_k.$$

This shows that the components of $\Delta \mathbf{r}$ are correlated when expressed in terms of the standard basis $\{\mathbf{e}_1, \ldots, \mathbf{e}_n\}$ for $\mathbb{R}^n$; however, they are uncorrelated when expressed in the alternative orthonormal basis $\{\mathbf{o}_1, \ldots, \mathbf{o}_n\}$ of $\mathbb{R}^n$.

If we study

$$\sum_{k=1}^{j} \lambda_k \bigg/ \sum_{k=1}^{n} \lambda_k$$

as a function of j, then we typically find that its value for $j = 1$ is close to one. This means that the variability of $\Delta \mathbf{r}$ is mainly in the direction of the first principal component $\mathbf{o}_1$ and that $\Delta \mathbf{r} \approx \mathrm{E}[\Delta \mathbf{r}] + \Delta r_1^* \mathbf{o}_1$ is a reasonably accurate approximation. To improve the accuracy of the approximation, we may pick a small j, $j = 1, 2$, or 3, say, and approximate $\Delta \mathbf{r}$ by

$$\mathrm{E}[\Delta \mathbf{r}] + \sum_{k=1}^{j} \Delta r_k^* \mathbf{o}_k.$$

For $\Delta \mathbf{r}$ corresponding to zero rate changes over a short time interval, we typically find that $\mathrm{E}[\Delta \mathbf{r}] \approx \mathbf{0}$ and that $\mathbf{o}_1 \approx \pm 1/\sqrt{n}$. In particular, the first principal component represents the most likely scenario corresponding to a parallel shift in the zero-rate curve. The second principal component $\mathbf{o}_2$ typically is a vector whose first components are positive and whose remaining components are negative (or vice versa). This eigenvector corresponds to an increase in zero rates for short maturity times and a decrease in zero rates for long maturity times (or vice versa). The arguments so far strongly indicate that a good choice for a hedging portfolio is obtained by solving

$$\sum_{k=1}^{m} h_k P_k(\mathbf{r}) = P(\mathbf{r}), \tag{3.14}$$

$$\sum_{k=1}^{m} h_k \nabla P_k(\mathbf{r})^{\mathrm{T}} \mathbf{o}_l = \nabla P(\mathbf{r})^{\mathrm{T}} \mathbf{o}_l \quad \text{for } l = 1, \ldots, q, \tag{3.15}$$

where $\mathbf{o}_k$ are the principal components and q is the number of considered principal components. It is often sufficient to take q equal to two or three.

Example 3.16 (Immunization for a nonlife insurer). Consider a nonlife insurer who faces the random cash flow $(C_1, \ldots, C_n)$ over the next n quarters from the settlement of claims due to events that have already occurred and events that will occur during the current year. The insurer has estimated the expected sizes of the payments for the next n quarters, $\mathrm{E}[C_1], \ldots, \mathrm{E}[C_n]$. Ideally, from a quadratic hedging perspective, the insurer should construct a hedge against the randomness in the cash flow $(C_1, \ldots, C_n)$ by purchasing a portfolio of $\mathrm{E}[C_k]$ k-quarter zero-coupon bonds (with face value 1) for $k = 1, \ldots, n$. However, from a practical perspective,

Table 3.2 Available bonds

Bond	#1	#2	#3	#4
Bond price	105.14	108.45	109.98	112.39
Maturity (quarters)	1	7	18	40
Annual coupon	5.25	5.5	4.5	5.0
Face value	100	100	100	100

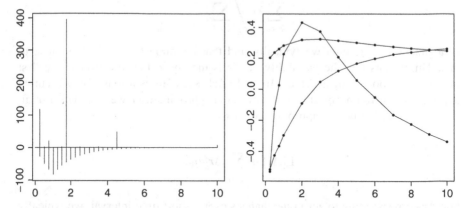

Fig. 3.2 *Left plot*: cash flow from bond portfolio (positive) and expected cash flow from claim settlement (negative). Time is on the x-axis and payment amounts are on the y-axis. *Right plot*: the first three principal components (parallel shifts, changes in slope, and changes in curvature) explain 85%, 13%, and 1%, respectively, of the variability (variance) in quarterly zero rate changes. The time to maturity is on the x-axis

the insurer has to hedge with whatever suitable hedging instruments are available. Here we assume that the available hedging instruments are the risk-free bonds, most of which are paying coupons, specified in Table 3.2. The aim of the insurer is therefore to buy a bond portfolio whose value closely matches the ideal quadratic hedging portfolio it cannot buy. To this end, we think of a portfolio consisting of $E[C_k]$ k-quarter zero-coupon bonds for $k = 1, \ldots, n$ as the liability and try to find a matching portfolio of available bonds that is immune to changes in the zero rates. The expected negative cash flow for the insurer is illustrated in the left-hand plot in Fig. 3.2 and numerically in Table 3.3. Data on quarterly changes in the zero rates for Swedish government bonds over more than a decade for the maturity times 3 months, 6 months, 9 months, and 1 up to 10 years were used in this example. The current zero rates are given in Table 3.4. The immunization approach is applied with three scenarios, o_1, o_2, o_3, that were taken to be the first three principal components extracted as eigenvectors of the empirical covariance matrix of quarterly zero rate changes. The scenarios o_1, o_2, o_3 are illustrated graphically in the right-hand plot in Fig. 3.2 and numerically in Table 3.4. The zero rates for the maturities not listed in Table 3.4 were obtained by linear interpolation between available zero rates.

A matching bond portfolio made up of the four bonds in Table 3.2 is found by solving the linear system of Eqs. (3.14) and (3.15), with $P_1(\mathbf{r}), \ldots, P_4(\mathbf{r})$ being the prices of the four available bonds. Computations similar to those in Example 3.14 lead to

Table 3.3 Expected cash flow of liability

Time	0.25	0.5	0.75	1.00	1.25	1.50	1.75	2.00	2.25	2.50
$E[C_k]$	27.20	49.47	67.70	82.63	67.65	55.39	45.35	37.13	30.40	24.89
Time	2.75	3.00	3.25	3.50	3.75	4.00	4.25	4.5	4.75	5.00
$E[C_k]$	20.38	16.68	13.66	11.18	9.16	7.50	6.14	5.02	4.11	3.37
Time	5.25	5.50	5.75	6.00	6.25	6.50	6.75	7.00	7.25	7.50
$E[C_k]$	2.76	2.26	1.85	1.51	1.24	1.01	0.83	0.68	0.56	0.46
Time	7.75	8.00	8.25	8.50	8.75	9.00	9.25	9.50	9.75	10.00
$E[C_k]$	0.37	0.31	0.25	0.20	0.17	0.14	0.11	0.09	0.08	0.06

First row: time to maturity (years) $k/4$; *second row*: expected payment $E[C_k]$, $k = 1, \ldots, 40$

$$P_1(\mathbf{r}) = 105.25\, e^{-r_1/4},$$

$$P_2(\mathbf{r}) = 5.5\, e^{-r_3 3/4} + 105.5\, e^{-r_7 7/4},$$

$$P_3(\mathbf{r}) = 4.5\, e^{-r_2 2/4} + \cdots + 4.5\, e^{-r_{14} 14/4} + 104.5\, e^{-r_{18} 18/4},$$

$$P_4(\mathbf{r}) = 5.0\, e^{-r_4 4/4} + \cdots + 5.0\, e^{-r_{36} 36/4} + 105.0\, e^{-r_{40} 40/4},$$

and, for $l = 1, \ldots, 4$,

$$\nabla P_1(\mathbf{r})^\mathsf{T} \mathbf{o}_l = -105.25 e^{-r_1/4} o_{l1},$$

$$\nabla P_2(\mathbf{r})^\mathsf{T} \mathbf{o}_l = -5.5 \frac{2}{4} e^{-r_3 3/4} o_{l1} - 105.5 \frac{7}{4} e^{-r_7 7/4} o_{j7},$$

$$\nabla P_3(\mathbf{r})^\mathsf{T} \mathbf{o}_l = -4.5 \frac{2}{4} e^{-r_2 2/4} o_{l1} - \cdots - 4.5 \frac{14}{4} e^{-r_{14} 14/4} o_{l14} - 104.5 \frac{18}{4} e^{-r_{18} 18/4} o_{l18},$$

$$\nabla P_4(\mathbf{r})^\mathsf{T} \mathbf{o}_l = -5.0 \frac{4}{4} e^{-r_4 4/4} o_{l1} - \cdots - 5.0 \frac{36}{4} e^{-r_{36} 36/4} o_{l36} - 105.0 \frac{40}{4} e^{-r_{40} 40/4} o_{l40}.$$

Similarly for the liability:

$$P(\mathbf{r}) = \sum_{k=1}^{40} E[C_k] e^{-r_k k/40},$$

$$\nabla P(\mathbf{r})^\mathsf{T} \mathbf{o}_l = -\sum_{k=1}^{40} \frac{k}{40} E[C_k] e^{-r_k k/40} o_{lk}, \quad l = 1, \ldots, 4.$$

The portfolio weights $h_1, \ldots, h_4$ can now be computed as the solution to (3.14) and (3.15), and the result is

$$h_1 = 1.12, \quad h_2 = 3.74, \quad h_3 = 0.46, \quad h_4 = 0.07.$$

Table 3.4 First three principal components o_1, o_2, o_3 of Swedish yield curve with time to maturity (years) and current zero rates (z. r.) as percentage (%)

Time	0.25	0.5	0.75	1	2	3	4	5	6	7	8	9	10
o_1	0.2036	0.2377	0.2609	0.2832	0.3207	0.3250	0.3147	0.3014	0.2876	0.2750	0.2652	0.2558	0.2477
o_2	−0.5163	−0.4252	−0.3655	−0.2961	−0.0907	0.0474	0.1185	0.1665	0.1978	0.2251	0.2395	0.2525	0.2613
o_3	−0.5289	−0.1257	0.0263	0.2278	0.4329	0.3745	0.2079	0.0578	−0.0517	−0.1642	−0.2238	−0.2894	−0.3380
z. r.	0.41	0.51	0.65	0.82	1.57	2.16	2.54	2.82	3.04	3.23	3.37	3.49	3.58

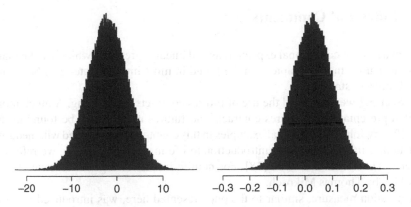

Fig. 3.3 *Left plot*: histogram of $P(\mathbf{r})e^{r_1/4} - P_{1/4}(\mathbf{r} + \Delta\mathbf{r})$ (without immunization). *Right plot*: histogram of difference between future values of matching bond portfolio and liability (with immunization)

The current value $P(\mathbf{r})$ of the liability, and therefore also that of the matching bond portfolio, was found to be 582.12. The positive cash flow generated from the bonds in the matching portfolio is illustrated in Fig. 3.2.

A small simulation study can be performed to evaluate the performance of the immunization technique. Changes $\Delta\mathbf{r}$ in the zero-rate curve were simulated 100,000 times from a multivariate normal distribution with mean $\boldsymbol{\mu}$ and covariance matrix $\boldsymbol{\Sigma}$ estimated from the data of quarterly zero rate changes. A vector $(X_1, \ldots, X_n)^{\mathrm{T}}$ with a normal distribution is drawn by first drawing independent standard normal variables $Z_1, \ldots, Z_n$ and then setting

$$\begin{pmatrix} X_1 \\ \vdots \\ X_n \end{pmatrix} = \begin{pmatrix} \mu_1 \\ \vdots \\ \mu_n \end{pmatrix} + \mathbf{A} \begin{pmatrix} Z_1 \\ \vdots \\ Z_n \end{pmatrix},$$

where $\mathbf{A}$ is a matrix satisfying $\boldsymbol{\Sigma} = \mathbf{A}\mathbf{A}^{\mathrm{T}}$. For each simulated zero-rate curve, $\mathbf{r} + \Delta\mathbf{r}$, the value of the liability and matching bond portfolio were recalculated, and the result is presented in Fig. 3.3. In the left-hand plot, the effect of investing the amount $P(\mathbf{r}) = 582.12$ in a one-quarter zero-coupon bond and hoping for the best is presented. The plot shows a histogram for the random variable $P(\mathbf{r})e^{r_1/4} - P_{1/4}(\mathbf{r} + \Delta\mathbf{r})$ representing the difference between the asset and liability values one quarter from now. The right-hand plot shows a histogram of the difference between the value of a matching bond portfolio, using the immunization approach, and that of the liability one quarter from now. Notice the difference in scale of the two probability distributions. The matching bond portfolio gives a vast reduction of the net value variability. Although the cash flows of the matching bond portfolio and the expected value of the liability are quite different (left-hand plot in Fig. 3.2), the values of the bond portfolio and the liability after one quarter are quite close.

3.7 Notes and Comments

More material on conditional expectations and linear regression, which are the main techniques used in this chapter, can be found in most intermediate-level books on probability and statistics.

In Sect. 3.2 we illustrated the use of futures contracts for hedging. A much more detailed presentation of futures contracts and futures markets can be found in the book [21] by John Hull. Several examples in this chapter are concerned with hedging of insurance liabilities. For an introduction to life insurance modeling we refer the reader to Hans Gerber's book [18]. For nonlife insurance a good reference is the book [34] by Thomas Mikosch.

A duration measure, similar to the one presented here, was introduced in 1938 by Frederick Macaulay and was originally thought of as an empirical measure of the length of a bond's cash flow. The insurance mathematician Frank Redington showed as early as 1952 how one could immunize a liability against value changes due to zero rate changes by applying a Taylor series expansion on the net value of a portfolio consisting of an asset portfolio and a liability. His rather general approach, similar to the immunization approach presented here, was not based on the duration concept. A historical account of the contributions of Macaulay and Redington is found in [27] by Geoffrey Poitras.

3.8 Exercises

In the exercises below, it is assumed, wherever applicable, that you can take positions corresponding to fractions of assets.

Exercise 3.1 (Annuity). Consider a life annuity contract that pays the holder a yearly fixed amount from a certain time until the death of the holder of the contract.

(a) Suppose the value today of the random cash flow C_k in k years is $E[C_k]e^{-r_k k}$, where r_k is the current k-year zero rate. All the current zero rates are assumed to be known. Determine an expression for the current value of an annuity whose holder is x years old, the annuity pays the yearly amount c starting from y years from today, and the current mortality rate is given by the Gompertz–Makeham formula $\mu_0(x) = A + Re^{\alpha x}$.

(b) What is the current value of an annuity that make a yearly payment of \$5,000 to an individual who is currently 65 years old if the k-year zero rate is 0.04 for all k, the parameters of the Gompertz–Makeham formula are $A = 0.002$, $R = e^{-12}$, and $\alpha = 0.12$, and the nearest yearly payment of the annuity is in 1 year.

Exercise 3.2 (Hedging with index futures). A bank has issued a European call option with strike price 110 on the value S_T of an index 1 year from today and wants to hedge the risk from its short position in the call option. The index is not traded, but there is a market of index futures contracts that can be used as hedging instruments, including a futures contract on the index value S_T at time T. Moreover, there is a zero-coupon bond available with price $B_0 = 97$ that matures in 1 year from today with a face value of 100 and a money market account that pays a daily interest rate of $r_{t-1,t}$ from day $t - 1$ to t (one unit of cash deposited on day $t - 1$ on the account has increased to $e^{r_{t-1,t}}$ units on day t).

The bank believes that the index value S_T can be modeled as $S_T = 100e^{0.035+0.1W}$, where W is standard normally distributed. The bank wants to use the futures strategy, with an initial cost equal to the current futures price, presented in Sect. 3.2 as a hedging instrument.

(a) Suppose the daily interest is deterministic and such that $r_{t-1,t} = r_{0,1}$ for $t = 1, \ldots, T$. Convince yourself that $r_{0,1} = -(1/365)\log(B_0/100)$, and determine the quadratic hedge of the short position in the call option with a zero-coupon bond and the futures strategy, i.e., determine the leverage of the futures strategy and the number of zero-coupon bonds to buy (fractions of units are allowed). Use the expressions computed in Exercise 1.4 to determine the standard deviation of the hedging error.

(b) Simulate from the standard normal distribution and make a histogram of the hedging error of the optimal portfolio in (a). Is the hedging error symmetric or skewed?

(c) Suppose that the daily interest rate is random and such that $r_{0,1} + \cdots + r_{T-1,T}$ is $N(0.0292, 0.0025)$-distributed and independent of W. The money market account can also be used as a hedging instrument to offset the additional randomness originating from the money market account in the futures strategy. Determine the quadratic hedge, i.e., the leverage of the futures strategy, the initial balance on the money market account, and the number of zero-coupon bonds to buy. Compare the result with the answer in (a).

(d) Simulate outcomes from two independent standard normally distributed random variables and make a histogram of the hedging error of the optimal portfolio in (c). Is the hedging error symmetric or skewed?

Exercise 3.3 (Leverage and margin calls). Suppose that a perceived arbitrage is identified on the oil market: there are futures contracts maturing in 2 months at a price of $99.95 per barrel of crude oil and the possibility of writing forward contracts with a forward price of $100 per barrel for delivery of crude oil in 2 months. Here it is assumed, to simplify the analysis, that the daily resettlement procedure for the futures contracts is replaced by a monthly resettlement procedure. Therefore, there are only three times $t = 0, 1, 2$ (months) to consider. It is also assumed that the interest rate on deposits on the margin account is zero.

(a) Determine an arbitrage portfolio.

The gain per barrel of crude oil on the arbitrage portfolio is small, and therefore the investor needs a highly leveraged position, corresponding to a large number of futures and forward contracts, to make a substantial amount of money on the arbitrage opportunity. However, a highly leveraged position may require a lot of

Table 3.5 Bond specifications

Bond	A	B	C	D
Bond price ($)	98.51	100.71	111.55	198.96
Maturity (years)	0.5	1	1.5	2
Annual coupon ($)	0	4	12	8
Face value ($)	100	100	100	200

Half of the annual coupon is paid every 6 months from today and including the time of maturity. The first coupon payment is in 6 months

cash to maintain the balance on the margin account in response to changes in the futures price and a subsequent margin call in 1 month. It is assumed that the investor has a buffer of $K = \$10,000$ at time $t = 1$, and if a margin call at $t = 1$ exceeds K, then the investor experiences great difficulties with respect to borrowing money. It is assumed that in this situation the investor may borrow money at a high interest rate of $R = 24\%$ per year.

(b) Express the value of the portfolio in (a) in 2 months from today as a function of the size of futures and forward contracts and a futures price of F_1 at time 1.

(c) Suppose that $F_1 = F_0 \exp\{-\sigma^2 \Delta/2 + \sigma\sqrt{\Delta} Z\}$, where Z is standard normally distributed, $\sigma = 0.6$, and $\Delta = 1/12$. Determine, numerically, the size h of the position in the futures contracts that maximizes the expected portfolio value in 2 months.

(d) Illustrate the distribution of the portfolio value in 2 months in a histogram for the portfolio in (c) that maximizes the expected payoff.

Exercise 3.4 (Immunization). Consider the bonds specified in Table 3.5. Half of the annual coupon is paid every 6 months, until and including the time of maturity, and the first coupon payment is in 6 months. Consider a company obliged to pay $\$100,000$ in 20 months. Form a bond portfolio consisting of only long positions in (some of or all of) bonds A–D that makes the net value of the bond portfolio and the liability immune to small parallel shifts in the zero-rate curve. The zero rates for arbitrary maturity times are determined by linear interpolation between zero rates for the maturity times of the four bonds.

Exercise 3.5 (Delta hedging with futures). Consider a futures contract and a forward contract on the value S_T of an asset at a future time T. Suppose that there is a known interest rate that applies to any loans and deposits until time T and that the cash flows of the futures and forward contracts will be delivered as contracted.

(a) Show that the current futures price equals the current forward price.

(b) Determine the delta hedge of a European call option on the future value of a stock market index in terms of positions in a risk-free bond and a futures contract on the future value of the stock market index, with the same time to maturity as the call option. Use Black's formula for call options to compute the option's delta.

Project 2 (Delta hedging of stock options). Find an exchange-traded European call option on the future value of a stock that does not pay dividends before the time of maturity of the option. Suggest a model for the daily returns of the stock and assume that the daily returns are independent and that the option's implied volatility does not change over time. Suggest a strategy for dynamic delta hedging of a short position in the call option, taking the bid–ask spread of the stock into account, and evaluate the distribution of the cost of the hedging strategy in a simulation study.

Project 3 (Delta hedging of index options). Find prices of traded European put and call options of the future value of a stock market index.

(a) Choose one European call option and investigate how to dynamically delta hedge a short position in the call option. Choose as hedging instrument an index futures contract with the same maturity as the call option. Suggest a model for the daily changes in the futures price; assume that the daily price changes are independent and that the option's implied volatility does not change over time. Evaluate the distribution of the cost of the hedging strategy in a simulation study.

(b) An issuer of an option typically benefits from issuing several options on the same asset because it can hedge the portfolio of options simultaneously at a lower cost than when hedging the options separately. Suggest a portfolio of short positions in European put and call options that you believe is less costly to hedge. Investigate the hedging cost for the option portfolio in a simulation study.

Chapter 4
Quadratic Investment Principles

In this chapter, we present investment principles solely based on means and variances of asset returns and budget restrictions. To begin with, we only consider risky assets in the sense that the variances of the returns are strictly positive. We will then consider the more interesting situation where we also have the possibility to invest (or deposit) money in a risk-free asset. We consider a fixed investment horizon. For convenience we measure time in units of the length of the investment horizon, and therefore $t_0 = 0$ denotes the current time and $t_1 = 1$ the end of the investment horizon. On the one hand, the investor wants to form a portfolio whose expected value $E[V_1]$ at time 1 is high. On the other hand, the investor wants the uncertainty in the future portfolio value V_1 to be as small as possible. Here the latter requirement means that the variance $\mathrm{Var}(V_1)$ should be as small as possible. The investor therefore needs to decide upon a suitable trade-off between maximizing $E[V_1]$ and minimizing $\mathrm{Var}(V_1)$.

We consider an investor with initial capital V_0 and a simple asset market with $n \geq 2$ risky assets with spot prices (the price for immediate delivery) S_t^k, where $t = 0, 1$ and $k = 1, \dots, n$. Note that S_0^k is known, whereas S_1^k is not (they are random variables as seen from today). In Sect. 4.2 we also allow positions in a risk-free zero-coupon bond that costs B_0 at time 0 and pays one unit of the chosen currency at time 1.

A position in the risky assets is represented by a vector $\mathbf{h} = (h_1, \dots, h_n)^{\mathrm{T}}$ in $\mathbb{R}^n$, where h_k is the number of units of asset number k held over the time period by the investor. When applicable, we let h_0 denote the position in the risk-free bond. Unless stated otherwise, we assume that short-selling is allowed. This means that h_k may take negative values. The prices or market values at time $t = 0$ and $t = 1$ of an affordable portfolio are

$$h_0 B_0 + \sum_{k=1}^{n} h_k S_0^k \leq V_0 \quad \text{and} \quad V_1 = h_0 + \sum_{k=1}^{n} h_k S_1^k,$$

H. Hult et al., *Risk and Portfolio Analysis: Principles and Methods*, Springer Series in Operations Research and Financial Engineering, DOI 10.1007/978-1-4614-4103-8_4, © Springer Science+Business Media New York 2012

respectively. If no risk-free bond is available, then we simply set $h_0 = 0$. It is often convenient to take the initial monetary value of the position in the kth asset, $w_k = h_k S_0^k$, as the kth portfolio weight instead of the size h_k of the kth position. Similarly, $w_0 = h_0 B_0$ is the monetary weight in the risk-free asset. With monetary portfolio weights the current and future portfolio values can be expressed as

$$w_0 + \sum_{k=1}^{n} w_k \leq V_0 \quad \text{and} \quad V_1 = w_0 \frac{1}{B_0} + \sum_{k=1}^{n} w_k \frac{S_1^k}{S_0^k}, \tag{4.1}$$

from which it is seen that determining the optimal allocation of the initial capital V_0 requires the knowledge of the expected value μ and covariance matrix Σ of the vector $\mathbf{R}$ of returns, where

$$\mathbf{R}^{\mathrm{T}} = \left(\frac{S_1^1}{S_0^1}, \ldots, \frac{S_1^n}{S_0^n} \right).$$

With $R_0 = 1/B_0$ and $\mathbf{w} = (w_1, \ldots, w_n)^{\mathrm{T}}$ we may write $V_1 = w_0 R_0 + \mathbf{w}^{\mathrm{T}} \mathbf{R}$, and therefore $\mathrm{E}[V_1] = w_0 R_0 + \mathbf{w}^{\mathrm{T}} \mu$ and $\mathrm{Var}(V_1) = \mathbf{w}^{\mathrm{T}} \Sigma \mathbf{w}$ (the latter identity was shown in Sect. 3.1.2 of Chap. 3).

Throughout this chapter we assume that the covariance matrix $\Sigma = \mathrm{Cov}(\mathbf{R}) = \mathrm{E}[(\mathbf{R} - \mu)(\mathbf{R} - \mu)^{\mathrm{T}}]$ is positive definite: $\mathbf{x}^{\mathrm{T}} \Sigma \mathbf{x} > 0$ for all $\mathbf{x} \neq \mathbf{0}$. By definition, any covariance matrix is symmetric and also positive-semidefinite: for any $\mathbf{x} \neq \mathbf{0}$

$$\mathbf{x}^{\mathrm{T}} \Sigma \mathbf{x} = \mathrm{Var}(\mathbf{x}^{\mathrm{T}} \mathbf{R}) \geq 0.$$

Therefore, assuming that Σ is positive definite is equivalent to assuming that Σ is invertible or, equivalently, that all the eigenvalues of Σ are positive (> 0).

This chapter is structured as follows. The first section considers investments without any risk-free borrowing and lending. The opportunity to make risk-free deposits is likely to be available to most investors, and risk-free borrowing is possible for many investors. Therefore, we focus mainly on analyzing the investment problem in this setting and consider variations on the investment problem with different sets of constraints. We then move on to consider investments in the presence of liabilities and find interesting connections to the hedging problems considered earlier. The next section considers investments when the number of available assets is large. In particular, we contrast the risk reduction from using quadratic hedging and exploiting dependencies among asset returns and from cancellation effects from holding large diversified positions. The chapter ends with a discussion of problems and pitfalls for investment principles that are based only on expected values and variances of portfolio returns.

4.1   Quadratic Investments Without a Risk-Free Asset

We assume that the investor wants to solve one of the following three closely connected optimization problems. The first version of the investment problem, called the trade-off problem, can be expressed as

$$\text{maximize } \text{E}[\sum_{k=1}^{n} h_k S_1^k] - \text{const} \cdot \text{Var}(\sum_{k=1}^{n} h_k S_1^k)/V_0$$
$$\text{subject to } \sum_{k=1}^{n} h_k S_0^k \leq V_0, \tag{4.2}$$

where const > 0 corresponds to the investor's choice of trade-off between maximizing $\text{E}[V_1]$ and minimizing $\text{Var}(V_1)$. The variance term in the objective function is normalized by dividing the variance by V_0 since we want the objective function to take values in units of our base currency and since the trade-off constant should be related to the available capital put at risk. The trade-off between risk and expected return does not only depend on the type of investor but also on the size of the investment.

The second version of the investment problem, called the maximization-of-expectation problem, can be expressed as

$$\text{maximize } \text{E}[\sum_{k=1}^{n} h_k S_1^k]$$
$$\text{subject to } \text{Var}(\sum_{k=1}^{n} h_k S_1^k) \leq \sigma_0^2 V_0^2$$
$$\sum_{k=1}^{n} h_k S_0^k \leq V_0.$$

A private investor may want to formulate the investment problem in this form. The third version of the investment problem, called the minimization-of-variance problem, can be expressed as

$$\text{minimize } \text{Var}(\sum_{k=1}^{n} h_k S_1^k)$$
$$\text{subject to } \text{E}[\sum_{k=1}^{n} h_k S_1^k] \geq \mu_0 V_0$$
$$\sum_{k=1}^{n} h_k S_0^k \leq V_0.$$

An institutional investor whose clients demand a good enough return may want to formulate the investment problem in this form. In a realistic investment situation, it is likely that more constraints, e.g., restrictions on short-selling, need to be considered. We will return to this issue later in this chapter.

We now express the investment problem (4.2) in terms of the vector $\mathbf{w}$ of (monetary) portfolio weights and note that (4.2) may be formulated as

$$\text{maximize } \mathbf{w}^T \boldsymbol{\mu} - \frac{c}{2V_0} \mathbf{w}^T \boldsymbol{\Sigma} \mathbf{w}$$
$$\text{subject to } \mathbf{w}^T \mathbf{1} \leq V_0, \tag{4.3}$$

where the dimensionless constant $c > 0$ here differs from the constant in (4.2) since the variance term here is multiplied by the factor $1/2$ (to produce a nicer looking solution to the investment problem). Investment problem (4.3) is a convex optimization problem, and we know from Proposition 2.3 that a solution $(\mathbf{w}, \lambda)$ to the system of equations

$$\nabla\left(\frac{c}{2V_0}\mathbf{w}^\mathsf{T}\boldsymbol{\Sigma}\mathbf{w} - \mathbf{w}^\mathsf{T}\boldsymbol{\mu}\right) + \lambda\nabla(\mathbf{w}^\mathsf{T}\mathbf{1}) = 0 \quad \text{and} \quad \mathbf{w}^\mathsf{T}\mathbf{1} = V_0 \qquad (4.4)$$

will determine the optimal solution if $\lambda > 0$. Computing partial derivatives (the gradients) shows that (4.4) is equal to the system of equations

$$\mathbf{w} = \frac{V_0}{c}\boldsymbol{\Sigma}^{-1}(\boldsymbol{\mu} - \lambda\mathbf{1}) \quad \text{and} \quad \mathbf{w}^\mathsf{T}\mathbf{1} = V_0,$$

which has the unique solution

$$\mathbf{w} = \frac{V_0}{c}\boldsymbol{\Sigma}^{-1}(\boldsymbol{\mu} - \lambda\mathbf{1}), \quad \lambda = (\mathbf{1}^\mathsf{T}\boldsymbol{\Sigma}^{-1}\boldsymbol{\mu} - c)/\mathbf{1}^\mathsf{T}\boldsymbol{\Sigma}^{-1}\mathbf{1}.$$

Since $\boldsymbol{\Sigma}$ is positive definite, also $\boldsymbol{\Sigma}^{-1}$ is positive definite. Therefore, $\mathbf{1}^\mathsf{T}\boldsymbol{\Sigma}^{-1}\mathbf{1} > 0$ and $\lambda \geq 0$ precisely when $\mathbf{1}^\mathsf{T}\boldsymbol{\Sigma}^{-1}\boldsymbol{\mu} \geq c$. If $\mathbf{1}^\mathsf{T}\boldsymbol{\Sigma}^{-1}\boldsymbol{\mu} < c$, then we cannot conclude from Proposition 2.3 that $\mathbf{w}$ above is not an optimal solution to (4.3). In this case, we may solve the unconstrained problem, which gives $\mathbf{w} = (V_0/c)\boldsymbol{\Sigma}^{-1}\boldsymbol{\mu}$; this is indeed an optimal solution to (4.3) since $\mathbf{w}^\mathsf{T}\mathbf{1} = (V_0/c)\mathbf{1}^\mathsf{T}\boldsymbol{\Sigma}^{-1}\boldsymbol{\mu} < V_0$. We observe that in this case it is not optimal to invest all the initial capital. The following proposition sums up the findings so far.

Proposition 4.1. *The optimal solution* $\mathbf{w}$ *to investment problem (4.3) is*

$$\mathbf{w} = \frac{V_0}{c}\boldsymbol{\Sigma}^{-1}\left(\boldsymbol{\mu} - \frac{(\mathbf{1}^\mathsf{T}\boldsymbol{\Sigma}^{-1}\boldsymbol{\mu} - c)_+}{\mathbf{1}^\mathsf{T}\boldsymbol{\Sigma}^{-1}\mathbf{1}}\mathbf{1}\right),$$

where $x_+ = \max(x, 0)$.

There is no straightforward way to choose the trade-off parameter c. A sensible way would be for the investor to determine two investment opportunities with returns R and $\widetilde{R}$ that are considered equally attractive investment opportunities for the capital V_0 and determine c as the solution to the equation

$$\mathrm{E}[V_0 R] - \frac{c}{2V_0}\mathrm{Var}(V_0 R) = \mathrm{E}[V_0\widetilde{R}] - \frac{c}{2V_0}\mathrm{Var}(V_0\widetilde{R}),$$

which gives $c = 2(\mathrm{E}[R] - \mathrm{E}[\widetilde{R}])/(\mathrm{Var}(R) - \mathrm{Var}(\widetilde{R}))$. On the other hand, it is not necessarily relevant to determine the value of the trade-off parameter c. If we believe that the investment problem in terms of means and variances is relevant but we find it difficult to specify the constant c, then we should evaluate the optimal solution $\mathbf{w}$ for several values of c and pick the solution $\mathbf{w}$ that we feel most comfortable with. Ideally, we should also assign a stochastic model to the vector of returns $\mathbf{R}$ and pick the solution $\mathbf{w}$ (corresponding to some c) that produces a histogram or density for the optimal portfolio value $V_1 = \mathbf{w}^T\mathbf{R}$ that we consider to be the most desirable among those for the optimal portfolio values.

Example 4.1 (Throwing away capital). The optimal solution to investment problem (4.3) may correspond to throwing away some or all of the initial capital. As an example we may take $n = 2$ and

$$\boldsymbol{\mu} = \begin{pmatrix} 1.5 \\ 1.05 \end{pmatrix} \quad \text{and} \quad \boldsymbol{\Sigma} = \begin{pmatrix} 0.3^2 & 0.3 \cdot 0.25 \cdot 0.99 \\ 0.3 \cdot 0.25 \cdot 0.99 & 0.25^2 \end{pmatrix}.$$

Proposition 4.1 gives the optimal solution

$$\mathbf{w} = \frac{V_0}{c}\boldsymbol{\Sigma}^{-1}\boldsymbol{\mu} = \frac{V_0}{c}\begin{pmatrix} 141.0385 \\ -150.7538 \end{pmatrix},$$

which means that, regardless of the trade-off parameter $c > 0$, nothing of the initial capital is used to take the optimal position in the two risky assets. The conclusion is that the investment problem here is not a good one unless a further safe asset is included in which we may invest the capital that is not invested in the risky assets.

Example 4.2 (Equally distributed returns). Consider the special case with risky assets whose returns are equally distributed, and suppose also that the linear correlation coefficient $\rho < 1$ is the same for any pair of returns. It is clear that all positions with the same initial cost will give the same expected portfolio return. However, the variance of the portfolio return is minimized by distributing the invested capital equally among the assets, and therefore this allocation should be the optimal solution to the investment problem. Let us now verify this guess.

The assumption here of equicorrelated and equally distributed returns means that

$$\boldsymbol{\mu} = \mu_0\mathbf{1} \quad \text{and} \quad \boldsymbol{\Sigma} = \sigma_0^2\{(1-\rho)\mathbf{I} + \rho\mathbf{1}\mathbf{1}^T\}.$$

The matrix identity $(\mathbf{A} + \mathbf{a}\mathbf{a}^T)^{-1} = \mathbf{A}^{-1} - \mathbf{A}^{-1}\mathbf{a}\mathbf{a}^T\mathbf{A}^{-1}/(1 + \mathbf{a}^T\mathbf{A}^{-1}\mathbf{a})$ gives

$$\boldsymbol{\Sigma}^{-1} = \frac{1}{\sigma_0^2(1-\rho)}\left(\mathbf{I} - \frac{\rho}{1+(n-1)\rho}\mathbf{1}\mathbf{1}^T\right), \quad \rho \in (-1/(n-1), 1).$$

Proposition 4.1 gives the optimal solution

$$\mathbf{w} = \frac{V_0\,\mu_0}{c}\frac{1}{\sigma_0^2}\frac{1}{1+(n-1)\rho}\mathbf{1} \quad \text{if} \quad \frac{\mu_0}{\sigma_0^2}\frac{n}{1+(n-1)\rho} < c,$$

which means that not all capital is invested and that the invested capital $(< V_0)$ is invested equally among the n assets. Otherwise, $(V_0/n)\mathbf{1}$, which means that all initial capital is invested equally among the n assets.

Example 4.3 (Uncorrelated returns). Consider the case with uncorrelated but not equally distributed risky assets. Suppose that the expected value and covariance matrix of the vector of returns are given by

$$
\mu = \begin{pmatrix} \mu_1 \\ \vdots \\ \mu_n \end{pmatrix} \quad \text{and} \quad \Sigma = \begin{pmatrix} \sigma_1^2 & 0 & \cdots \\ 0 & \ddots & \\ \vdots & & \sigma_n^2 \end{pmatrix}.
$$

Assume also that $\lambda = (1^T \Sigma^{-1} \mu - c)/1^T \Sigma^{-1} 1 > 0$, which implies that all capital is invested. We find that the optimal solution to the investment problem is given by

$$
w_k = \frac{V_0}{c\sigma_k^2} \left(\mu_k - \frac{\sum_{j=1}^n \mu_j \sigma_j^{-2} - c}{\sum_{j=1}^n \sigma_j^{-2}} \right).
$$

In the special case with $\mu_k = \mu_0$ for all k, we get

$$
w_k = V_0 \frac{1}{\sigma_k^2} \bigg/ \sum_{j=1}^n \frac{1}{\sigma_j^2},
$$

which means that the capital is distributed among the assets proportional to the reciprocal of the variance of the return.

In the special case with $\sigma_k = \sigma_0$ for all k, we get

$$
w_k = \frac{V_0}{n} + c\frac{\mu_k - \bar{\mu}}{\sigma_0^2}, \quad \bar{\mu} = \frac{1}{n}\sum_{j=1}^n \mu_j,
$$

which means that the capital is first evenly split among the assets and then adjusted so that more of the capital is invested in the assets with above-average expected returns.

Example 4.4 (Minimum variance portfolio). If the investor is forced to invest all initial capital (corresponding to the binding budget constraint $1^T w = V_0$) and if the investor cares only about minimizing variance (corresponding to $c \to \infty$), then the optimal solution to the modified investment problem is $w = V_0 \Sigma^{-1} 1 / 1^T \Sigma^{-1} 1$. The portfolio w is called the minimum variance portfolio and is sometimes advocated as being a sensible choice. However, there is no trade-off parameter c for which w is a solution to the investment problem (4.3) with a not necessarily binding budget constraint. Therefore, a mean–variance investor would rather throw away money than invest in this portfolio!

Empirical studies suggest that if the estimates of means and covariances are based only on historical price data, then the minimum variance portfolio is not necessarily a bad choice. Estimators of the mean are inaccurate, even on observations of independent and identically distributed random variables. Moreover, the strong performance of an asset in the past is not necessarily a good prediction of its future performance. On the other hand, estimates of covariances between asset returns may vary less than estimates of means over time. Therefore, it may be reasonable to select a set of assets that we believe will perform well—although we do not feel comfortable with assigning them expected values—and allocate our capital to the linear combination of these assets that gives the smallest variance of the future portfolio value. Therefore, the minimum variance portfolio can be seen as a sensible portfolio choice when the value of the parameter μ is uncertain.

Consider the covariance matrix (expressed as a product of standard deviations and linear correlations)

$$
\Sigma = \begin{pmatrix} 0.3 & 0 & 0 & 0 \\ 0 & 0.25 & 0 & 0 \\ 0 & 0 & 0.2 & 0 \\ 0 & 0 & 0 & 0.15 \end{pmatrix} \begin{pmatrix} 1 & 0.6 & 0.5 & 0.4 \\ 0.6 & 1 & -0.1 & 0 \\ 0.5 & -0.1 & 1 & 0.6 \\ 0.4 & 0 & 0.6 & 1 \end{pmatrix} \begin{pmatrix} 0.3 & 0 & 0 & 0 \\ 0 & 0.25 & 0 & 0 \\ 0 & 0 & 0.2 & 0 \\ 0 & 0 & 0 & 0.15 \end{pmatrix}
\tag{4.5}
$$

of yearly returns, say. The minimum variance portfolio weights are

$$
\mathbf{w} = V_0 \frac{\Sigma^{-1}\mathbf{1}}{\mathbf{1}^{\mathrm{T}}\Sigma^{-1}\mathbf{1}} \approx V_0 \begin{pmatrix} -0.31 \\ 0.45 \\ 0.37 \\ 0.49 \end{pmatrix}.
\tag{4.6}
$$

If short sales are not allowed, then some of the minimum variance portfolio weights will be zero since the optimal solution without the restriction to long-only positions is no longer a feasible solution. Judging from (4.6), it seems plausible that $w_1 = 0$ and $w_k > 0$ for $k = 2, 3, 4$ in this case. The short position in (4.6) is a consequence of the higher variance of the first asset return and the high correlation between the first asset return and the second (and third) asset return. Indeed, if short sales are not allowed, then the minimum variance portfolio weights are given by, with $\mathbf{b} = (0, 1, 1, 1)^{\mathrm{T}}$,

$$
\mathbf{w} = V_0 \frac{\Sigma^{-1}\mathbf{b}}{\mathbf{b}^{\mathrm{T}}\Sigma^{-1}\mathbf{b}} \approx V_0 \begin{pmatrix} 0 \\ 0.27 \\ 0.17 \\ 0.56 \end{pmatrix}.
$$

We saw in Proposition 4.1 and subsequent examples that the optimal solution to investment problem (4.3) may correspond to throwing away some of the initial capital. This seems rather strange but makes sense when considering that in the current setting there is no risk-free savings account or zero-coupon bond in which to put the initial capital that is not used to take positions in risky assets. The conclusion is that formulation (4.3) of the investment problem is unrealistic. One should at least be allowed to put money aside for use later (corresponding to a long position in a risk-free asset). To make the investment problem more realistic, we now consider the investment problem when a risk-free asset has been included.

4.2   Quadratic Investments with a Risk-Free Asset

Suppose now that we are able to invest in a risk-free asset with return R_0. The return $R_0 = 0$ would correspond to throwing away money, and the return $R_0 = 1$ would correspond to storing cash in a safe if we ignore inflation. We take the risk-free asset to be a zero-coupon bond maturing at time 1 with face value 1 and write B_0 for the current bond price. Note that $R_0 = 1/B_0$ is the return on the bond. From (4.1) we know that the portfolio values at times 0 and 1 are

$$w_0 + \mathbf{w}^\mathsf{T}\mathbf{1} \le V_0 \quad \text{and} \quad V_1 = w_0 R_0 + \mathbf{w}^\mathsf{T}\mathbf{R}.$$

Therefore, the expected value and variance of the future portfolio value are given by $E[\mathbf{w}^\mathsf{T}\mathbf{R} + w_0 R_0] = \mathbf{w}^\mathsf{T}\boldsymbol{\mu} + w_0 R_0$ and $\mathrm{Var}(\mathbf{w}^\mathsf{T}\mathbf{R}) = \mathbf{w}^\mathsf{T}\boldsymbol{\Sigma}\mathbf{w}$, respectively.

4.2.1   The Trade-Off Problem

The investment problem (4.3) modified by including a risk-free asset may now be formulated as

$$\text{maximize } w_0 R_0 + \mathbf{w}^\mathsf{T}\boldsymbol{\mu} - \frac{c}{2V_0}\mathbf{w}^\mathsf{T}\boldsymbol{\Sigma}\mathbf{w}$$
$$\text{subject to } w_0 + \mathbf{w}^\mathsf{T}\mathbf{1} \le V_0. \tag{4.7}$$

A straightforward application of Proposition 2.3 gives the solution to (4.7).

Proposition 4.2. *The optimal solution to investment problem* (4.7) *is given by* $(w_0, \mathbf{w})$, *where*

$$\mathbf{w} = \frac{V_0}{c}\boldsymbol{\Sigma}^{-1}(\boldsymbol{\mu} - R_0\mathbf{1}) \quad \text{and} \quad w_0 = V_0 - \mathbf{w}^\mathsf{T}\mathbf{1}. \tag{4.8}$$

Proof. Condition (1) of Proposition 2.3 here reads

$$\frac{c}{V_0}\boldsymbol{\Sigma}\mathbf{w} - \boldsymbol{\mu} + \lambda\mathbf{1} = \mathbf{0} \quad \text{and} \quad -R_0 + \lambda = 0.$$

Since $\lambda = R_0 > 0$, Condition (4) of Proposition 2.3 gives $w_0 + \mathbf{w}^\mathsf{T}\mathbf{1} = V_0$, from which we find that (4.8) is indeed the optimal solution to (4.7). $\square$

Remark 4.1. (i) It is not surprising that not to invest all initial capital is suboptimal here. Suppose, falsely, that $(w_0, \mathbf{w})$ is an optimal solution to (4.7) for which it holds that $w_0 + \mathbf{w}^T \mathbf{1} < V_0$. Then we can increase w_0, which increases the value of the objective function without violating the budget constraint—a contradiction.

(ii) From expression (4.8) for the optimal position in the risky assets we find that if we are given the optimal solution $\mathbf{w}^*$ for one value c^* of the trade-off parameter c, then we know the optimal solution for all trade-off parameters (they are just scalar multiples of the optimal solution $\mathbf{w}^*$). In particular, any optimal solution $(w_0, \mathbf{w})$ is a combination of a position in the portfolio $\mathbf{w}^*$ and the risk-free asset. This fact is often referred to as the one-fund theorem.

(iii) Note that if the optimal position in the risky assets has a positive initial value $\mathbf{w}^T \mathbf{1} = (V_0/c)\mathbf{1}^T \boldsymbol{\Sigma}^{-1}(\boldsymbol{\mu} - R_0 \mathbf{1})$ for some (and therefore for all) values of c, then there is precisely one value for the risk–reward trade-off parameter c that gives an optimal portfolio fully invested in the risky assets only. Solving the equation $\mathbf{w}^T \mathbf{1} = V_0$ for c gives $c = \mathbf{1}^T \boldsymbol{\Sigma}^{-1}(\boldsymbol{\mu} - R_0 \mathbf{1})$.

(iv) An often-used risk-adjusted performance measure for an investment with return $R = V_1/V_0$ is the Sharpe ratio $(\mathrm{E}[R] - R_0)/\sqrt{\mathrm{Var}(R)}$. For any value of the trade-off parameter $c > 0$, the optimal solution to investment problem (4.7) gives the Sharpe ratio

$$\sqrt{(\boldsymbol{\mu} - R_0 \mathbf{1})^T \boldsymbol{\Sigma}^{-1}(\boldsymbol{\mu} - R_0 \mathbf{1})}.$$

In particular, the Sharpe ratio for optimal portfolios does not depend on c and V_0. All feasible but suboptimal solutions to the investment problem yield lower Sharpe ratios.

Example 4.5 (Uncorrelated returns). Suppose that all the risky assets have expected returns that are greater than the risk-free return and that the returns of the risky assets are uncorrelated. In this case, $w_k = (V_0/c)(\mathrm{E}[R_k] - R_0)/\mathrm{Var}(R_k)$ for $k = 1, \ldots, n$, and the solution has a natural interpretation: a return with a high expected value and a small variance is an attractive investment opportunity. If necessary (depending on the initial capital and the trade-off parameter c), money is borrowed to afford the long positions in the risky assets.

Example 4.6 (Portfolio densities). Consider optimal investment, solutions to (4.7), in four risky assets and a risk-free asset. Here we illustrate the probability distribution of the random portfolio value V_1 at the end of the investment period for which

$$\mathrm{E}[V_1] = V_0\left(R_0 + \frac{1}{c}(\boldsymbol{\mu} - R_0 \mathbf{1})^T \boldsymbol{\Sigma}^{-1}(\boldsymbol{\mu} - R_0 \mathbf{1})\right),$$

$$\mathrm{Var}(V_1) = \frac{V_0^2}{c^2}(\boldsymbol{\mu} - R_0 \mathbf{1})^T \boldsymbol{\Sigma}^{-1}(\boldsymbol{\mu} - R_0 \mathbf{1}).$$

Fig. 4.1 Density functions
for optimal portfolio value V_1
in Example 4.6 with
$E[V_1] = 1 + 0.29c$ and
$Var(V_1) = 0.29c^2$ for three
choices of trade-off parameter
($c \approx 5.82, 2.91, 0.97$) under
the assumption that V_1 is
normally distributed

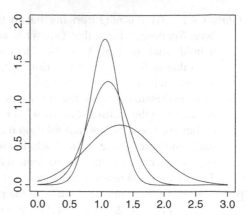

In particular, we can achieve an arbitrary high expected value $E[V_1]$, but at the price of a high variance $Var(V_1)$. The trade-off parameter c specifies the investor's trade-off between maximizing $E[V_1]$ and minimizing $Var(V_1)$.

Suppose that the returns have expected values 1.025, 1.075, 1.1, and 1.15, standard deviations 0.1, 0.2, 0.3, and 0.4, and a pairwise linear correlation coefficient of 0.2. Suppose further that the return is 1 for the risk-free asset and that $V_0 = 1$. These values imply that $(\mu - R_0 \mathbf{1})^T \Sigma^{-1} (\mu - R_0 \mathbf{1}) \approx 0.29$ and that the optimal allocation in the risky assets is given by $\mathbf{w}^T \approx c^{-1}(1.04, 1.30, 0.69, 0.65)$. For instance, $E[V_1] = 1.05$ corresponds to $c \approx 5.82$ and $\mathbf{w}^T\mathbf{1} \approx 0.63$ (implying a long position in the risk-free asset), $E[V_1] = 1.1$ corresponds to $c \approx 2.91$ and $\mathbf{w}^T\mathbf{1} \approx 1.27$ (implying a short position in the risk-free asset), $E[V_1] = 1.3$ corresponds to $c \approx 0.97$ and $\mathbf{w}^T\mathbf{1} \approx 3.81$ (implying a large short position in the risk-free asset). Figure 4.1 shows the density functions of V_1 for the three different trade-off parameters under the assumption that the joint distribution of the returns is a normal distribution.

Example 4.7 (Parameter uncertainty). Consider a 1-year investment problem with two risky assets and a risk-free bond. Here we illustrate the risks that misspecified parameters can lead to and the effect of adding a long-positions-only constraint.

Suppose that the linear correlation coefficient for the two 1-year returns of the risky assets is $\rho > 0$. From Proposition 4.2 we find that the solution to the investment problem is

$$\mathbf{w} = \frac{V_0}{c(1 - \rho^2)} \begin{pmatrix} \frac{\mu_1 - R_0}{\sigma_1^2} - \rho \frac{\mu_2 - R_0}{\sigma_1 \sigma_2} \\ \frac{\mu_2 - R_0}{\sigma_2^2} - \rho \frac{\mu_1 - R_0}{\sigma_1 \sigma_2} \end{pmatrix}.$$

Now suppose that we are rather confident in the values we assign to (μ_1, σ_1) but less sure about (μ_2, σ_2), although our best guess would be that $(\mu_2, \sigma_2) = (\mu_1, \sigma_1)$. One way to handle the difference in quality in our probability beliefs would be to assign a higher value to σ_2. With $\mu_1 = \mu_2 > R_0$ this gives

$$\mathbf{w} = \frac{V_0(\mu_1 - R_0)}{c(1 - \rho^2)} \begin{pmatrix} \frac{1}{\sigma_1}\left(\frac{1}{\sigma_1} - \frac{\rho}{\sigma_2}\right) \\ \frac{1}{\sigma_2}\left(\frac{1}{\sigma_2} - \frac{\rho}{\sigma_1}\right) \end{pmatrix}.$$

The higher value we assigned to σ_2, which was intended to reflect the uncertainty in the mean–variance characteristics of the second asset, could lead to $w_2 < 0$, i.e., a short position in the second asset. It certainly does not seem wise to short-sell an asset just because we are unsure about its mean–variance characteristics. Short-selling is often risky and should be based on good information.

If we add the constraint $w_1, w_2 \geq 0$ to the investment problem, thereby ruling out short sales, then the solution behaves better. In this case, we cannot capitalize on the positive correlation between the two asset returns. If both w_1 and w_2 above are positive, then the solution stays the same. However, if the higher σ_2 produced a negative w_2 above, then allowing only long positions gives the optimal solution $(w_0, w_1, w_2) = (V_0 - w_1, w_1, 0)$, where $w_1 \geq 0$ (or $w_1 \in [0, V_0]$) maximizes

$$V_0 R_0 + w_1(\mu_1 - R_0) - \frac{c}{2V_0}w_1^2\sigma_1^2.$$

Example 4.8 (Efficient frontiers). Consider four assets whose vector of returns has the covariance matrix $\mathbf{\Sigma}$ in (4.5) and expected value $\boldsymbol{\mu} = (1.05, 1.15, 1.1, 1.1)^{\mathrm{T}}$. Recall from Proposition 4.1 that the optimal solution to the investment problem (4.3) without a risk-free asset is given by

$$\mathbf{w} = \frac{V_0}{c}\mathbf{\Sigma}^{-1}\left(\boldsymbol{\mu} - \frac{(\mathbf{1}^{\mathrm{T}}\mathbf{\Sigma}^{-1}\boldsymbol{\mu} - c)_+}{\mathbf{1}^{\mathrm{T}}\mathbf{\Sigma}^{-1}\mathbf{1}}\mathbf{1}\right).$$

For $c > 0$, the pairs $(\sigma_p(c), \mu_p(c))$, where $\sigma_p(c) = (\mathbf{w}^{\mathrm{T}}\mathbf{\Sigma}\mathbf{w})^{1/2}$ and $\mu_p(c) = \mathbf{w}^{\mathrm{T}}\boldsymbol{\mu}$ $(\mathbf{w} = \mathbf{w}(c))$, constitute the so-called efficient frontier. The solid curve in Fig. 4.2 shows the efficient frontier for the values of $\boldsymbol{\mu}$ and $\mathbf{\Sigma}$ considered here. For any feasible nonoptimal solution to (4.3) the corresponding pair of standard deviation and expected value of the future portfolio value will be a point below the efficient frontier. For $c > \mathbf{1}^{\mathrm{T}}\mathbf{\Sigma}^{-1}\boldsymbol{\mu}$ the efficient frontier is a straight line (for these values of c a fraction of the initial capital is discarded). For $c \leq \mathbf{1}^{\mathrm{T}}\mathbf{\Sigma}^{-1}\boldsymbol{\mu}$ the efficient frontier curves down since here $\sigma_p^2(c)$ is a second-degree polynomial in $\mu_p(c)$ with nonzero coefficients of the terms of orders $0, 1, 2$.

Recall from Proposition 4.2 that the optimal solution to investment problem (4.7) with a risk-free asset is given by

$$\mathbf{w} = \frac{V_0}{c}\mathbf{\Sigma}^{-1}(\boldsymbol{\mu} - R_0\mathbf{1}) \quad \text{and} \quad w_0 = V_0 - \mathbf{w}^{\mathrm{T}}\mathbf{1}.$$

The pairs $(\sigma_p(c), \mu_p(c))$ of the efficient frontier are given by $\sigma_p(c) = (\mathbf{w}^{\mathrm{T}}\mathbf{\Sigma}\mathbf{w})^{1/2}$ and $\mu_p(c) = w_0 R_0 + \mathbf{w}^{\mathrm{T}}\boldsymbol{\mu}$, and the efficient frontier is a straight line. The dashed line in Fig. 4.2 shows the efficient frontier for $R_0 = 1.05$ and the values of $\boldsymbol{\mu}$ and $\mathbf{\Sigma}$ considered here. It is not surprising that the opportunity to take positions in the

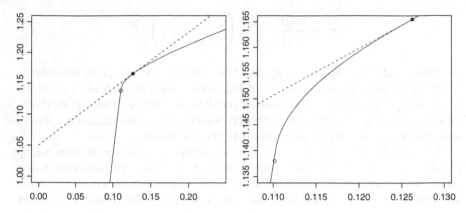

Fig. 4.2 Efficient frontiers for investment problems (4.3) and (4.7). The efficient frontiers are made up of pairs $(\sigma_p(c), \mu_p(c))$ of standard deviations and expected values of the future optimal portfolio values. The *straight line* is the efficient frontier for (4.7). The points *circle* and *filled circle* correspond to the minimum variance portfolio and the tangent portfolio, respectively

risk-free asset leads to an efficient frontier that dominates that of the investment problem without the risk-free asset. There is one point of tangency of the two efficient frontiers. The corresponding portfolio is called the tangent portfolio, and it is an optimal solution to both investment problems. Setting the two expressions for the optimal solutions to the two investment problems to be equal and solving the equation for the trade-off parameter c gives $c = \mathbf{1}^T \boldsymbol{\Sigma}^{-1}(\boldsymbol{\mu} - R_0\mathbf{1}) < \mathbf{1}^T \boldsymbol{\Sigma}^{-1}\boldsymbol{\mu}$. The standard deviation and expected value for the tangent portfolio are shown as the point on the efficient frontiers marked by the filled circle in Fig. 4.2.

The point corresponding to the minimum variance portfolio is marked by the circle in Fig. 4.2. The plot to the right shows that the minimum variance portfolio is a feasible but nonoptimal solution (the point is close to but not on the efficient frontier).

4.2.2   Maximization of Expectation and Minimization of Variance

We now turn to the two alternative versions of the investment problem. Here we assume that $\boldsymbol{\mu} \neq R_0\mathbf{1}$ since this rules out an unrealistic degenerate form of the investment problem. In the first version, the maximization-of-expectation problem, one seeks to maximize the expected portfolio value given the risk (variance) constraint

$$
\begin{aligned}
\text{maximize} \quad & w_0 R_0 + \mathbf{w}^T \boldsymbol{\mu} \\
\text{subject to} \quad & \mathbf{w}^T \boldsymbol{\Sigma} \mathbf{w} \leq \sigma_0^2 V_0^2 \\
& w_0 + \mathbf{w}^T \mathbf{1} \leq V_0.
\end{aligned}
\tag{4.9}
$$

The second version of the investment problem, the minimization-of-variance problem, seeks the portfolio that has minimum risk (variance) given a lower bound on the expected value:

$$
\begin{aligned}
&\text{minimize} \quad \tfrac{1}{2}\mathbf{w}^{\mathrm{T}}\boldsymbol{\Sigma}\mathbf{w} \\
&\text{subject to} \quad w_0 R_0 + \mathbf{w}^{\mathrm{T}}\boldsymbol{\mu} \geq \mu_0 V_0 \\
&\qquad\qquad\quad w_0 + \mathbf{w}^{\mathrm{T}}\mathbf{1} \leq V_0.
\end{aligned}
\tag{4.10}
$$

Both investment problems (4.9) and (4.10) are convex optimization problems. This is easily seen by minimizing -1 times the objective function in the former and rewriting the constraint $\mathbf{w}^{\mathrm{T}}\boldsymbol{\mu} + w_0 R_0 \geq \mu_0 V_0$ as $-\mathbf{w}^{\mathrm{T}}\boldsymbol{\mu} - w_0 R_0 \leq -\mu_0 V_0$ in the latter optimization problem.

Proposition 4.3. *The solution to* (4.9) *is given by*

$$
\mathbf{w} = \sigma_0 V_0 \frac{\boldsymbol{\Sigma}^{-1}(\boldsymbol{\mu} - R_0\mathbf{1})}{\sqrt{(\boldsymbol{\mu} - R_0\mathbf{1})^{\mathrm{T}}\boldsymbol{\Sigma}^{-1}(\boldsymbol{\mu} - R_0\mathbf{1})}}.
\tag{4.11}
$$

The solution to (4.10) *is given by*

$$
\mathbf{w} = V_0(\mu_0 - R_0)\frac{\boldsymbol{\Sigma}^{-1}(\boldsymbol{\mu} - R_0\mathbf{1})}{(\boldsymbol{\mu} - R_0\mathbf{1})^{\mathrm{T}}\boldsymbol{\Sigma}^{-1}(\boldsymbol{\mu} - R_0\mathbf{1})}
\tag{4.12}
$$

if $\mu_0 > R_0$, *and* $\mathbf{w} = \mathbf{0}$ *otherwise.*

Proof. The two problems (4.9) and (4.10) can be solved using Propositions 2.3 and 2.1, respectively. We begin with (4.9). From Proposition 2.3 we have the following system of equations:

$$
-\boldsymbol{\mu} + 2\lambda_1\boldsymbol{\Sigma}\mathbf{w} + \lambda_2\mathbf{1} = \mathbf{0},
$$

$$
-R_0 + \lambda_2 = 0,
$$

$$
\mathbf{w}^{\mathrm{T}}\boldsymbol{\Sigma}\mathbf{w} = \sigma_0^2 V_0^2,
$$

$$
\mathbf{w}^{\mathrm{T}}\mathbf{1} + w_0 = V_0.
$$

Combining the first two equations yields $\mathbf{w} = (2\lambda_1)^{-1}\boldsymbol{\Sigma}^{-1}(\boldsymbol{\mu} - R_0\mathbf{1})$ and $\lambda_2 = R_0$. Inserting the expression for $\mathbf{w}$ in the third and fourth equations gives

$$
w_0 = V_0(1 - (2\lambda_1)^{-1}\mathbf{1}^{\mathrm{T}}\boldsymbol{\Sigma}^{-1}(\boldsymbol{\mu} - R_0\mathbf{1})),
$$

$$
\lambda_1 = \frac{\sqrt{(\boldsymbol{\mu} - R_0\mathbf{1})^{\mathrm{T}}\boldsymbol{\Sigma}^{-1}(\boldsymbol{\mu} - R_0\mathbf{1})}}{2\sigma_0 V_0}.
$$

In particular, the optimal solution to (4.9) is (4.11).

Now we turn to (4.10). From Proposition 2.1 we have the following system of linear equations:

$$\boldsymbol{\Sigma}\mathbf{w} - \lambda_1 \boldsymbol{\mu} + \lambda_2 \mathbf{1} = \mathbf{0},$$

$$-\lambda_1 R_0 + \lambda_2 = 0,$$

$$\mathbf{w}^{\mathsf{T}}\boldsymbol{\mu} + w_0 R_0 = \mu_0 V_0,$$

$$\mathbf{w}^{\mathsf{T}}\mathbf{1} + w_0 = V_0.$$

Combining the first two equations yields $\mathbf{w} = \lambda_1 \boldsymbol{\Sigma}^{-1}(\boldsymbol{\mu} - R_0 \mathbf{1})$ and $\lambda_2 = \lambda_1 R_0$. Inserting the expression for $\mathbf{w}$ in the third and fourth equations, multiplying the fourth equation by $-R_0$, adding the two equations, and solving for λ_1 give

$$\lambda_1 = \frac{\mu_0 - R_0}{(\boldsymbol{\mu} - R_0 \mathbf{1})^{\mathsf{T}} \boldsymbol{\Sigma}^{-1}(\boldsymbol{\mu} - R_0 \mathbf{1})}.$$

Thus, if $\mu_0 - R_0 > 0$, then $\lambda_1 > 0$, and the optimal solution to (4.10) is

$$\mathbf{w} = V_0(\mu_0 - R_0)\frac{\boldsymbol{\Sigma}^{-1}(\boldsymbol{\mu} - R_0 \mathbf{1})}{(\boldsymbol{\mu} - R_0 \mathbf{1})^{\mathsf{T}} \boldsymbol{\Sigma}^{-1}(\boldsymbol{\mu} - R_0 \mathbf{1})}.$$

If $\mu_0 - R_0 \leq 0$, then a quick look at (4.10) leads to the conclusion that the optimal solution is to take $\mathbf{w} = \mathbf{0}$ and w_0 to be any number less than or equal to V_0. If we seek to minimize the risk and are satisfied with an expected return that is smaller than the risk-free return, then clearly we should take a position in the risk-free asset and stay away from the risky assets.   □

Remark 4.2. (i) For the maximization-of-expectation investment problem (4.9) we observe that, for fixed parameters $\boldsymbol{\mu}$, $\boldsymbol{\Sigma}$, and R_0, if $\mathbf{w}$ is the optimal solution for σ_0^2, then $(\widetilde{\sigma}_0/\sigma_0)\mathbf{w}$ is the solution for $\widetilde{\sigma}_0^2$. We also observe that, for fixed parameters $\boldsymbol{\Sigma}$ and σ_0^2, if $\mathbf{w}$ is the optimal solution for $\boldsymbol{\mu}$ and R_0, then $\mathbf{w}$ is also the optimal solution for $\widetilde{\boldsymbol{\mu}}$ and $\widetilde{R}_0$ such that $\boldsymbol{\mu} - R_0 \mathbf{1} = c(\widetilde{\boldsymbol{\mu}} - \widetilde{R}_0 \mathbf{1})$ for some constant $c \neq 0$.

 (ii) For the minimization-of-variance investment problem (4.10) we observe that, for fixed parameters $\boldsymbol{\mu}$, $\boldsymbol{\Sigma}$, and R_0, if $\mathbf{w}$ is the optimal solution for μ_0, then $(\widetilde{\mu}_0 - R_0)/(\mu_0 - R_0)\mathbf{w}$ is the solution for $\widetilde{\mu}_0$.

(iii) Comparing the solutions to (4.9) and (4.10) shows that if $\mu_0 - R_0 > 0$, then the solutions coincide precisely when

$$\frac{\mu_0 - R_0}{\sigma_0} = \sqrt{(\boldsymbol{\mu} - R_0 \mathbf{1})^{\mathsf{T}} \boldsymbol{\Sigma}^{-1}(\boldsymbol{\mu} - R_0 \mathbf{1})}.$$

4.2.3 Evaluating the Methods on Simulated Data

Propositions 4.2 and 4.3 give the optimal solutions to the three versions of the investment problem with n risky and one risk-free asset. However, the solutions assume that we know the expected value μ and the covariance matrix Σ of the vector $\mathbf{R}$ of returns on the risky assets. In reality, we can never be sure that the values we assign to μ and Σ are the right ones. If we have no reliable information about the probability distribution of $\mathbf{R}$ but believe strongly that the observed historical returns on assets, over time periods of the same length as the one for $\mathbf{R}$, can be seen as a sample from the distribution of $\mathbf{R}$, then the sample mean $\widehat{\mu}$ and sample covariance matrix $\widehat{\Sigma}$ can be used as proxies for μ and Σ. Here we investigate how the optimal solutions $\widehat{\mathbf{w}}$, for the three versions of the investment problem, based on estimated parameters $\widehat{\mu}$ and $\widehat{\Sigma}$ compares to the theoretical (but unknown) solution.

Consider a vector $\mathbf{R}$ of returns for two risky asset whose mean vector and covariance matrix are given by

$$\mu = \begin{pmatrix} 1.025 \\ 1.075 \end{pmatrix} \quad \text{and} \quad \Sigma = \begin{pmatrix} \sigma_1^2 & \sigma_1\sigma_2 0.5 \\ \sigma_1\sigma_2 0.5 & \sigma_2^2 \end{pmatrix}, \qquad (4.13)$$

where $\sigma_1 = 0.3$ and $\sigma_2 = 0.5$. Suppose that there also exists a risk-free asset with return $R_0 = 1$. We will compare the three versions of the investment problem under the assumption that μ and Σ are not known but can be estimated on simulated samples from the distribution of $\mathbf{R}$, which we take to be the bivariate normal distribution $N_2(\mu, \Sigma)$. We draw a $N_2(\mu, \Sigma)$-distributed vector by first drawing two independent standard normal variables Z_1 and Z_2 and then setting

$$\begin{pmatrix} R_1 \\ R_2 \end{pmatrix} = \begin{pmatrix} \mu_1 \\ \mu_2 \end{pmatrix} + \mathbf{A} \begin{pmatrix} Z_1 \\ Z_2 \end{pmatrix},$$

where $\mathbf{A}$ is a matrix satisfying $\mathbf{A}\mathbf{A}^{\mathrm{T}} = \Sigma$. The multivariate normal distribution is presented in more detail in Sect. 4.5.

Suppose that we want to invest according to the solution to one of the three versions (4.7), (4.9), and (4.10) of the investment problem that we have solved analytically. For the minimization-of-variance problem we set $\sigma_0 = 0.3$, and for the trade-off problem and the maximization-of-expectation problem we set the parameters $c \approx 0.50$ and $\mu_0 \approx 1.05$, so that the optimal solutions to the three investment problems coincide. Without loss of generality we set $V_0 = 1$, which means that the solution $\mathbf{w}$ is the position in the risky assets per unit of initial capital. The common theoretical solution $\mathbf{w}$ to the investment problems (trade-off, maximization-of-expectation, and minimization-of-variance) is given by

$$\mathbf{w} = \frac{V_0}{c} \boldsymbol{\Sigma}^{-1}(\boldsymbol{\mu} - R_0 \mathbf{1})$$

$$= \sigma_0 V_0 \frac{\boldsymbol{\Sigma}^{-1}(\boldsymbol{\mu} - R_0 \mathbf{1})}{\sqrt{(\boldsymbol{\mu} - R_0 \mathbf{1})^{\mathrm{T}} \boldsymbol{\Sigma}^{-1}(\boldsymbol{\mu} - R_0 \mathbf{1})}}$$

$$= V_0(\mu_0 - R_0) \frac{\boldsymbol{\Sigma}^{-1}(\boldsymbol{\mu} - R_0 \mathbf{1})}{(\boldsymbol{\mu} - R_0 \mathbf{1})^{\mathrm{T}} \boldsymbol{\Sigma}^{-1}(\boldsymbol{\mu} - R_0 \mathbf{1})}$$

$$\approx \begin{pmatrix} 0.074 \\ 0.577 \end{pmatrix}$$

and $w_0 = 1 - \mathbf{w}^{\mathrm{T}} \mathbf{1} \approx 0.349$. However, here we consider the situation where we do not know $\boldsymbol{\mu}$ and $\boldsymbol{\Sigma}$ and, hence, do not know the optimal solution $(w_0, \mathbf{w})$.

Suppose that we have observed 200 outcomes of independent copies of $\mathbf{R}$, which is normally distributed with mean $\boldsymbol{\mu}$ and covariance matrix $\boldsymbol{\Sigma}$. From these observations we can compute estimates $\widehat{\boldsymbol{\mu}}$ and $\widehat{\boldsymbol{\Sigma}}$ and obtain estimates $(\widehat{w}_0, \widehat{\mathbf{w}})$ by replacing $\boldsymbol{\mu}$ and $\boldsymbol{\Sigma}$ with $\widehat{\boldsymbol{\mu}}$ and $\widehat{\boldsymbol{\Sigma}}$ in the preceding expressions for the solutions to the three versions of the investment problem. Note that although the true optimal solutions coincide here, the random variables $(\widehat{w}_0, \widehat{\mathbf{w}})$ for the three cases do not coincide: the effect of the random parameters $\widehat{\boldsymbol{\mu}}$ and $\widehat{\boldsymbol{\Sigma}}$ are not identical for the three cases.

To determine the accuracy of the estimates $(\widehat{w}_0, \widehat{\mathbf{w}})$ of optimal portfolio weights, we repeat this scheme 3,000 times and plot the estimated weights $\widehat{\mathbf{w}}$ for the solutions to the three investment problems. For each of the 3,000 repetitions, $\widehat{\boldsymbol{\mu}}$ and $\widehat{\boldsymbol{\Sigma}}$ are estimated on a sample of size 200.

The upper left plot in Fig. 4.3 is a scatter plot of the 3,000 portfolio weights in the risky assets for the trade-off problem. The upper right plot in Fig. 4.3 is a scatter plot of the 3,000 points

$$(\sigma(\widehat{\mathbf{w}}), \mu(\widehat{\mathbf{w}})) = (\sqrt{\widehat{\mathbf{w}}^{\mathrm{T}} \boldsymbol{\Sigma} \widehat{\mathbf{w}}}, (1 - \widehat{\mathbf{w}}^{\mathrm{T}} \mathbf{1}) R_0 + \widehat{\mathbf{w}}^{\mathrm{T}} \boldsymbol{\mu}) \qquad (4.14)$$

for the trade-off problem. Similarly, the two scatter plots in the middle of Fig. 4.3 are the corresponding plots for the maximization-of-expectation problem. The two scatter plots at the bottom of Fig. 4.3 are the corresponding plots for the minimization-of-variance problem.

For the maximization-of-expectation problem the risk constraint and the estimate $\widehat{\boldsymbol{\Sigma}}$ force the solution $\widehat{\mathbf{w}}$ to be a point on the ellipse $\widehat{\mathbf{w}}^{\mathrm{T}} \widehat{\boldsymbol{\Sigma}} \widehat{\mathbf{w}} = \sigma_0^2$. Since the estimates $\widehat{\boldsymbol{\Sigma}}$ vary across the 3,000 samples, the points $\widehat{\mathbf{w}}$ of the scatter plot form a point cloud that is concentrated near the ellipse $\mathbf{w}^{\mathrm{T}} \boldsymbol{\Sigma} \mathbf{w} = \sigma_0^2$. The points of the scatter plots for the trade-off problem and the minimization-of-variance problem are more spread out, especially for the latter. In particular, many of the the solutions $\widehat{\mathbf{w}}$ for the minimization-of-variance problem based on the simulated samples are very far from the theoretical solution $\mathbf{w} = (0.074, 0.577)^{\mathrm{T}}$. The reason for the poor accuracy is that many of the estimated values $\widehat{\boldsymbol{\mu}} - R_0 \mathbf{1}$ are very close to $\mathbf{0}$,

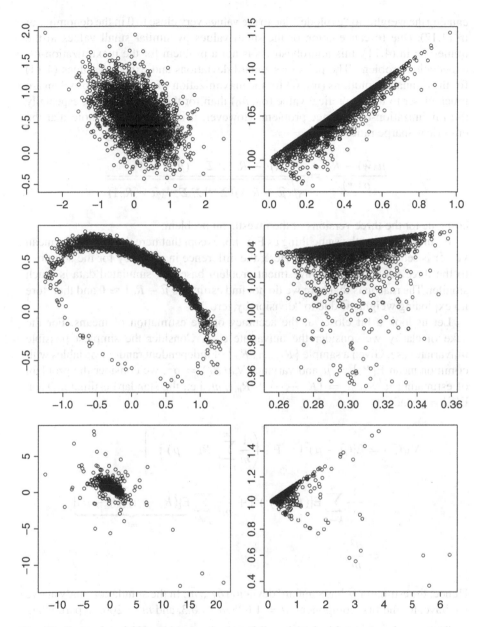

Fig. 4.3 *Upper plots*: 3,000 empirical optimal portfolio weights in risky assets and corresponding standard deviation–mean pairs, based on samples of size 200, for trade-off problem; *middle plots*: maximization-of-expectation problem; *lower plots*: minimization-of-variance problem. Parameters μ and Σ used in the simulations are those in (4.13)

causing the weights to "explode" due to the values very close to 0 in the denominator
in (4.12). Due to cancellation of the small values by similar small values in the
numerator in (4.11), this nonrobustness is not a problem for the maximization-of-
expectation problem. The pairs of standard deviations and expected values (4.14)
for the estimated solutions $(\widehat{w}_0, \widehat{w})$ for the maximization-of-expectation problem are
much closer to the theoretical value (σ_0, μ_0) than for the other versions, especially
the minimization-of-variance problem. However, it is interesting to note that the
empirical Sharpe ratios

$$\frac{\mu(\widehat{w}) - R_0}{\sigma(\widehat{w})} = \frac{(\mu - R_0 \mathbf{1})^{\mathrm{T}} \widehat{\boldsymbol{\Sigma}}^{-1} (\widehat{\mu} - R_0 \mathbf{1})}{\sqrt{(\widehat{\mu} - R_0 \mathbf{1})^{\mathrm{T}} \widehat{\boldsymbol{\Sigma}}^{-1} \boldsymbol{\Sigma} \widehat{\boldsymbol{\Sigma}}^{-1} (\widehat{\mu} - R_0 \mathbf{1})}}$$

coincide for the three versions of the investment problem.

Figure 4.4 shows the same thing as Fig. 4.3, except that here the theoretical mean
vector is set to $\mu = (1.1, 1.2)^{\mathrm{T}}$. Here, the difference in accuracy for the solutions
to the three versions of the investment problem based on simulated data is much
smaller. The reason is that here we do not find estimates $\widehat{\mu} - R_0 \mathbf{1} \approx \mathbf{0}$ and therefore
no exploding weights $\widehat{w}$ due to "division by zero."

Let us look a bit closer at the accuracy of the estimation of means. For the
sake of clarity we consider the univariate case. Consider the simplest possible
univariate case. Given a sample $\{R_1, \ldots, R_m\}$ of independent random variables with
common mean $E[R_k] = \mu$ and variance $\text{Var}(R_k) = \sigma^2$, we consider the problem
of estimating μ. Set $\widehat{\mu} = (R_1 + \cdots + R_m)/m$, i.e., the standard estimator. Then
$E[\widehat{\mu}] = \mu$ and

$$\text{Var}(\widehat{\mu}) = E[(\widehat{\mu} - \mu)^2] = E\left[\left(\frac{1}{m}\sum_{k=1}^{m}(R_k - \mu)\right)^2\right]$$

$$= \frac{1}{m^2}\sum_{k=1}^{m}E[(R_k - \mu)^2] + \frac{1}{m^2}\sum_{j \neq k}\underbrace{E[(R_j - \mu)(R_k - \mu)]}_{=0}$$

$$= \frac{\sigma^2}{m}.$$

Hence, the estimator $\widehat{\mu}$ has standard deviation $\sigma/\sqrt{m}$. In the simulation study above
we have, for the first component, $\mu = 1.025$, $\sigma = 0.3$, and $m = 200$. In particular,

$$\mu - R_0 = 0.025 \quad \text{and} \quad \sqrt{\text{Var}(\widehat{\mu} - R_0)} = \sigma/\sqrt{m} \approx 0.021,$$

from which we see that it is likely that $\widehat{\mu} - R_0$ has outcomes very close to zero.

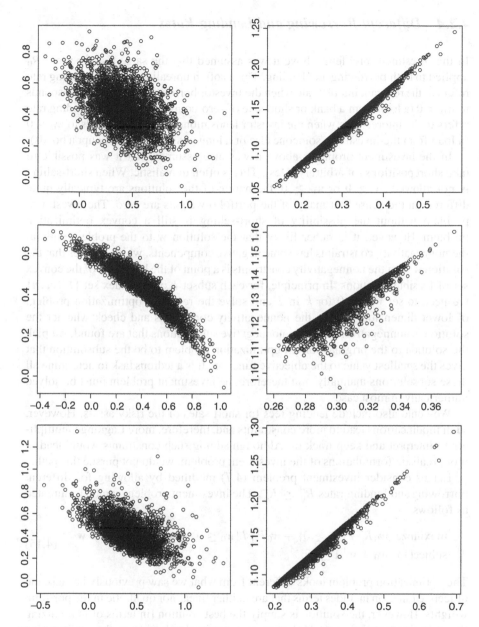

Fig. 4.4 Same plots as in Fig. 4.3 except that parameter μ used in the simulation is $\mu = (1.1, 1.2)^{\mathrm{T}}$ here

4.2.4   Different Borrowing and Lending Rates

In the investment problems above it was assumed that the same risk-free rate R_0 applied to both borrowing and lending. This is often unrealistic. The borrowing rate refers to the relevant interest rate when the investor borrows money and corresponds to $w_0 < 0$ (a loan from a bank or short sale of zero-coupon bonds). The lending rate refers to the interest rate when the investor lends money and corresponds to $w_0 > 0$ (a loan from the investor to someone else or a long position in zero-coupon bonds).

In the investment problems above it was also assumed that it was possible to take short positions (of arbitrary sizes). This is often unrealistic. When short-selling is not allowed, $\mathbf{w} \geq \mathbf{0}$ or $w_0 \geq 0$, the nature of the solutions are typically quite different in the sense that many of the portfolio weights are zero. The investment problem without the possibility of short-selling is still a convex optimization problem. However, it is rather likely that the solution $\mathbf{w}$ to the problem without the nonnegativity constraints has some negative components, which implies that the solution $\mathbf{w}$ with the nonnegativity constraints is a point of the boundary of the convex set of feasible solutions. In principle, for each subset I of the index set $\{1, \ldots, n\}$ we need to set $w_k = 0$ for k in I and solve the resulting optimization problem of lower dimension without the nonnegativity constraints and check whether the solution is nonnegative. For those nonnegative subsolutions that are found, we pick the solution to the original convex optimization problem to be the subsolution that gives the smallest value to the objective function. It is a tedious task to determine all these subsolutions manually, and therefore the investment problem must be solved numerically in most cases.

We could also include lending fees for short sales of the risky assets. However, such modifications lead to more constraints and, therefore, more Lagrange multipliers to interpret and keep track of. Although adding such constraints would lead to more realistic formulations of the investment problem, we do not pursue this path.

Let us consider investment problem (4.7) modified by allowing for different borrowing and lending rates $R_0^+ \leq R_0^-$. The investment problem can be formulated as follows:

$$\text{maximize} \quad w_0 R_0^+ I\{w_0 \geq 0\} + w_0 R_0^- I\{w_0 \leq 0\} + \mathbf{w}^\mathsf{T} \boldsymbol{\mu} - \frac{c}{2V_0} \mathbf{w}^\mathsf{T} \boldsymbol{\Sigma} \mathbf{w} \tag{4.15}$$
$$\text{subject to} \quad w_0 + \mathbf{w}^\mathsf{T} \mathbf{1} \leq V_0.$$

The optimization problem looks different from what we saw previously because the objective function includes terms that are neither linear nor quadratic in the portfolio weights. However, the solution is simply the best solution [in terms of maximizing the objective function in (4.15)] of the two optimal solutions to the optimization problems

$$\text{maximize} \quad w_0 R_0^+ + \mathbf{w}^\mathsf{T} \boldsymbol{\mu} - \frac{c}{2V_0} \mathbf{w}^\mathsf{T} \boldsymbol{\Sigma} \mathbf{w}$$
$$\text{subject to} \quad w_0 + \mathbf{w}^\mathsf{T} \mathbf{1} \leq V_0 \tag{4.16}$$
$$w_0 \geq 0$$

and

$$\text{maximize } w_0 R_0^- + \mathbf{w}^T \boldsymbol{\mu} - \frac{c}{2V_0} \mathbf{w}^T \boldsymbol{\Sigma} \mathbf{w}$$
$$\text{subject to } w_0 + \mathbf{w}^T \mathbf{1} \le V_0 \qquad (4.17)$$
$$w_0 \le 0.$$

Both (4.16) and (4.17) are convex optimization problems (write $-w_0 \le 0$ instead of $w_0 \ge 0$) and (4.16) is an investment problem without the possibility of risk-free borrowing for the investor.

Proposition 4.4. *The optimal solution to* (4.16) *is given by*

$$\mathbf{w}^+ = \frac{V_0}{c} \boldsymbol{\Sigma}^{-1}(\boldsymbol{\mu} - R_0^+ \mathbf{1}) - \frac{V_0}{c} \frac{(\mathbf{1}^T \boldsymbol{\Sigma}^{-1}(\boldsymbol{\mu} - R_0^+ \mathbf{1}) - c)_+}{\mathbf{1}^T \boldsymbol{\Sigma}^{-1} \mathbf{1}} \boldsymbol{\Sigma}^{-1} \mathbf{1}.$$

The optimal solution to (4.17) *is given by*

$$\mathbf{w}^- = \frac{V_0}{c} \boldsymbol{\Sigma}^{-1}(\boldsymbol{\mu} - R_0^- \mathbf{1}) + \frac{V_0}{c} \frac{(c - \mathbf{1}^T \boldsymbol{\Sigma}^{-1}(\boldsymbol{\mu} - R_0^- \mathbf{1}))_+}{\mathbf{1}^T \boldsymbol{\Sigma}^{-1} \mathbf{1}} \boldsymbol{\Sigma}^{-1} \mathbf{1}.$$

The optimal solution to (4.15) *is* $\mathbf{w}^+$ *or* $\mathbf{w}^-$, *depending on which of the two maximizes the objective function in* (4.15).

Proof. Problem (4.16) is a convex optimization problem, and the conditions for an optimal solution $(w_0, \mathbf{w})$ are

$$(c/V_0) \boldsymbol{\Sigma} \mathbf{w} - \boldsymbol{\mu} + \lambda_1 \mathbf{1} = \mathbf{0},$$

$$- R_0^+ + \lambda_1 - \lambda_2 = 0,$$

$$\lambda_1 (\mathbf{w}^T \mathbf{1} + w_0 - V_0) = 0,$$

$$\lambda_2 w_0 = 0,$$

$$\lambda_1, \lambda_2, w_0 \ge 0,$$

$$\mathbf{w}^T \mathbf{1} + w_0 \le V_0.$$

We first note that $\mathbf{w}^T \mathbf{1} + w_0 = V_0$ since if $(w_0, \mathbf{w})$ were an optimal solution satisfying $\mathbf{w}^T \mathbf{1} + w_0 < V_0$, then $(\mathbf{w}, w_0 + \delta)$ with $\delta = V_0 - \mathbf{w}^T \mathbf{1} - w_0$ would be a better feasible solution [therefore contradicting the assumption that $(w_0, \mathbf{w})$ is an optimal solution]. Combining the first two equations above gives

$$\mathbf{w} = \frac{V_0}{c} \boldsymbol{\Sigma}^{-1}(\boldsymbol{\mu} - R_0^+ \mathbf{1}) - \lambda_2 \frac{V_0}{c} \boldsymbol{\Sigma}^{-1} \mathbf{1}, \quad \mathbf{w}^T \mathbf{1} + w_0 = V_0.$$

If $\lambda_2 = 0$, then the optimal solution $\mathbf{w} = (V_0/c)\boldsymbol{\Sigma}^{-1}(\boldsymbol{\mu} - R_0^+\mathbf{1})$ to (4.7) turns out to be affordable without any risk-free borrowing and, therefore, also the optimal solution to (4.16). If $\lambda_2 > 0$, then $w_0 = 0$, and we get the optimal solution

$$\mathbf{w} = \frac{V_0}{c}\boldsymbol{\Sigma}^{-1}(\boldsymbol{\mu} - R_0^+\mathbf{1}) - \lambda_2\frac{V_0}{c}\boldsymbol{\Sigma}^{-1}\mathbf{1}, \quad \mathbf{w}^{\mathrm{T}}\mathbf{1} = V_0.$$

The optimal solution to (4.17) is determined by repeating the preceding arguments with minor modifications.   □

Remark 4.3. The optimal solutions $\mathbf{w}^+$ to (4.16) and $\mathbf{w}^-$ to (4.17) in Proposition 4.4 merit some explanations.

The solution $\mathbf{w}^+$ to (4.16) is the solution $\mathbf{w}$ to (4.7) (with $R_0 = R_0^+$) if $V_0 - \mathbf{w}^{\mathrm{T}}\mathbf{1} \geq 0$ (the portfolio is affordable without any risk-free borrowing). Otherwise, $\mathbf{w}^+$ is the position obtained after withdrawing money from the risky assets until no risk-free borrowing is needed.

The solution $\mathbf{w}^-$ to (4.17) is the solution $\mathbf{w}$ to (4.7) (with $R_0 = R_0^-$) if $V_0 - \mathbf{w}^{\mathrm{T}}\mathbf{1} \leq 0$ (the portfolio is affordable only with risk-free borrowing). Otherwise, $\mathbf{w}^-$ is the position obtained after adding money to the risky assets until risk-free borrowing is needed.

4.3 Investments in the Presence of Liabilities

We now consider optimal investments in the presence of liabilities. Let $w_0 R_0 + \mathbf{w}^{\mathrm{T}}\mathbf{R}$ and L denote the future values of assets and liabilities, respectively, and consider the investment problem

$$\begin{aligned} &\text{maximize} \quad \mathrm{E}[w_0 R_0 + \mathbf{w}^{\mathrm{T}}\mathbf{R} - L] - \tfrac{c}{2V_0}\,\mathrm{Var}(w_0 R_0 + \mathbf{w}^{\mathrm{T}}\mathbf{R} - L) \\ &\text{subject to} \quad w_0 + \mathbf{w}^{\mathrm{T}}\mathbf{1} \leq V_0. \end{aligned} \tag{4.18}$$

The investment problem is clearly related to both the corresponding investment problem (4.7) without a liability and the minimum variance hedging problem in Proposition 3.3 in Chap. 3. The following proposition states that the optimal solution to (4.18) is an easily interpreted combination of optimal investments and optimal minimum variance hedging.

Proposition 4.5. *The solution to investment problem* (4.18) *is given by*

$$\mathbf{w} = \frac{V_0}{c}\boldsymbol{\Sigma}^{-1}(\boldsymbol{\mu} - R_0\mathbf{1}) + \boldsymbol{\Sigma}^{-1}\boldsymbol{\Sigma}_{L,\mathbf{R}} \quad and \quad w_0 = V_0 - \mathbf{w}^{\mathrm{T}}\mathbf{1}. \tag{4.19}$$

If risk-free borrowing is not allowed, i.e., $w_0 \geq 0$, then the solution changes to

$$\mathbf{w} = \frac{V_0}{c}\mathbf{\Sigma}^{-1}(\boldsymbol{\mu} - R_0\mathbf{1}) + \mathbf{\Sigma}^{-1}\mathbf{\Sigma}_{L,\mathbf{R}} - \lambda_2\frac{V_0}{c}\mathbf{\Sigma}^{-1}\mathbf{1} \quad and \quad \mathbf{w}^T\mathbf{1} = V_0, \quad (4.20)$$

where

$$\lambda_2 = \frac{1}{\mathbf{1}^T\mathbf{\Sigma}^{-1}\mathbf{1}}\left(\mathbf{1}^T\mathbf{\Sigma}^{-1}(\boldsymbol{\mu} - R_0\mathbf{1}) + \frac{c}{V_0}\mathbf{1}^T\mathbf{\Sigma}^{-1}\mathbf{\Sigma}_{L,\mathbf{R}} - c\right)_+.$$

We observe that the solution (4.19) to (4.18) corresponds to taking both the optimal investment position without a liability, the solution to (4.7), and the minimum variance hedge position. If the initial capital V_0 is insufficient to take this position, then more capital is raised by risk-free borrowing (a short position in the risk-free bond).

If risk-free borrowing is not allowed and if the optimal solution without borrowing restriction (4.19) is too expensive, then taking $\lambda_2 > 0$ large enough and subtracting $\lambda_2(V_0/c)\mathbf{\Sigma}^{-1}\mathbf{1}$ from the monetary portfolio weights gives a modified position that is affordable without borrowed money: in this case the position (4.19) is adjusted so that the adjusted position (4.20) costs precisely V_0.

Proof of Proposition 4.5. We formulate (4.18) as the convex optimization problem

$$\text{minimize} \quad \frac{c}{2V_0}(\mathbf{w}^T\mathbf{\Sigma}\mathbf{w} + \sigma_L^2 - 2\mathbf{w}^T\mathbf{\Sigma}_{L,\mathbf{R}}) - (w_0R_0 + \mathbf{w}^T\boldsymbol{\mu} - \mu_L)$$
$$\text{subject to} \quad w_0 + \mathbf{w}^T\mathbf{1} \leq V_0.$$

Proposition 2.1 gives the sufficient conditions for an optimal solution:

$$\mathbf{w} = (V_0/c)\mathbf{\Sigma}^{-1}(\boldsymbol{\mu} - \lambda_1\mathbf{1}) + \mathbf{\Sigma}^{-1}\mathbf{\Sigma}_{L,\mathbf{R}},$$

$$\lambda_1 - R_0 = 0,$$

$$\lambda_1(w_0 + \mathbf{w}^T\mathbf{1} - V_0) = 0,$$

$$w_0 + \mathbf{w}^T\mathbf{1} \leq V_0.$$

Therefore, the optimal solution is given by

$$\mathbf{w} = \frac{V_0}{c}\mathbf{\Sigma}^{-1}(\boldsymbol{\mu} - R_0\mathbf{1}) + \mathbf{\Sigma}^{-1}\mathbf{\Sigma}_{L,\mathbf{R}} \quad and \quad w_0 = V_0 - \mathbf{w}^T\mathbf{1}.$$

We now turn to (4.18) modified by including the constraint $w_0 \geq 0$ (or, equivalently, $-w_0 \leq 0$). The sufficient conditions for an optimal solution are

$$\mathbf{w} = (V_0/c)\mathbf{\Sigma}^{-1}(\boldsymbol{\mu} - \lambda_1\mathbf{1}) + \mathbf{\Sigma}^{-1}\mathbf{\Sigma}_{L,\mathbf{R}},$$

$$\lambda_1 - R_0 - \lambda_2 = 0,$$

$$\lambda_1(w_0 + \mathbf{w}^T\mathbf{1} - V_0) = 0,$$

$$\lambda_2 w_0 = 0,$$

$$\lambda_1, \lambda_2, w_0 \geq 0,$$

$$w_0 + \mathbf{w}^\mathsf{T}\mathbf{1} \leq V_0.$$

If $\lambda_2 = 0$, then we have the solution above, with $w_0 \geq 0$. Therefore, the interesting case is where $\lambda_2 > 0$, which implies $w_0 = 0$. In this case, $\mathbf{w}^\mathsf{T}\mathbf{1} = V_0$ since $\mathbf{w}^\mathsf{T}\mathbf{1} < V_0$ would correspond to throwing away money rather than investing it in a risk-free bond, which is clearly suboptimal. The optimal solution is

$$\mathbf{w} = \frac{V_0}{c}\boldsymbol{\Sigma}^{-1}(\boldsymbol{\mu} - R_0\mathbf{1}) - \lambda_2\frac{V_0}{c}\boldsymbol{\Sigma}^{-1}\mathbf{1} + \boldsymbol{\Sigma}^{-1}\boldsymbol{\Sigma}_{L,\mathbf{R}} \quad \text{and} \quad \mathbf{w}^\mathsf{T}\mathbf{1} = V_0.$$

Combining the two equations gives

$$(V_0/c)\mathbf{1}^\mathsf{T}\boldsymbol{\Sigma}^{-1}(\boldsymbol{\mu} - R_0\mathbf{1}) - \lambda_2(V_0/c)\mathbf{1}^\mathsf{T}\boldsymbol{\Sigma}^{-1}\mathbf{1} + \mathbf{1}^\mathsf{T}\boldsymbol{\Sigma}^{-1}\boldsymbol{\Sigma}_{L,\mathbf{R}} = V_0,$$

which can be solved for λ_2. □

Example 4.9. Suppose that risky assets can be divided into a set of hedging instruments (e.g., bonds) and a set of pure investment assets (e.g., stocks), where the values of the assets of the latter kind are uncorrelated with the liability. Suppose further that the vectors of returns for both sets of assets have invertible covariance matrices. Write $\mathbf{w}_i$ and $\mathbf{w}_h$ for the vectors of monetary weights for the investment and hedging assets, respectively. Therefore,

$$\boldsymbol{\Sigma} = \begin{pmatrix} \boldsymbol{\Sigma}_i & 0 \\ 0 & \boldsymbol{\Sigma}_h \end{pmatrix} \text{ and } \boldsymbol{\Sigma}^{-1} = \begin{pmatrix} \boldsymbol{\Sigma}_i^{-1} & 0 \\ 0 & \boldsymbol{\Sigma}_h^{-1} \end{pmatrix},$$

and the solution to the optimal investment problem with a liability and no risk-free borrowing reads

$$\mathbf{w}_i = \frac{V_0}{c}\boldsymbol{\Sigma}_i^{-1}(\boldsymbol{\mu}_i - R_0\mathbf{1}) - \lambda_2\frac{V_0}{c}\boldsymbol{\Sigma}_i^{-1}\mathbf{1},$$

$$\mathbf{w}_h = \frac{V_0}{c}\boldsymbol{\Sigma}_h^{-1}(\boldsymbol{\mu}_h - R_0\mathbf{1}) - \lambda_2\frac{V_0}{c}\boldsymbol{\Sigma}_h^{-1}\mathbf{1} + \boldsymbol{\Sigma}_h^{-1}\boldsymbol{\Sigma}_{L,\mathbf{R}_h},$$

where $\lambda_2 > 0$ is the smallest number such that $\mathbf{w}_i^\mathsf{T}\mathbf{1} + \mathbf{w}_h^\mathsf{T}\mathbf{1} = V_0$; if there is no such $\lambda_2 > 0$, then $\lambda_2 = 0$, $w_0 > 0$, and $w_0 + \mathbf{w}_i^\mathsf{T}\mathbf{1} + \mathbf{w}_h^\mathsf{T}\mathbf{1} = V_0$. We observe that here we can divide the original investment problem with a liability into two simpler problems, one with the liability and only the hedging instruments as risky assets, and one without a liability and only the pure investment assets as risky assets.

Example 4.10 (Pension arrangement). Here we consider a pension arrangement between a company and its employees. The company has promised to pay $L = V_0 \max(1, R_1)$, where R_1 is the return on some basket of exchange-traded stocks, on a future date, to its employees. The company has decided to invest the capital kV_0 in such a way that it provides a hedge against an increase in the value L of its liability to its employees.

Suppose that the company considers three investment and hedging instruments: a position in a risk-free asset, a position in the basket of stocks, and a long position in a European call option with strike price V_0 on the basket of stocks. The call option is not a traded asset, but it can be issued and sold to the company by a bank if the bank receives a good enough price C_0. The company decides to invest the capital kV_0 according to the solution to the investment problem

$$\text{maximize } E[w_0 R_0 + w_1 R_1 + w_2(V_0/C_0)(R_1 - 1)_+ - L]$$
$$- \frac{c}{2V_0} \text{Var}(w_0 R_0 + w_1 R_1 + w_2(V_0/C_0)(R_1 - 1)_+ - L)$$
$$\text{subject to } w_0 + w_1 + w_2 \leq kV_0$$
$$w_2 \geq 0. \tag{4.21}$$

Notice that $(w_0, w_1, w_2) = (V_0/R_0, 0, C_0)$ gives

$$w_0 R_0 + w_1 R_1 + w_2(V_0/C_0)(R_1 - 1)_+ - L = 0,$$

which means a perfect hedge of the liability. However, here we assume that $V_0/R_0 + C_0 \geq kV_0$, so that the company cannot hedge the liability perfectly and make a profit just by buying the call option from a bank.

We now consider a numerical example. The company invests $1.1V_0$ and is obliged to pay $V_0 \max(1, R_1) = V_0 + V_0(R_1 - 1)_+$, i.e., the amount V_0 plus the payoff of the amount V_0 invested in an at-the-money call option on the basket of stocks, in 1 year from now. The risk-free return is $R_0 = 1/0.97$. The company considers the return R_1 on the basket of stocks to be $\text{LN}(\mu, \sigma^2)$-distributed with $\mu = 0.05$ and $\sigma = 0.3$.

The company can purchase a call option with payoff $V_0(R_1 - 1)_+$ from a bank, but the price $C_0 = 0.17V_0$ is high, corresponding to an implied volatility of approximately 0.4. In particular, $1.1V_0 < V_0/R_0 + C_0 = 1.14V_0$, so there is no risk-free profit to be made.

The expected value and variance of the return R_1 are given by

$$E[R_1] = e^{\mu + \sigma^2/2} \quad \text{and} \quad \text{Var}(R_1) = e^{2\mu + \sigma^2}(e^{\sigma^2} - 1).$$

The expected value and variance of the return $R_2 = (V_0/C_0)(R_1 - 1)_+$ can be computed as follows. Since R_1 has a lognormal distribution, it has the same distribution as $\exp\{\mu + \sigma Z\}$, where Z has a standard normal distribution. Then

$$E[(R_1 - 1)_+] = E[R_1 I\{R_1 > 1\}] - P(R_1 > 1)$$
$$= E[e^{\mu + \sigma Z} I\{Z > -\mu/\sigma\}] - P(Z > -\mu/\sigma)$$

$$= e^{\mu} \int_{-\mu/\sigma}^{\infty} \frac{1}{\sqrt{2\pi}} e^{-\frac{1}{2}(z-\sigma)^2 + \frac{1}{2}\sigma^2} dz - \Phi(\mu/\sigma)$$

$$= e^{\mu+\sigma^2/2} \, \mathrm{P}(Z - \sigma > -\mu/\sigma) - \Phi(\mu/\sigma)$$

$$= e^{\mu+\sigma^2/2} \Phi(\sigma + \mu/\sigma) - \Phi(\mu/\sigma).$$

Multiplying by V_0/C_0 gives an expression for $\mathrm{E}[R_2]$. Similarly, the variance of R_2 can be computed from

$$\mathrm{Var}((R_1 - 1)_+) = \mathrm{E}[(R_1 - 1)^2 I\{R_1 > 1\}] - \mathrm{E}[(R_1 - 1)_+]^2$$

$$= e^{2\mu+2\sigma^2} \Phi(2\sigma + \mu/\sigma) - 2e^{\mu+\sigma^2/2} \Phi(\sigma + \mu/\sigma) + \Phi(\mu/\sigma)$$

$$- \mathrm{E}[(R_1 - 1)_+]^2.$$

Finally, the covariances $\mathrm{Cov}(R_1, R_2)$ and $\mathrm{Cov}(L, R_1)$ can be computed from

$$\mathrm{Cov}(R_1, (R_1 - 1)_+) = \mathrm{E}[R_1(R_1 - 1)I\{R_1 > 1\}] - \mathrm{E}[R_1]\,\mathrm{E}[(R_1 - 1)_+]$$

$$= e^{2\mu+2\sigma^2} \Phi(2\sigma + \mu/\sigma) - e^{\mu+\sigma^2/2} \Phi(\sigma + \mu/\sigma)$$

$$- \mathrm{E}[R_1]\,\mathrm{E}[(R_1 - 1)_+].$$

Upon inserting numerical values we find that

$$\mu \approx \begin{pmatrix} 1.0997 \\ 1.0658 \end{pmatrix}, \quad \Sigma \approx \begin{pmatrix} 0.1139 & 0.4899 \\ 0.4899 & 2.3704 \end{pmatrix}, \quad \text{and} \quad \Sigma_{L,\mathbf{R}} \approx V_0 \begin{pmatrix} 0.0832 \\ 0.4030 \end{pmatrix}.$$

From Proposition 4.5 we know that if there are no restrictions on short-selling, then the optimal solution (in the risky assets) to (4.21) is

$$\Sigma^{-1}\left(\frac{1.1V_0}{c}(\mu - R_0\mathbf{1}) + \Sigma_{L,\mathbf{R}}\right) \approx V_0\left(\frac{1}{c}\begin{pmatrix} 5.3503 \\ -1.0894 \end{pmatrix} + \begin{pmatrix} 0 \\ 0.1700 \end{pmatrix}\right). \quad (4.22)$$

Therefore, $w_1, w_2 \geq 0$ for $c \geq c_0 \approx 6.4087$, and $w_1 > 0$ and $w_2 < 0$ otherwise. The interpretation is that to hedge the liability, the company should buy the call option, but for an investment it should short-sell the call option and take a long position in the basket of stocks. However, the company cannot short the call option. In particular, the optimal solution to (4.21) is for $c < c_0$ not given by the expression in Proposition 4.5. We conclude that in the case $c < c_0$, the solution is either of the form $(1.1V_0 - w_2, 0, w_2)$ with $w_2 > 0$ or of the form $(1.1V_0 - w_1, w_1, 0)$ with $w_1 > 0$.

If we take c very large ($c \to \infty$), then the optimal solution is approximately $w_0 = 1.1V_0 - C_0$, $w_1 = 0$, and $w_2 = C_0$. This solution gives a zero-variance hedge but also a certain loss:

$$w_0 R_0 + w_1 R_1 + w_2 R_2 - L = V_0((1.1 - 0.17)/0.97 - 1) \approx -0.04V_0.$$

It can be shown by computing the values of the objective function, which is plausible given what we have seen so far, that if $c \leq c_0$, then the optimal solution is of the form $(1.1V_0 - w_1, w_1, 0)$, with $w_1 > 0$. More precisely, if $c \leq c_0$, then the optimal solution to (4.21) is given by $w_0 = 1.1V_0 - w_1$,

$$w_1 = \frac{1.1V_0}{c}\frac{\mu_1 - R_0}{\mathrm{Var}(R_1)} + \frac{\mathrm{Cov}(L, R_1)}{\mathrm{Var}(R_1)} \approx V_0(0.6639/c + 0.7313) \quad \text{and} \quad w_2 = 0.$$

We see that the position w_1 is the sum of the position for speculation and the position for hedging the liability, where the former depends on the trade-off parameter c. Notice that for $c = c_0 \approx 6.4087$ the solution here coincides with the solution in (4.22). Also notice that for $c \leq c_0$ the future net value of the company's position is a random variable distributed as

$$V_0\frac{1}{0.97}\left(1.1 - \frac{a}{c} - b\right) + V_0\left(\frac{a}{c} + b\right)e^{0.05+0.3Z} - V_0\max(1, e^{0.05+0.3Z}), \quad (4.23)$$

where $a \approx 0.6639$, $b \approx 0.7313$, and Z is $N(0, 1)$-distributed. Histograms illustrating the probability distribution of the net value in (4.23) are shown in Fig. 4.5.

For $c = 1$ (the upper left plot) and $c = 2$ (the upper right plot) the company wants to capitalize on its positive view of the performance of the stocks and invests $w_1 > V_0$ in the stocks. The positions are profitable if the stocks perform well, $R_1 > 1$, but can be rather costly if they do not. For $c = 1$ the net result for the company is approximately

$$V_0(0.39R_1 - 0.29) \text{ if } R_1 > 1,$$
$$V_0(1.39R_1 - 1.29) \text{ if } R_1 \leq 1.$$

In this case, the company is short the risk-free asset to finance speculation on the basket of stocks. For $c = 2$ the net result for the company is approximately

$$V_0(0.06R_1 + 0.04) \text{ if } R_1 > 1,$$
$$V_0(1.06R_1 - 0.96) \text{ if } R_1 \leq 1.$$

For $c = 3$ (the lower left plot) and $c = 6$ (the lower right plot) the situation is different. Here the company invests $w_1 < V_0$ in the stocks and the rest $1.1V_0 - w_1$ in the risk-free bond. For $c = 3$ the net result for the company is approximately

$$V_0(0.15 - 0.05R_1) \text{ if } R_1 > 1,$$
$$V_0(0.95R_1 - 0.85) \text{ if } R_1 \leq 1,$$

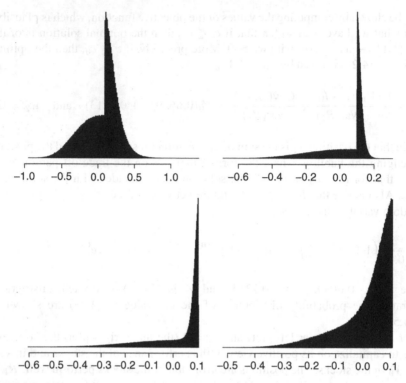

Fig. 4.5 Distributions of net value in (4.23) for $c = 1$ (*upper left*), $c = 2$ (*upper right*), $c = 3$ (*lower left*), and $c = 6$ (*lower right*)

and for $c = 6$ the net result for the company is

$$V_0(0.26 - 0.16R_1) \text{ if } R_1 > 1,$$
$$V_0(0.84R_1 - 0.76) \text{ if } R_1 \le 1.$$

In both cases, the best outcome for the company corresponds to $R_1 \approx 1$ and gives a net result of approximately $0.1V_0$.

4.4   Large Portfolios

Consider the random variables $R_1, \ldots, R_n$ that represent returns for some assets over some future time period. For any random variables $Z_1, \ldots, Z_m$, called factors, we write

$$R_k = h_{k,0} + h_{k,1}Z_1 + \cdots + h_{k,m}Z_m + W_k, \quad k = 1, \ldots, n,$$

by simply setting $W_k = R_k - h_{k,0} - h_{k,1}Z_1 - \cdots - h_{k,m}Z_m$. We want as much as possible of the variances of the R_ks to be explained by linear combinations of the factors. If there are good reasons to believe that R_k depends nonlinearly on one of the factors (e.g., e^Z), then we should transform this factor into a new factor by applying a nonlinear function to the original factor (e.g., $\log e^Z$). Moreover, we want the number of factors m to be substantially smaller than n in order to obtain a model for the high-dimensional vector of R_k that is easily expressed in the model for the low-dimensional vector of Z_k. However, this can only be achieved if we consider a large enough number of factors so that the residual noise terms W_k can be considered independent and also independent of the factors Z_k. We know from Proposition 3.2 that the choice of $h_{k,0}$ and $\mathbf{h}_k = (h_{k,1}, \ldots, h_{k,m})^{\mathrm{T}}$ given by

$$\mathbf{h}_k = \boldsymbol{\Sigma}_{\mathbf{Z}}^{-1}\boldsymbol{\Sigma}_{R_k,\mathbf{Z}} \quad \text{and} \quad h_{k,0} = \mathrm{E}[R_k] - \mathbf{h}_k^{\mathrm{T}}\mathrm{E}[\mathbf{Z}]$$

minimizes $\mathrm{Var}(W_k)$ and makes $\mathrm{E}[W_k] = 0$ and $\mathrm{Cor}(Z_j, W_k) = 0$ for all j.

Suppose R_k represents returns on potentially rewarding investment opportunities for which the investor does not have sufficient information to construct an n-dimensional model. Suppose, however, that the investor feels reasonably comfortable with assigning values to the regression coefficients of R_k onto factors $Z_1, \ldots, Z_m$ whose covariance structure can be inferred from available historical data. The factors $Z_1, \ldots, Z_m$ are selected because there are good reasons, e.g., economic, to believe that they should explain a substantial part of the variance of the R_ks. That is,

$$R_k = h_{k,0} + h_{k,1}Z_1 + \cdots + h_{k,m}Z_m + W_k, \quad k = 1, \ldots, n,$$

where the regression coefficients $h_{k,1}, \ldots, h_{k,m}$ are considered to be known. If w_k denotes the amount invested in the kth investment opportunity, then the future portfolio value can be expressed as

$$\sum_{k=1}^{n} w_k R_k = \sum_{j=1}^{m}\left(\sum_{k=1}^{n} w_k h_{k,j}\right)Z_j + \sum_{k=1}^{n} w_k h_{k,0} + \sum_{k=1}^{n} w_k W_k.$$

For the investor the aim is to select w_k such that $\sum_{k=1}^{n} w_k R_k$ is likely to be large while at the same time keeping the variance

$$\mathrm{Var}\left(\sum_{k=1}^{n} w_k R_k\right) = \sum_{j=1}^{m}\left(\sum_{k=1}^{n} w_k h_{k,j}\right)^2 \mathrm{Var}(Z_j)$$

$$+ \sum_{i=1}^{m}\sum_{j=1}^{m}\left(\sum_{k=1}^{n} w_k h_{k,i}\right)\left(\sum_{k=1}^{n} w_k h_{k,j}\right)\mathrm{Cov}(Z_i, Z_j)$$

$$+ \mathrm{Var}\left(\sum_{k=1}^{n} w_k W_k\right)$$

small. If the factors are well chosen so that W_k are uncorrelated, then the distribution of $\sum_{k=1}^{n} w_k W_k$ is probably well approximated by a normal distribution with zero mean and variance

$$\mathrm{Var}\left(\sum_{k=1}^{n} w_k W_k\right) = \sum_{k=1}^{n} w_k^2 \, \mathrm{Var}(W_k).$$

In particular, from these expressions the investor can easily assess whether a modification in portfolio weight w_k makes sense in terms of maintaining a good balance between risk and potential return.

The following two examples illustrate that for a large homogeneous portfolio and a correctly specified factor model, the systematic risk factors explain most of the randomness in the future portfolio value.

Example 4.11 (Homogeneous portfolios I). Write $V_1 = V_0(R_1 + \cdots + R_n)/n$ for the future value of a portfolio with capital V_0 invested in equal amounts in n assets. Note that

$$V_1 = \frac{V_0}{n} \sum_{k=1}^{n} h_{k,0} + \frac{V_0}{n} \sum_{j=1}^{m} \left(\sum_{k=1}^{n} h_{k,j}\right) Z_j + \frac{V_0}{n} \sum_{k=1}^{n} W_k,$$

and therefore Chebyshev's inequality $(\mathrm{P}(|X| > a) \leq \mathrm{E}[X^2]/a^2)$ yields

$$\mathrm{P}\left(\left|\frac{V_1}{V_0} - \frac{\sum_{k=1}^{n} h_{k,0}}{n} - \sum_{j=1}^{m} \frac{\sum_{k=1}^{n} h_{k,j}}{n} Z_j\right| > \varepsilon\right) \leq \frac{\mathrm{E}[(\sum_{k=1}^{n} W_k)^2]}{\varepsilon^2 n^2}.$$

If W_k are also uncorrelated, then $\mathrm{E}[(\sum_{k=1}^{n} W_k)^2] = \sum_{k=1}^{n} \mathrm{E}[W_k^2] = \sum_{k=1}^{n} \mathrm{Var}(W_k)$, and we may write

$$\mathrm{P}\left(\left|\frac{V_1}{V_0} - \overline{h_0} - \sum_{j=1}^{m} \overline{h_j} Z_j\right| > \varepsilon\right) \leq \frac{\overline{\sigma}^2}{\varepsilon^2 n},$$

where $\overline{h}_j$ is the average jth factor loading among $R_1, \ldots, R_n$, for $j = 0, 1, \ldots, m$, and $\overline{\sigma}$ is the average standard deviation among the n residuals.

Example 4.12 (Homogeneous portfolios II). Suppose that for all k, $R_k = h_0 + h_1 Z_1 + \cdots + h_m Z_m + W_k$, where W_k are uncorrelated and uncorrelated with Z_j. Suppose also that $\mathrm{Var}(W_k)$ does not vary with k. Write $\sigma_R^2 = \mathrm{Var}(R_k)$, $\sigma_Z^2 = \mathrm{Var}(h_1 Z_1 + \cdots + h_m Z_m)$, and $\sigma_W^2 = \mathrm{Var}(W_k)$, and let λ be the fraction of the variance of each R_k that is explained by the factors Z_k, i.e.,

$$\lambda = \frac{\sigma_Z^2}{\sigma_R^2} = \frac{\sigma_Z^2}{\sigma_Z^2 + \sigma_W^2}.$$

With $\sigma_V^2 = \text{Var}(V_1/V_0)$, $V_1 = V_0(R_1 + \cdots + R_n)/n$, and λ_n the fraction of the variance of V_1/V_0 that is explained by the factors Z_k, we find that

$$\lambda_n = \frac{\sigma_Z^2}{\sigma_V^2} = \frac{\sigma_Z^2}{\sigma_Z^2 + \sigma_W^2/n} = \frac{\lambda n}{\lambda(n-1)+1}.$$

In particular, if $n = 1{,}000$ and only 5% of the variance of each R_k is explained by the factors ($\lambda = 0.05$), then approximately 98% of the variance of V_1/V_0 is explained by the factors ($\lambda_n \approx 0.98$).

The following two examples consider two homogeneous groups of assets that can be expressed in terms of a one-factor model, the same factor for both asset types, where each asset of the first type is a more attractive investment opportunity than each asset of the second type. The examples illustrate that if the number of assets is large and if the assets of the first type have a stronger dependence on the common factor, then the optimal investment is to invest only in the assets of the second type. The reason for this result is simply that the benefits from diversification in large portfolios can be substantial if the assets are weakly correlated.

Example 4.13 (Two homogeneous groups I). Here we analyze optimal investments, long positions only, in a large set of assets whose returns can be divided into two homogeneous groups:

$$R_k = h_0 + h_1 Z + W_k, \quad k = 1, \ldots, n,$$
$$\widetilde{R}_k = \widetilde{h}_0 + \widetilde{h}_1 Z + \widetilde{W}_k, \quad k = 1, \ldots, n,$$

where the set of all W_k and $\widetilde{W}_k$ are uncorrelated with mean zero and all of them are uncorrelated with Z. We set $\mu_Z = E[Z]$ and assume that all the W_k have the variance $\sigma^2 = \text{Var}(W_1)$, and all the $\widetilde{W}_k$ have the variance $\widetilde{\sigma}^2 = \text{Var}(\widetilde{W}_1)$. Each asset of the first type is more attractive than any asset of the second type of asset from a mean–variance perspective in the sense that

$$E[R_k] = h_0 + h_1\mu_Z > \widetilde{h}_0 + \widetilde{h}_1\mu_Z = E[\widetilde{R}_k] \quad \text{and}$$
$$\text{Var}(R_k) = h_1^2\sigma_Z^2 + \sigma^2 < \widetilde{h}_1^2\sigma_Z^2 + \widetilde{\sigma}^2 = \text{Var}(\widetilde{R}_k).$$

However, we assume that $h_1 > \widetilde{h}_1$, i.e., that the returns of the first type have a stronger dependence on the systematic risk factor Z.

To determine how to optimally invest the available capital V_0 in the $2n$ assets we only need to determine how the capital is divided optimally between the to asset types. For each of the two assets types, the optimal allocation of the capital λV_0, where $\lambda \in [0, 1]$ is a constant to be determined, among these assets is clearly to allocate $\lambda V_0/n$ to each of the n assets. Therefore, the investment problem may be formulated as follows:

maximize $\mathrm{E}\left[\sum_{k=1}^{n} \frac{w}{n} R_k + \sum_{k=1}^{n} \frac{\widetilde{w}}{n} \widetilde{R}_k\right] - \frac{c}{2V_0} \operatorname{Var}\left(\sum_{k=1}^{n} \frac{w}{n} R_k + \sum_{k=1}^{n} \frac{\widetilde{w}}{n} \widetilde{R}_k\right)$

subject to $w + \widetilde{w} = V_0$

$\quad\quad\quad w, \widetilde{w} \geq 0.$

If $(w, \widetilde{w}) = (V_0, 0)$, then the value of the objective function is

$$V_0(h_0 + h_1 \mu_Z) - \frac{cV_0}{2}\left(h_1^2 \sigma_Z^2 + \frac{\sigma^2}{n}\right). \tag{4.24}$$

The corresponding value for $(w, \widetilde{w}) = (0, V_0)$ is

$$V_0(\widetilde{h}_0 + \widetilde{h}_1 \mu_Z) - \frac{cV_0}{2}\left(\widetilde{h}_1^2 \sigma_Z^2 + \frac{\widetilde{\sigma}^2}{n}\right). \tag{4.25}$$

The expected return is higher when the capital is invested in the first type of assets. However, if n is sufficiently large, then an investment in these assets only also gives a higher variance for the portfolio return since the higher systematic risk ($h_1 > \widetilde{h}_1$) becomes much more important than the idiosyncratic risk, which is essentially diversified away. Therefore, it seems plausible that if both c and n are large enough, then the assets of the second type, although not attractive on a standalone basis compared to an asset of the first type, will form the most attractive portfolio return. We may verify this guess by first noting that $\widetilde{w} = V_0 - w$, so the solution to the investment problem is found by first setting to 0 the derivative of the objective function with respect to w and solving it for w. If the resulting value for w is in the interval $(0, V_0)$, then the solution to the original problem has been found. Otherwise, the solution is $(w, \widetilde{w}) = (0, V_0)$ or $(V_0, 0)$, and we can easily determine which one by comparing the corresponding values of the objective function. Setting to 0 the derivative of the objective function, with $\widetilde{w} = V_0 - w$, with respect to w and solving it for w gives

$$w = \left(\frac{V_0}{c}\right) \frac{h_0 - \widetilde{h}_0 + (h_1 - \widetilde{h}_1)\mu_Z - c\widetilde{h}_1(h_1 - \widetilde{h}_1)\sigma_Z^2 + c\widetilde{\sigma}^2/n}{(h_1 - \widetilde{h}_1)^2 \sigma_Z^2 + \sigma^2/n + \widetilde{\sigma}^2/n}. \tag{4.26}$$

Example 4.14 (Two homogeneous groups II). Consider the situation in Example (4.13) with

$$h_0 = \widetilde{h}_0 = 1/2, \, h_1 = 1, \, \widetilde{h}_1 = 1/2, \, \sigma_Z = \sigma = 0.25, \, \widetilde{\sigma} = \sqrt{2}\sigma, \, \text{and} \, \mu_Z = 1.1.$$

Suppose that we are indifferent to investing all capital V_0 in an index fund with return Z or all capital in an optimally diversified portfolio of assets of the first type. That is,

$$V_0\left(\mu_Z - \frac{c}{2}\sigma_Z^2\right) = V_0\left(h_0 + h_1\mu_Z - \frac{c}{2}\left(h_1^2\sigma_Z^2 + \frac{\sigma^2}{n}\right)\right),$$

which, after inserting the foregoing parameter values, gives $c = n/\sigma^2$. We now investigate the value of w in (4.26) and find that $w = V_0(2.55 - n/4)/(3 + n/4)$, which is negative for values of n corresponding to a large portfolio. We conclude that a mix of assets from the two asset types is not optimal and therefore compare the values of (4.24) and (4.25):

$$(4.25) = V_0\left(\frac{1}{2} + \frac{\mu_Z}{2} - \frac{n}{2}\left(\frac{1}{4} + \frac{2}{n}\right)\right) = V_0\left(\frac{1}{2} + \mu_Z - \frac{n}{2}\left(1 + \frac{1}{n}\,\frac{3}{4} + \frac{1 + \mu_Z}{n}\right)\right)$$

$$> V_0\left(\frac{1}{2} + \mu_Z - \frac{n}{2}\left(1 + \frac{1}{n}\right)\right) = (4.24)$$

for n in the range we are considering. In particular, the optimal allocation is a diversified position in the assets of the second type only.

4.5   Problems with Mean–Variance Analysis

The mean–variance approach to comparing investment opportunities is a cornerstone in portfolio theory. However, many portfolios have future values $V_1 = V_0 R$ whose probability distributions are not well summarized by means and variances. Figure 4.6 shows the density functions of four probability distributions for R with $E[R] = 1.1$ and $Var(R) = 0.3^2$. Therefore, they are identical from a mean–variance perspective, although the density functions are quite different. Location (mean) and dispersion (variance) are reasonable measures of likely reward and risk, respectively, if R is approximately normally distributed (upper left plot in Fig. 4.6). The variance $Var(R) = E[(R - E[R])^2]$ quantifies a range of likely deviations from the mean. Whereas deviations of R below the mean are bad for the long holder of the asset, deviations above the mean are good. For a probability distribution with a unimodal density function that is symmetric around the mean, the outcomes of R are likely to take values close to the mean; the variance quantifies what is meant by close and also quantifies reasonably well both deviations above and below the mean. In particular, the variance can be used to measure the riskiness of the position.

For a bimodal probability density that is symmetric around the mean such as that shown in the lower left plot in Fig. 4.6, the mean and variance do not give a good summary of the probability distribution. The lower left plot in Fig. 4.6 shows the density function of a random variable with stochastic representation

$$I(1.1 - h + \sigma Y_1) + (1 - I)(1.1 + h + \sigma Y_2), \tag{4.27}$$

where $h = \sqrt{0.3^2 - \sigma^2}$ and $\sigma = 0.1$ and where I is $Be(1/2)$-distributed (Bernoulli) and Y_1, Y_2 are independent and standard normally distributed. Decreasing σ pushes

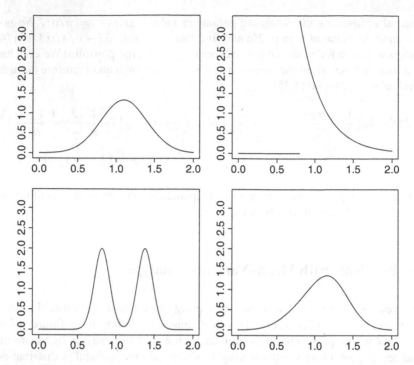

Fig. 4.6 Density functions for random variables R with $E[R] = 1.1$ and $\text{Var}(R) = 0.3^2$. *Upper left*: R is $N(1.1, 0.3^2)$-distributed. *Upper right*: $R = 0.8 + Z$ with Z $\text{Exp}(1/0.3)$-distributed. *Lower left*: $R = IZ_1 + (1 - I)Z_2$, where I is $\text{Be}(1/2)$-distributed and Z_1, Z_2 are independent with Z_1 $N(1 - \sqrt{0.08}, 0.1^2)$-distributed and Z_2 $N(1 + \sqrt{0.08}, 0.1^2)$-distributed. *Lower right*: $R = IZ_1 + (1 - I)Z_2$, where I is $\text{Be}(1/4)$-distributed and Z_1, Z_2 are independent with Z_1 $N(0.8, 0.06)$-distributed and Z_2 $N(1.2, 0.06)$-distributed

the two scaled normal density curves apart and makes them more narrow while preserving the overall mean and variance. The outcomes of the random variable are not necessarily likely to take values close to the overall mean. Although each of the probability distributions with stochastic representation (4.27) is determined by two parameters, the mean and variance are here not the right parameters.

The asymmetric density function shown in the upper right plot in Fig. 4.6 corresponds to a translated exponentially distributed random variable, $0.8 + Z$, where Z is $\text{Exp}(1/0.3)$-distributed (exponential). Each translated exponentially distributed random variable has a probability distribution that is determined by two parameters, but again here the mean and variance are not the right parameters.

The lower right plot in Fig. 4.6 shows the density function of a random variable with stochastic representation

$$I(\mu_1 + \sigma_1 Y_1) + (1 - I)(\mu_2 + \sigma_2 Y_2), \tag{4.28}$$

where I is Be$(1/4)$-distributed and Y_1, Y_2 are independent and standard normally distributed, and where $\mu_1 = 0.8$, $\mu_2 = 1.2$, and $\sigma_1^2 = \sigma_2^2 = 0.06$. In particular, from the asymmetric unimodal density function we cannot see that it comes from a two-point mixture of normal distributions.

In the mean–variance approach to portfolio optimization presented in this chapter, we have chosen an allocation of the available capital to obtain the optimal trade-off between a high mean and a low variance for the portfolio return $R_p = w_0 R_0 + \mathbf{w}^T \mathbf{R}$, where $\mathbf{R}$ is the vector of returns. The approach makes perfect sense if R_p is normally distributed but, as we have just seen, may be rather inappropriate if R_p is far from normally distributed. However, $R_p = w_0 R_0 + \mathbf{w}^T \mathbf{R}$ is normally distributed for any vector $\mathbf{w}$ if and only if $\mathbf{R}$ has a multivariate normal distribution. In particular, for R_p to be normally distributed for any $\mathbf{w}$ it is necessary but not sufficient that $R_1, \ldots, R_n$ be normally distributed. We now present the multivariate normal distribution, which is parameterized by the mean vector E$[\mathbf{R}]$ and the covariance matrix Cov$(\mathbf{R})$. In Chap. 9 we will study the normal distribution in more detail and also find that the mean–variance approach is appropriate for a wider set of multivariate distributions called elliptical distributions.

The random vector $\mathbf{Z}$ has a standard normal distribution if the components $Z_1, \ldots, Z_n$ of $\mathbf{Z}$ are independent and standard normally distributed. We denote this multivariate distribution by N$_n(\mathbf{0}, \mathbf{I})$, where $\mathbf{I}$ is the $n \times n$ identity matrix. The random vector $\mathbf{X}$ has a normal distribution with mean $\boldsymbol{\mu}$ and covariance matrix $\boldsymbol{\Sigma}$, written N$_n(\boldsymbol{\mu}, \boldsymbol{\Sigma})$, if there exists a N$_m(\mathbf{0}, \mathbf{I})$-distributed random vector $\mathbf{Z}$, for $m \leq n$, and an $n \times m$ matrix $\mathbf{A}$ satisfying $\mathbf{A}\mathbf{A}^T = \boldsymbol{\Sigma}$ such that

$$\begin{pmatrix} X_1 \\ \vdots \\ X_n \end{pmatrix} \stackrel{\mathrm{d}}{=} \begin{pmatrix} \mu_1 \\ \vdots \\ \mu_n \end{pmatrix} + \mathbf{A} \begin{pmatrix} Z_1 \\ \vdots \\ Z_m \end{pmatrix}.$$

If $m < n$, then $\boldsymbol{\Sigma} = \mathbf{A}\mathbf{A}^T$ does not have full rank. In particular, $\boldsymbol{\Sigma}$ is not invertible. If $\boldsymbol{\Sigma}$ is invertible, which is typically the more interesting case, then we can always take $m = n$ and $\mathbf{A}$ to be invertible.

Remark 4.4. $\boldsymbol{\Sigma}$ is a covariance matrix if and only if it is symmetric and positive semidefinite. Since $\boldsymbol{\Sigma}$ is symmetric, it can be expressed as $\boldsymbol{\Sigma} = \mathbf{O}\mathbf{D}\mathbf{O}^T$, where $\mathbf{D}$ is a diagonal matrix with the eigenvalues of $\boldsymbol{\Sigma}$ as diagonal elements and $\mathbf{O}$ is an orthogonal matrix (meaning that $\mathbf{O}^{-1} = \mathbf{O}^T$) with the eigenvectors of $\boldsymbol{\Sigma}$ as columns. Since $\boldsymbol{\Sigma}$ is positive semidefinite, it has nonnegative eigenvalues. If $\boldsymbol{\Sigma}$ is positive definite, then it has strictly positive eigenvalues, so $\mathbf{D}$ is invertible. In particular, a covariance matrix is positive definite if and only if it is invertible. In this case, we may take $\mathbf{A} = \mathbf{O}\mathbf{D}^{1/2}$, which is an invertible $n \times n$ matrix. However, often $\mathbf{A}$ is chosen as the so-called Cholesky decomposition of $\boldsymbol{\Sigma}$.

Any linear combination $\mathbf{w}^T \mathbf{X} = w_1 X_1 + \cdots + w_n X_n$ of the components of $\mathbf{X}$ is normally distributed. The mean and variance are given by $\mathbf{w}^T \boldsymbol{\mu}$ and $\mathbf{w}^T \boldsymbol{\Sigma} \mathbf{w}$.

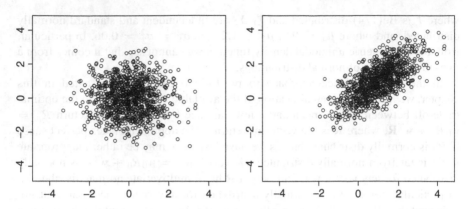

Fig. 4.7 Simulated samples from bivariate normal distributions

To verify this claim, we recall that any sum of independent normally distributed random variables is again normally distributed. With $\mathbf{v} = \mathbf{A}^T\mathbf{w}$ (a vector in $\mathbb{R}^m$) we have

$$\mathbf{w}^T\mathbf{X} \overset{\mathrm{d}}{=} \mathbf{w}^T(\boldsymbol{\mu} + \mathbf{A}\mathbf{Z}) = \mathbf{w}^T\boldsymbol{\mu} + (\mathbf{A}^T\mathbf{w})^T\mathbf{Z} = \mathbf{w}^T\boldsymbol{\mu} + \sum_{k=1}^{m} v_k Z_k,$$

i.e., a constant plus a sum of independent normally distributed random variables. Moreover, we see that $E[\mathbf{w}^T\mathbf{X}] = \mathbf{w}^T\boldsymbol{\mu}$ and

$$\mathrm{Var}(\mathbf{w}^T\mathbf{X}) = \sum_{k=1}^{m} v_k^2 = \mathbf{v}^T\mathbf{v} = (\mathbf{A}^T\mathbf{w})^T\mathbf{A}^T\mathbf{w} = \mathbf{w}^T\boldsymbol{\Sigma}\mathbf{w}.$$

In particular, $\mathbf{w}^T\mathbf{X}$ and $\mathbf{w}^T\boldsymbol{\mu} + (\mathbf{w}^T\boldsymbol{\Sigma}\mathbf{w})^{1/2}Z_1$ have the same (normal) distribution.

Figure 4.7 shows the result of simulations from two bivariate normal distributions. The left plot shows a simulated sample of size 1,000 from a bivariate standard normal distribution. Each of the points (z_1, z_2) in the left plot in Fig. 4.7 generates a point (x_1, x_2) in the right plot given by

$$\begin{pmatrix} x_1 \\ x_2 \end{pmatrix} = \begin{pmatrix} 1 \\ 1 \end{pmatrix} + \begin{pmatrix} 1 & 0 \\ 0.7 & \sqrt{1 - 0.7^2} \end{pmatrix} \begin{pmatrix} z_1 \\ z_2 \end{pmatrix}.$$

Therefore, the points shown in the right plot is a sample of size 1,000 from $N_2(\boldsymbol{\mu}, \boldsymbol{\Sigma})$, where $\mu_1 = \mu_2 = 1$, $\Sigma_{1,1} = \Sigma_{2,2} = 1$, and $\Sigma_{1,2} = \Sigma_{2,1} = 0.7$.

Example 4.15. Consider an investor who is a mean–variance optimizer and wants to make a 1-year investment. Suppose that the investor may invest in either an index that does not pay dividends or in a risk-free zero-coupon bond and call options on the index. Suppose further that the 1-year return on the index has mean $\mu = 1.05$ and standard deviation $\sigma = 0.2$, and suppose that the index has value 1 today. The zero-coupon bond maturing in 1 year with face value 1 costs $B_0 = e^{-0.03}$ today.

Buying one share of the index means buying a portfolio with the random value S_1 of the index in 1 year as payoff. The investor's view of buying a share of the index is determined by the mean–standard deviation pair $(\mu, \sigma) = (1.05, 0.2)$ of its return. However, the portfolio, consisting of a suitable mix of a long position of h_0 zero-coupon bonds and a short position of h_1 call options on S_1, gives the same mean–standard deviation pair for the portfolio return and has the same price 1 today. This portfolio with payoff $g(S_1) = h_0 - h_1(S_1 - K)_+$ is determined by solving the system of equations

$$1 = h_0 B_0 - h_1 C_0(K),$$

$$\mu = E[h_0 - h_1(S_1 - K)_+] = h_0 - h_1 E[(S_1 - K)_+],$$

$$\sigma^2 = \mathrm{Var}(h_0 - h_1(S_1 - K)_+) = h_1^2 \mathrm{Var}((S_1 - K)_+),$$

where $C_0(K)$ is the price of a call option with strike price K. For simplicity we assume that the call option prices are given by the Black–Scholes formula (1.7) with implied volatility $\sigma = 0.2$. We find that

$$h_0 = \mu + \sigma \frac{E[(S_1 - K)_+]}{\sqrt{\mathrm{Var}((S_1 - K)_+)}},$$

$$h_1 = \frac{\sigma}{\sqrt{\mathrm{Var}((S_1 - K)_+)}},$$

where K solves

$$B_0 \left(\mu + \sigma \frac{E[(S_1 - K)_+]}{\sqrt{\mathrm{Var}((S_1 - K)_+)}} \right) - C_0(K) \frac{\sigma}{\sqrt{\mathrm{Var}((S_1 - K)_+)}} = 1.$$

If the investor considers S_1 to be $N(\mu, \sigma^2)$-distributed, then the solution (h_0, h_1, K) to the system of equations is (rounded off to two decimals) $h_0 = 1.10$, $h_1 = 5.01$, and $K = 1.30$. If the investor considers S_1 to be lognormally distributed, then the solution is (rounded off to two decimals) $h_0 = 1.08$, $h_1 = 3.54$, and $K = 1.30$. The distribution function of the portfolio value $g(S_1)$ is

$$P(h_0 - h_1(S_1 - K)_+ \le x) = \begin{cases} 1, & x \ge h_0, \\ P(S_1 \ge K + (h_0 - x)/h_1), & x < h_0. \end{cases}$$

The distribution functions of S_1 and $g(S_1)$ are shown in Fig. 4.8 (for normally and lognormally distributed S_1). Note that at the point $h_0 \approx 1.1$, the distribution function of $g(S_1)$ jumps to value 1. In particular, we see that portfolios with the same price today but very different payoffs can be equivalent from a mean–variance perspective.

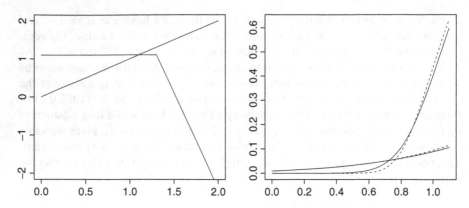

Fig. 4.8 *Left plot*: payoff functions $f(s) = s$ and $g(s) = h_0 - h_1(s-K)_+$. *Right plot*: distribution functions of payoffs S_1 and $g(S_1)$. *Solid curves* correspond to a normally distributed return; *dashed curves* correspond to a lognormally distributed return

4.6 Notes and Comments

The investment principles presented here originate from the classic work of Harry Markowitz on portfolio theory in [29,30]. Markowitz focused on linear portfolios of assets rather than on individual assets and explained how the use of diversification could make a portfolio of assets with average risk-reward characteristics more attractive than individual assets with more attractive risk-reward characteristics than those making up the portfolio. Although the mathematical basis of the quadratic investment principles is very simple, essentially the formula for the variance of a sum of correlated random variables, this approach to portfolio analysis has had an enormous effect on research in economics and finance and the way investment opportunities are analyzed in practice.

In Sect. 4.5 we pointed out some potential problems with the quadratic approach to portfolio selection. For more details on problems and pitfalls in connection with quadratic investment problems we refer the reader to Sect. 6.5 in Attilio Meucci's book [33]. The Sharpe ratio is due to William Sharpe, one of the key contributors to modern portfolio theory, see e.g. [44]

4.7 Exercises

In the exercises below it is assumed, whenever applicable, that you can take positions corresponding to fractions of assets.

Exercise 4.1 (Efficient frontiers). Consider an investor with initial capital $10,000 who wants to make a 1-year investment in two risky assets. The mean μ and covariance matrix Σ of the vector of returns on the risky assets are given by

$$\mu = \begin{pmatrix} 1.05 \\ 1.12 \end{pmatrix} \quad \text{and} \quad \Sigma = \begin{pmatrix} 0.04 \; 0.03 \\ 0.03 \; 0.09 \end{pmatrix}.$$

(a) Determine and plot the efficient frontier when the initial capital is invested fully in risky assets. Both long and short positions are allowed.

(b) Determine and plot the efficient frontier when the initial capital is invested fully in risky assets. Only long positions are allowed. Compare the efficient frontier to the one in (a).

(c) Determine and plot the efficient frontier when the initial capital is invested in the risky assets and in a risk-free asset with return 1.03. Both long and short positions are allowed. Compare the efficient frontier to the one in (a).

(d) Suppose that, today, the interest rate for lending money is 3% and that the interest rate for borrowing money is 5%. Determine and plot the efficient frontier when the initial capital is invested in risky assets, both long and short positions are allowed, and both lending and borrowing is allowed. Compare the efficient frontier to the one in (c).

Exercise 4.2 (Sports betting). The odds offered by a bookmaker on a Premier League game between Chelsea and Liverpool are: "Chelsea": 2.50, "draw": 3.25, and "Liverpool": 2.70. A gambler who believes that all outcomes are equally likely has 100 British pounds to place on bets on the outcome of the game. Determine the efficient frontier.

Exercise 4.3 (Uncorrelated returns). An investor with capital amounting to $10,000 considers a repeated 1-day investment in five risky assets and a risk-free asset. The 1-day return on the risk-free asset is 1, and the one-day returns on the risky assets are uncorrelated with expected values μ_k and standard deviations σ_k given by

$$\mu_1 = 1.01, \; \sigma_1 = 0.02,$$
$$\mu_2 = 1.02, \; \sigma_2 = 0.04,$$
$$\mu_3 = 1.03, \; \sigma_3 = 0.06,$$
$$\mu_4 = 1.04, \; \sigma_4 = 0.08,$$
$$\mu_5 = 1.05, \; \sigma_5 = 0.10.$$

Determine the amount to be invested today in each asset to maximize the expected portfolio value tomorrow when the standard deviation of the portfolio value tomorrow is not allowed to exceed $30. Short-selling is allowed.

Exercise 4.4 (Hedging a zero-coupon bond). A bank has written a contract that requires the bank to pay $10,000 in 6 months from today. In return, the bank receives $9,700 today and wants to invest this amount to manage the liability. There are two investment opportunities available: a long position in a 9-month zero-coupon bond and a deposit in an account that does not pay interest. A 9-month zero-coupon bond with a face value of $10,000 costs $9,510 today. The bank believes that the 3-month zero rate, per year, in 6 months from today is normally distributed with mean 6% and standard deviation 1.5%.

(a) Determine the portfolio, among those whose initial value does not exceed
    $9,700, that minimizes the variance of the value of the assets minus that of
    your liability 6 months from today, subject to the constraint that the expectation
    of the portfolio value in 6 months from today is nonnegative. Determine the
    expected value and the standard deviation of the value in 6 months of the
    optimal portfolio.
(b) Determine and plot the efficient frontier.

Exercise 4.5 (Hedging stocks with options). Consider an investor liable to deliver
1,000 shares of a stock 1 year from now. The share price S_1 at that time is modeled
as $S_1 = S_0 e^{\mu - \sigma^2/2 + \sigma Z}$, where Z is standard normally distributed. The investor
believes the share price will go down and that $\mu = -0.05$ and $\sigma = 0.2$. Therefore,
the investor is reluctant to hedge the liability by buying shares. Instead, the investor
wants to hedge by buying at-the-money call options with payoff $(S_1 - S_0)_+$ 1 year
from now and investing in a risk-free 1-year zero-coupon bond with return 1.05.
The current share price is $S_0 = \$87$ and the current price of the call option is
$C_0 = \$9.04$.

(a) Determine the quadratic hedge.
(b) Determine the solution to the trade-off problem when the initial capital is
    $87,000 and the trade-off parameter $c = 5$.

Exercise 4.6 (Credit rating migration). Consider two corporate bonds with
different issuers. Both bonds have a face value of $100 and mature in 2 years. The
values of the bonds after 1 year depend on the 1-year risk-free zero rate at that
time, the credit ratings of the issuers, and the credit spreads of the issuers' rating
classes. Suppose that the possible credit ratings are "Excellent," "Good," "Poor,"
and "Default." The credit spreads corresponding to the rating classes are 0.5, 2.0,
9.0, and 80%. For instance, if an issuer after 1 year has a rating of "Good," then the
associated bond's 1-year zero rate is the risk-free zero rate at that time plus 2.0%.

The 1-year risk-free zero rate in 1 year is assumed to be normally distributed
with mean 6% and standard deviation 1.2%. Table 4.1 shows the probabilities of the
pairs of credit ratings of the two bond issuers in 1 year (the probability that the first
issuer has the rating "Good" and the second issuer has the rating "Excellent" is 0.6,
etc.). The 1-year risk-free zero rate in 1 year is assumed to be independent of the
credit ratings at that time. The current price of the bond issued by the first issuer is
$83.68, and the price of the bond issued by the second issuer is $87.50.

Consider an investment of $10,000 fully invested in positions in the two bonds.

(a) Determine and plot the efficient frontier.
(b) Select two efficient portfolios, one rather risky and one conservative. For both
    portfolios, illustrate the distribution of the portfolio value in 1 year, for instance
    by simulating from the distribution of the portfolio value, and construct a
    histogram.

Table 4.1 Probabilities for pair of credit ratings in 1 year for two issuers in Exercise 4.6

	Excellent	Good	Poor	Default	Sum
Excellent	0.098736	0.001056	0.000050	0	0.099842
Good	0.632718	0.097799	0.019020	0.000921	0.750458
Poor	0.064579	0.043802	0.019078	0.002226	0.129703
Default	0.004582	0.007243	0.006436	0.001736	0.019997
Sum	0.80063	0.149900	0.044584	0.004883	

The first issuer's rating corresponds to the rows, whereas the second issuer's rating corresponds to the columns

Exercise 4.7 (Insurer's asset allocation). Let L be the value of an insurer's liabilities 1 year from today. The liability value L represents the aggregate claim amount and is positively correlated with the return on a price index measuring inflation. The insurer wants to choose an asset portfolio to hedge the liability and to generate a good return on the assets. The investor considers three assets. The first is an inflation-linked asset with return R_1. The second is a pure investment asset with return R_2. The third is a bond portfolio with return R_3. The expected values and covariances are given by

$$
E\begin{bmatrix} R_1 \\ R_2 \\ R_3 \\ L \end{bmatrix} = \begin{pmatrix} 1.05 \\ 1.15 \\ 1.05 \\ 10^6 \end{pmatrix} \quad \text{and} \quad \text{Cov}\begin{pmatrix} R_1 \\ R_2 \\ R_3 \\ L \end{pmatrix} = \begin{pmatrix} 0.2^2 & 0 & 0 & 3{,}000 \\ 0 & 0.25^2 & 0 & 0 \\ 0 & 0 & 0.05^2 & 0 \\ 3{,}000 & 0 & 0 & 10^{10} \end{pmatrix}.
$$

The insurer invests according to the solution to the investment problem

$$
\begin{aligned}
\text{minimize} \quad & \text{Var}(w_1 R_1 + w_2 R_2 + w_3 R_3 - L) \\
\text{subject to} \quad & E[w_1 R_1 + w_2 R_2 + w_3 R_3] \geq 1.3\,E[L] \\
& w_1 + w_2 + w_3 = 1.2\,E[L] \\
& w_1, w_2, w_3 \geq 0.
\end{aligned}
$$

Determine the solution (w_1, w_2, w_3). Compare the solution to the corresponding solution in the case where the first asset is uncorrelated with the liability.

Project 4 (Repeated investments). Consider a pair of assets whose vectors of weekly log returns $(\log(S^1_{t+1}/S^1_t), \log(S^2_{t+1}/S^2_t))^T$ are assumed to be independent and $N_2(\mu, \Sigma)$-distributed, where

$$
\mu = \begin{pmatrix} 1 \\ 1 \end{pmatrix} \cdot 10^{-3} \quad \text{and} \quad \Sigma = \begin{pmatrix} 9 & 3 \\ 3 & 9 \end{pmatrix} \cdot 10^{-4}.
$$

An investor with initial capital of \$1,000,000 considers the following investment strategies with weekly rebalancing of the portfolio for 50 weeks.

(a) Rebalance weekly to hold half of the portfolio value in each asset at the beginning of each week.
(b) Rebalance weekly to reinvest the portfolio value according to the minimum-variance portfolio based on the computed mean vector and covariance matrix for the asset return vector based on estimates of μ and Σ from the 50 most recent log-return vectors.
(c) Rebalance weekly to reinvest the portfolio value fully in the two assets according to the solution to the trade-off problem without a risk-free asset and with trade-off parameter $c = 1$, with parameters estimated as in (b).

Simulate n samples of 100 weekly log returns, where the first 50 log-return vectors are considered as historical data and the remaining 50 log-return vectors are considered as future data within the investment period. Evaluate the distributions of the 50-week returns for investment strategies (a)–(c) by generating histograms based on n simulated 50-week returns. Choose n sufficiently large and use the same simulated samples for the evaluation of strategies (a)–(c).

Suppose that the investor is allowed, at the start of each week, to invest some of the portfolio value in a 1-week risk-free bond that pays no interest. Suggest and evaluate an investment strategy that includes the risk-free bonds as investment opportunities with the aim of generating an attractive distribution for the 50-week return.

Chapter 5
Utility-Based Investment Principles

In the previous chapter we measured the quality of an investment in terms of the expected value $E[V_1]$ and the variance $Var(V_1)$ of the future portfolio value V_1 and determined portfolio weights (subject to constraints) that maximize a suitable trade-off $E[V_1] - c\,Var(V_1)/(2V_0)$ between a large expected value and a small variance. Attractive features of this approach are that the probability distribution of V_1 does not have to be specified in detail and that explicit expressions for the optimal portfolio weights are found that have intuitive interpretations. We saw that this approach makes perfect sense if we consider portfolio values V_1 that can be expressed as linear combinations of asset returns whose joint distribution is a multivariate normal distribution. However, unless there are good reasons to assume a multivariate normal distribution (or, more generally, as will be made clear in Chap. 9, an elliptical distribution), solutions provided by the quadratic investment principles can be rather misleading. Here we want to allow for a probability distribution of any kind, and this calls for more general investment principles that are not only based on the variance and expected value of V_1.

This chapter consists of three sections. In Sect. 5.1, we introduce concepts such as subjective expected utility and risk aversion and derive a flexible parametric family of utility function. We also formulate investment problems in terms of maximization of expected utility of future portfolio values and analyze the consequences of this approach in a series of examples. In Sect. 5.2, we take a closer look at a special case, corresponding to a horse race with a given set of odds, of the general investment problem. For this special case explicit computations are possible, and from the explicit solutions we arrive at important conclusions about investment approaches may be drawn. In Sect. 5.3 we consider the future value of a single asset but assume that we may purchase any derivative contract on this value, as long as the derivative price does not violate our budget constraints. We show how to determine the optimal derivative contract given a utility function capturing our attitude toward risk, subjective probability beliefs, and market prices. The problem considered in Sect. 5.3 can be viewed as a limiting case of the problem considered in Sect. 5.2 as

H. Hult et al., *Risk and Portfolio Analysis: Principles and Methods*, Springer Series in Operations Research and Financial Engineering, DOI 10.1007/978-1-4614-4103-8_5, © Springer Science+Business Media New York 2012

the number of horses tends to infinity and the bookmaker's odds take the form of a
forward probability density determined by the market.

5.1 Maximization of Expected Utility

We consider an investor who at time 0 has capital V_0 and invests it until time 1 by
taking positions in a risk-free asset with value 1 at time 1 and in n risky assets with
future random values $S_1^1, \ldots, S_1^n$. At the end of the investment horizon the aggregate
value of the investor's positions is

$$V_1 = h_0 + h_1 S_1^1 + \cdots + h_n S_1^n - L,$$

where $h_0, h_1, \ldots, h_n$ are the positions in the $n + 1$ assets and L is the future value of
the investor's liabilities. Often we will consider the case $L = 0$, which corresponds
to a pure investment problem. If the current values $B_0, S_0^1, \ldots, S_0^n$ of the risk-free
and risky assets are positive, then we may consider monetary portfolio weights $w_0 = h_0 B_0$ and $w_k = h_k S_0^k$ and returns $R_0 = 1/B_0$ and $R_k = S_1^k/S_0^k$ and write $V_1 = w_0 R_0 + \mathbf{w}^T \mathbf{R} - L$, where $\mathbf{w} = (w_1, \ldots, w_n)^T$ and $\mathbf{R} = (R_1, \ldots, R_n)^T$.

A good approach for an investor to measuring the quality of a portfolio with
future value V_1 would be to plot the distribution function or density function of
V_1, compare it to the corresponding plots for alternative investments, and consider
how the shape of the plots relate to the investor's attitude toward the riskiness and
potential reward of the investment. However, it is often necessary to simplify the
decision process by summarizing the quality of a future portfolio value by a single
number. In the previous chapter, the investor summarized the quality of the portfolio
value V_1 by the number $E[V_1] - c \, \text{Var}(V_1)/(2V_0)$ and compared this number to the
corresponding values for alternative portfolios. A more general approach that can
handle possible deviations from the normality of V_1 is obtained by considering the
number $E[u(V_1)]$ for an appropriate choice of the function u. The function u is called
a utility function and should measure the utility of the random portfolio value V_1
from the investor's perspective. The number $E[u(V_1)]$ is the investor's subjective
expected utility of the portfolio value V_1 at time 1. For two attainable future portfolio
values V_1 and V_1', the investor prefers V_1 to V_1' if $E[u(V_1)] > E[u(V_1')]$. The investor's
investment principle here is the maximization of expected utility of the future
portfolio value. Since more money is preferred to less, u is an increasing function.
The increase in utility from an additional monetary unit is typically assumed to
decrease with increasing wealth. Therefore, we assume that u is a concave function:

$$u(\lambda x + (1 - \lambda)y) \geq \lambda u(x) + (1 - \lambda)u(y), \quad \lambda \in [0, 1].$$

If u is concave and twice differentiable, then $u'(x) \geq 0$ and $u''(x) \leq 0$.

That maximization of expected utility is a sound investment principle is supported by many classical mathematical results that form the basis of the well-
established theory of decision under uncertainty. The famous theorems of Leonard

Savage, for example (stripped of their details and in their simplest form), state that as long as an investor ranks portfolios according to certain axioms of rational behavior, there exist a bounded utility function u and a subjective probability distribution for $(S_1^1, \ldots, S_1^n, L)$ such that the investor's preferences are consistent with expected utility maximization. In principle, from the rational investor's ranking of all available portfolios one can determine the subjective probability distribution and the utility function u up to positive affine transformations ($au + b$ for $a > 0$). In this chapter, we assume that utility functions are concave. This assumption makes sense from an economic point of view and leads to natural parametric families of utility functions and to convex optimization problems. Although concavity is a convenient assumption, it should be stressed that it is not a consequence of the axioms of rational behavior proposed by Savage.

An investor with a utility function u and a subjective probability distribution assigned to the vector $\mathbf{R}$ of returns seeks to determine the optimal solution $(w_0, \mathbf{w})$ to the investment problem

$$
\begin{aligned}
&\text{maximize } \mathrm{E}\left[u\left(w_0 R_0 + \mathbf{w}^\mathrm{T}\mathbf{R} - L\right)\right] \\
&\text{subject to } w_0 + \mathbf{w}^\mathrm{T}\mathbf{1} \leq V_0,
\end{aligned}
\tag{5.1}
$$

where $\mathbf{w} = (w_1, \ldots, w_n)^\mathrm{T}$ and $\mathbf{R} = (R_1, \ldots, R_n)^\mathrm{T}$. Depending on the particular investment problem considered, it is likely that more constraints, such as restrictions on short-selling, have to be included. It is assumed that the utility function u and the probability distribution assigned to $(\mathbf{R}, L)$ are chosen such that the expected utility in (5.1) exists finitely, at least for some subset of portfolio weights $(w_0, \mathbf{w})$. Since u is assumed to be concave, the investment problem (5.1) is a convex optimization problem of the kind considered in Chap. 2. Indeed, it follows from Lemma 3.1 that the objective function $\mathrm{E}[u(w_0 R_0 + \mathbf{w}^\mathrm{T}\mathbf{R} - L))]$ is a concave function in $(w_0, \mathbf{w})$ since u is concave. Therefore, Proposition 2.3 applies, and it follows that if we find a number w_0, a vector $\mathbf{w}$, and a nonnegative number λ such that

$$
\begin{aligned}
\mathrm{E}\left[u'\left(w_0 R_0 + \mathbf{w}^\mathrm{T}\mathbf{R} - L\right) R_0\right] &= \lambda, \\
\mathrm{E}\left[u'\left(w_0 R_0 + \mathbf{w}^\mathrm{T}\mathbf{R} - L\right) \mathbf{R}\right] &= \lambda\mathbf{1}, \\
w_0 + \mathbf{w}^\mathrm{T}\mathbf{1} &= V_0,
\end{aligned}
\tag{5.2}
$$

then $(w_0, \mathbf{w})$ is an optimal solution to investment problem (5.1). If a risk-free asset is not available, then we set $w_0 = 0$ and omit the second equation in (5.2).

Remark 5.1. For investment problem (5.1) to make sense, an obvious requirement is that the expected utility must exist finitely. If u is bounded on an interval that contains all the values that the random variables $w_0 R_0 + \mathbf{w}^\mathrm{T}\mathbf{R} - L$ can take, then the finiteness of the expected value is guaranteed. However, if u is concave and a probability distribution is assigned to $(\mathbf{R}, L)$ that allows $w_0 R_0 + \mathbf{w}^\mathrm{T}\mathbf{R} - L$ to take any value on the real line, then the expected value in (5.1) may fail to exist finitely. A (nontrivial) concave function defined on the entire real line is necessarily unbounded from below, and many of the parametric families of concave utility function defined on the positive real line that are tractable for analytical computations are unbounded from above. The standard textbook probability distributions that seem to be reasonable model choices for the components of vector $\mathbf{R}$ and for L are defined on the entire positive real line, which may cause problems here. For instance, if $u(x) = -\tau e^{-x/\tau}$ and L is lognormally distributed, then the expected value in (5.1) does not exist finitely.

If $L = 0$ and only long position (nonnegative portfolio weights) in the risky assets are allowed, then the situation is better. However, if u is unbounded from above or below, then the probability mass that the return distribution assigns to intervals of the form (x, ∞) or $(0, x)$ must decay sufficiently fast as $x \to \infty$ or $x \to 0$ to ensure that the expected value in (5.1) exists finitely.

Summing up, we may run into technical difficulties when combining unbounded utility functions with typical textbook probability distributions that allow random variables to take arbitrary large values. Moreover, it is not easy to find examples that demonstrate the necessity of allowing for strictly positive probabilities of arbitrary large utility values (with plus or minus sign). However, from a pragmatic point of view parametric models that are computationally tractable but give rise to unbounded utilities may be acceptable if they give rise to useful procedures for decision making.

Example 5.1 (Arbitrage). If the utility function u is strictly increasing, then a necessary condition for the existence of an optimal solution to (5.1) is the absence of arbitrage opportunities. We now verify this claim. In the current setting, an arbitrage opportunity is a position $(w_0, \mathbf{w})$ such that

$$w_0 + \mathbf{w}^\mathsf{T}\mathbf{1} = 0, \quad w_0 R_0 + \mathbf{w}^\mathsf{T}\mathbf{R} \geq 0, \quad \text{and } w_0 R_0 + \mathbf{w}^\mathsf{T}\mathbf{R} > 0 \text{ with positive probability.}$$

The claim follows if we show that the existence of a solution to problem (5.1) implies the absence of arbitrage opportunities. Suppose that $(w_0^*, \mathbf{w}^*)$ is an optimal solution to (5.1) and that an arbitrage opportunity $(w_0, \mathbf{w})$ exists. Then also $(w_0^* + w_0, \mathbf{w}^* + \mathbf{w})$ is a feasible solution since $w_0 + \mathbf{w}^\mathsf{T}\mathbf{1} = 0$. However,

$$\mathrm{E}\left[u\left((w_0^* + w_0)R_0 + (\mathbf{w}^* + \mathbf{w})^\mathsf{T}\mathbf{R} - L\right)\right] > \mathrm{E}\left[u\left(w_0^* R_0 + \mathbf{w}^{*\mathsf{T}}\mathbf{R} - L\right)\right],$$

which contradicts the claim that $(w_0^*, \mathbf{w}^*)$ is an optimal solution to (5.1).

The axioms of rational behavior underlying expected utility maximization have been criticized for not reflecting many peoples' observed preferences among choices. One such illustration of irrational preferences is called the Allais paradox and is presented below.

Example 5.2 (Allais paradox). Consider four contracts with payoffs, in millions of dollars given by

$$A = 1, \quad B = \begin{cases} 1 \text{ w. p. } 0.89 \\ 0 \text{ w. p. } 0.01 \,, \\ 5 \text{ w. p. } 0.10 \end{cases} \quad C = \begin{cases} 0 \text{ w. p. } 0.89 \\ 1 \text{ w. p. } 0.11 \end{cases}, \quad D = \begin{cases} 0 \text{ w. p. } 0.90 \\ 5 \text{ w. p. } 0.10 \end{cases},$$

where w. p. stands for "with probability." Suppose that the payoffs (contracts) A and B have the same price and that the payoffs (contracts) C and D have the same price. Empirical investigations show that many people tend to prefer A to B and also D to C. If a person with these preferences is maximizing expected utility, then there must exist a utility function u such that

$$u(1) > 0.10u(5) + 0.89u(1) + 0.01u(0), \tag{5.3}$$

$$0.10u(5) + 0.90u(0) > 0.11u(1) + 0.89u(0). \tag{5.4}$$

Combining (5.3) and (5.4) gives $0.10u(5) + 0.90u(0) > 0.10u(5) + 0.90u(0)$, from which we conclude that preferring A to B and also D to C is inconsistent with expected utility maximization. The conclusion is, however, likely to be perceived as rather irrelevant to someone who has these preferences and does not care much about mathematical theories. To demonstrate more clearly the irrationality of these preferences, consider payoff F, which takes the value 0 with probability $1/11$ and the value 5 with probability $10/11$ and has the same price as A (and B). Suppose that A is preferred to F (the argument below is identical if the converse is true). Consider an indicator variable I that is independent of F and takes the value 1 with probability 0.89. We notice that

$$IA + (1 - I)A = A,$$

$$IA + (1 - I)F = B,$$

$$I0 + (1 - I)A = C,$$

$$I0 + (1 - I)F = D.$$

One of the axioms of rational behavior, called the independence axiom or the sure-thing principle, says that if A is preferred to F, then, for any probability p, receiving A with probability p and a fixed amount otherwise must be preferred to receiving F with probability p and the same fixed amount otherwise. Most people would agree that preferences violating the independence axiom are irrational. Here, preferring A to B and also D to C is clearly a violation of the independence axiom. It seems likely that once confronted with the consequences of preferring A to B and also D to C, a person that initially held these preferences would modify them into preferences that are consistent with expected utility maximization.

The Allais paradox is not really a paradox but rather an illustration of the fact that people do not always make rational decisions when the alternatives are somewhat complicated. However, here we are not focusing on describing observations of preferences but rather on presenting sound decision tools that lead to preferences between complicated alternatives that are not modified after analyzing the alternatives in more detail.

Consider an asset, portfolio, or financial contract that has the random value V_1 at time 1. For each investor there should be some fixed amount C of money at time 1 such that the investor is indifferent to receiving the fixed amount C or the random amount V_1 at time 1. For an expected-utility-maximizing investor the amount C is given by

$$u(C) = \mathrm{E}[u(V_1)]$$

and is called the investor's certainty equivalent of random value V_1. The certainty equivalent could be seen as an upper bound on the forward price of V_1 that would make the investor want to enter into a forward contract with a counterparty that promises to deliver V_1 at time 1. What can be said about C? If the utility function u is strictly increasing and continuous, then $C = u^{-1}(\mathrm{E}[u(V_1)])$, where the inverse u^{-1} is strictly increasing. In particular, maximizing the certainty equivalent is tantamount to maximizing the expected utility. The utility functions that we consider are concave, and therefore Proposition 2.2 in Chap. 2 implies that $u(\mathrm{E}[V_1]) \geq \mathrm{E}[u(V_1)]$ (this inequality is called Jensen's inequality). Since u is also increasing, the relation

$$u(C) = \mathrm{E}[u(V_1)] \leq u(\mathrm{E}[V_1])$$

tells us that $C \leq \mathrm{E}[V_1]$. The investor is said to be risk averse since the certain amount C, which is preferred to investing and receiving the random value V_1, is less than or equal to the expected future portfolio value $\mathrm{E}[V_1]$. A risk-averse expected-utility maximizer may find it rational to pay a risk premium to an insurer.

The certainty equivalent is illustrated in Fig. 5.1. Along the x-axis, the density of V_1 is illustrated along with the graph of the utility function. The density on the y-axis is the density of $u(V_1)$. It is obtained by reflecting the density of V_1 in the graph of the utility function, $\{(x, u(x)), x > 0\}$. The concavity of the utility function leads to a skew to the right in the density of $u(V_1)$. The expected value $\mathrm{E}[V_1]$ is illustrated by the solid line segment orthogonal to the x-axis. The value $u(\mathrm{E}[V_1])$ is illustrated by the solid line segment orthogonal to the y-axis. The expected utility $\mathrm{E}[u(V_1)]$ is illustrated by the dashed line segment orthogonal to the y-axis. Note that, because of the concavity of u, we have the inequality $\mathrm{E}[u(V_1)] \leq u(\mathrm{E}[V_1])$. The certainty equivalent $C = u^{-1}(\mathrm{E}[u(V_1)]$ is illustrated by the dashed line segment orthogonal to the x-axis. Notice that $C \leq \mathrm{E}[V_1]$.

Closely related to the certainty equivalent is the absolute risk premium, which is the fixed amount $\pi[V_1]$ of money at time 1 such that

$$\mathrm{E}[u(V_1)] = u(\mathrm{E}[V_1] - \pi[V_1]). \tag{5.5}$$

We see that the risk premium is the certain amount that the investor requires as compensation to become indifferent to the choice between the risky investment and its expected value. Note also that $\mathrm{E}[V_1] - \pi[V_1]$ is the certainty equivalent of V_1 and that $\pi[V_1] \geq 0$ since u is concave.

The degree of risk aversion for a twice-differentiable utility function may be measured in terms of the Arrow–Pratt absolute risk aversion coefficient, which is given by

Fig. 5.1 Density of portfolio value V_1 (x-axis) and density of $u(V_1)$ (y-axis). The *solid line* segment orthogonal to the x-axis represents $E[V_1]$, whereas that orthogonal to the y-axis represents $u(E[V_1])$. The *dashed line* segment orthogonal to the y-axis represents $E[u(V_1)]$, whereas that orthogonal to the x-axis represents the certainty equivalent C

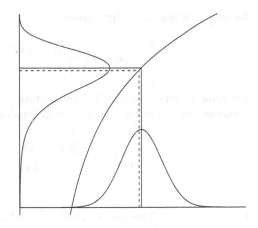

$$A(x) = -\frac{u''(x)}{u'(x)}.$$

It measures the investor's risk aversion locally as a function of the wealth level and does not depend on the investor's probability beliefs. Clearly, the absolute risk aversion coefficient A is defined only for strictly increasing utility functions u, $u'(x) > 0$. Moreover, the concavity of u implies that $A(x) \geq 0$. The reason for looking at this coefficient rather than just u'' is that the measure A is invariant under a positive affine transformation of u ($au + b$ with $a > 0$).

There are infinitely many potentially useful parametric families of utility functions. To understand how properties of an investor's utility function affect the investor's portfolio choice, we consider only one parametric family. The popular and flexible family of utility functions called HARA (hyperbolic absolute risk aversion) utility functions follows from a natural parameterization of the Arrow–Pratt absolute risk aversion coefficient A. If x and h have the dimension money, then it follows from the definitions $u'(x) = \lim_{h \to 0}(u(x + h) - u(x))/h$ and $u''(x) = \lim_{h \to 0}(u'(x + h) - u'(x))/h$ that $A(x)$ has the dimension money to the power -1. Therefore, a natural parameterization is

$$A(x) = -\frac{u''(x)}{u'(x)} = \frac{1}{\tau + \gamma x}, \quad \tau + \gamma x > 0,$$

where γ is a dimensionless constant and the constant τ has the dimension money. The rightmost foregoing equality is a differential equation for u that can be solved as follows. With $y(x) = u'(x)$ we solve the linear homogeneous equation $y'(x) = -y(x)/(\tau + \gamma x)$ by separation of variables:

$$\int \frac{dy}{y} = -\int \frac{dx}{\tau + \gamma x} + C_1$$

for some constant C_1, which gives

$$y = C_2 \cdot \begin{cases} (\tau + \gamma x)^{-1/\gamma}, & \gamma \neq 0, \\ e^{-x/\tau}, & \gamma = 0, \end{cases}$$

for some positive constant C_2. Integrating $y(x) = u'(x)$ with respect to x gives solutions that are positive affine transformations of

$$u(x) = \begin{cases} \frac{1}{\gamma-1}(\tau + \gamma x)^{1-1/\gamma}, & \gamma \neq 1, 0; \\ \log(\tau + x), & \gamma = 1; \\ -\tau e^{-x/\tau}, & \gamma = 0 \end{cases} \tag{5.6}$$

for $\tau + \gamma x > 0$. This utility function is called a HARA utility function. The three cases correspond to power, logarithmic, and (negative) exponential utility functions. As a limit of the HARA utility function as $\gamma \to \infty$ we obtain the linear utility function

$$\lim_{\gamma \to \infty} u(x) = \lim_{\gamma \to \infty} \frac{1}{\gamma - 1}(\tau + \gamma x)^{1-1/\gamma} = x.$$

The HARA utility function with $\gamma = -1$ is the quadratic utility function

$$u(x) = -\frac{1}{2}(\tau - x)^2, \quad x \leq \tau,$$

which is seen to be strictly increasing and strictly concave for $x < \tau$. The expected utility of a future portfolio value bounded from above by τ here takes the form

$$\mathrm{E}[u(V_1)] = \tau \, \mathrm{E}[V_1] - \frac{1}{2}\left(\mathrm{Var}(V_1) + \mathrm{E}[V_1]^2 \right) - \frac{\tau^2}{2},$$

which indicates that maximizing the expected quadratic utility is closely connected with solving the mean–variance trade-off problem (4.7) [or (4.3)]. Example 5.4 below shows that this is indeed true.

So which utility function is the right utility function for a rational investor? This question is not particularly relevant. It is quite possible that the decisions over time of a rational decision maker are such that each of the decisions is consistent with expected utility maximization but also such that no utility function is consistent with all of the decisions: the degree of risk aversion and the subjective probabilities assigned to future events are likely to vary over time.

Example 5.3 (Fire insurance). Consider a company that is contemplating buying fire insurance on a factory for the following year. For simplicity we assume that the company will have the known net wealth V 1 year from now if there is no fire and $V - fV$, for some $f \in (0, 1)$, if there is a fire. Therefore, without fire insurance the random net wealth for the company in 1 year is $V_1 = V(1 - If)$, where I takes the value 1 if there is a fire during the year and 0 otherwise. The company estimates that the probability of a fire in its factory during the following year is p and that the company's attitude toward risk can be captured by a HARA utility function u with $\gamma \neq 0, 1$ and $\tau = 0$, i.e.,

$$u(x) = \frac{1}{\gamma - 1} \gamma^{1-1/\gamma} x^{1-1/\gamma}, \quad x > 0.$$

Therefore, the certainty equivalent C of V_1 is

$$C = V \left(p(1 - f)^{(\gamma-1)/\gamma} + 1 - p \right)^{\gamma/(\gamma-1)}.$$

If we assume that buying fire insurance makes the company immune to any financial consequences of a fire and only affects the company's future net wealth by reducing it from V to $V - cV$, where cV discounted to money today can be interpreted as the fire insurance premium, then the number c, given by $E[u(V(1-If))] = u(V(1-c))$, that makes the company indifferent to buying fire insurance is

$$c(f, p, \gamma) = 1 - \left(p(1 - f)^{(\gamma-1)/\gamma} + 1 - p \right)^{\gamma/(\gamma-1)}.$$

Computing the partial derivative of c with respect to γ gives

$$-\frac{1}{\gamma(\gamma - 1)} \left(p(1 - f)^{(\gamma-1)/\gamma} + 1 - p \right)^{1/(\gamma-1)} p(1 - f)^{(\gamma-1)/\gamma} \log(1 - f) < 0,$$

so the premium that the company is willing to pay for fire insurance is decreasing in γ: the more risk averse the company is, the higher the fire insurance premium it is willing to accept. Notice that $\lim_{\gamma \to \infty} u(x) = x$, which is the utility function of a risk-neutral decision maker, and that $\lim_{\gamma \to \infty} c(f, p, \gamma) = pf$, which is the expected fire loss. The case $\gamma \to \infty$ corresponds to setting the expected net wealth without insurance equal to the net wealth with insurance and solving the equation for c. We now compute the limits for $c(p, f, \gamma)$ as γ tends to 1 and 0, respectively.

$$\lim_{\gamma \to 1} c(f, p, \gamma) = 1 - \exp \left(-\lim_{x \to 0} \frac{1}{x} \log(p(1 - f)^{-x} + 1 - p) \right)$$

$$= \left\{ \text{l'Hôpital's rule}, \quad da^{-x}/dx = -a^{-x} \log a \right\}$$

$$= 1 - \exp \left(\lim_{x \to 0} \frac{(1 - f)^{-x} p \log(1 - f)}{p(1 - f)^{-x} + 1 - p} \right)$$

$$= 1 - (1 - f)^p.$$

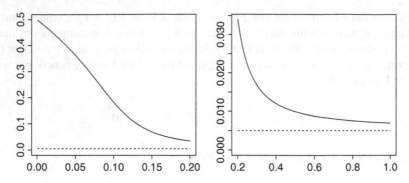

Fig. 5.2 Graph of $c(f, p, \gamma)$ as a function of γ for $(f, p) = (0.5, 0.01)$ (*solid curves*) and the level $pf = 0.005$ (*dashed line*)

We also find that

$$
\lim_{\gamma \to 0} c(f, p, \gamma) = 1 - \lim_{x \to -\infty} (p(1-f)^x + 1 - p)^{1/x}
$$

$$
= 1 - \lim_{x \to -\infty} (p(1-f)^x)^{1/x}
$$

$$
= f,
$$

which is extremely risk averse and does not take the (probably rather small) probability p into account. Figure 5.2 shows a graph of $c(f, p, \gamma)$ as a function of γ for $(f, p) = (0.5, 0.01)$.

Example 5.4 (Quadratic utility). Consider an investor with a quadratic utility function $u(x) = \tau x - x^2/2$ who wants to find the optimal solution to the investment problem

$$
\text{maximize } E[u(w_0 R_0 + \mathbf{w}^T \mathbf{R})]
$$
$$
\text{subject to } w_0 + \mathbf{w}^T \mathbf{1} \le V_0.
$$

Since $u(x)$ is increasing and concave only for $x \le \tau$, the possible outcome of the future portfolio value $w_0 R_0 + \mathbf{w}^T \mathbf{R}$ should be bounded from above by τ. Here we ignore this restriction.

The sufficient conditions for an optimal solution in (5.2) here take the form

$$
E[\tau \mathbf{R} - (w_0 R_0 + \mathbf{w}^T \mathbf{R}) \mathbf{R}] = \lambda \mathbf{1},
$$

$$
E[\tau R_0 - (w_0 R_0 + \mathbf{w}^T \mathbf{R}) R_0] = \lambda,
$$

$$
w_0 + \mathbf{w}^T \mathbf{1} = V_0,
$$

or, equivalently, since $E[\mathbf{R}\mathbf{R}^T] = \text{Cov}(\mathbf{R}) + E[\mathbf{R}] E[\mathbf{R}]^T$,

$$(\tau - w_0 R_0)\mu - (\Sigma + \mu\mu^\mathsf{T})\mathbf{w} = \lambda\mathbf{1},$$

$$(\tau - w_0 R_0)R_0 - \mathbf{w}^\mathsf{T}\mu R_0 = \lambda,$$

$$w_0 + \mathbf{w}^\mathsf{T}\mathbf{1} = V_0.$$

Inserting the expression of λ from the second equation into the first equation leads to

$$-(\Sigma + \mu\mu^\mathsf{T})\mathbf{w} = [(\tau - w_0 R_0)R_0 - \mathbf{w}^\mathsf{T}\mu R_0]\mathbf{1} - (\tau - w_0 R_0)\mu.$$

Since $w_0 = V_0 - \mathbf{w}^\mathsf{T}\mathbf{1}$, the equation in the last display may be rewritten in the form

$$(\Sigma + (\mu - R_0\mathbf{1})(\mu - R_0\mathbf{1})^\mathsf{T})\mathbf{w} = (\mu - R_0\mathbf{1})(\tau - V_0 R_0).$$

The identity

$$(\Sigma + \mathbf{v}\mathbf{v}^\mathsf{T})^{-1} = \Sigma^{-1} - \frac{\Sigma^{-1}\mathbf{v}\mathbf{v}^\mathsf{T}\Sigma^{-1}}{1 + \mathbf{v}^\mathsf{T}\Sigma^{-1}\mathbf{v}}$$

now leads to the solution

$$\mathbf{w} = \frac{\tau - V_0 R_0}{1 + (\mu - R_0\mathbf{1})^\mathsf{T}\Sigma^{-1}(\mu - R_0\mathbf{1})}\Sigma^{-1}(\mu - R_0\mathbf{1}).$$

In particular, the solution to the maximization of the expected quadratic utility is the solution to the trade-off version (4.7) of the quadratic investment problems studied in Chap. 4 with trade-off parameter

$$c = \frac{1 + (\mu - R_0\mathbf{1})^\mathsf{T}\Sigma^{-1}(\mu - R_0\mathbf{1})}{\tau/V_0 - R_0}.$$

Example 5.5 (Exponential utility and multivariate normal returns). Suppose that $\mathbf{R}$ is $N_n(\mu, \Sigma)$-distributed, which implies that $w_0 R_0 + \mathbf{w}^\mathsf{T}\mathbf{R}$ is $N(w_0 R_0 + \mathbf{w}^\mathsf{T}\mu, \mathbf{w}^\mathsf{T}\Sigma\mathbf{w})$-distributed. In particular, we may write

$$E\left[u\left(w_0 R_0 + \mathbf{w}^\mathsf{T}\mathbf{R}\right)\right] = E\left[u\left(w_0 R_0 + \mathbf{w}^\mathsf{T}\mu + \sqrt{\mathbf{w}^\mathsf{T}\Sigma\mathbf{w}}Z\right)\right]$$

for a standard normally distributed Z. If $u(x) = -\tau e^{-x/\tau}$, then $u(w_0 R_0 + \mathbf{w}^\mathsf{T}\mathbf{R})$ is lognormally distributed and

$$E[u(w_0 R_0 + \mathbf{w}^\mathsf{T}\mathbf{R})] = -\tau e^{-(w_0 R_0 + \mathbf{w}^\mathsf{T}\mu)/\tau + \mathbf{w}^\mathsf{T}\Sigma\mathbf{w}/(2\tau^2)}. \qquad (5.7)$$

Since $-\tau \log(-x/\tau)$ is an increasing function of x for $x < 0$, maximizing (5.7) over all $\mathbf{w}$ satisfying the budget constraint $w_0 + \mathbf{w}^\mathsf{T}\mathbf{1} \leq V_0$ is equivalent to solving

$$\text{maximize } w_0 R_0 + \mathbf{w}^{\mathsf{T}} \boldsymbol{\mu} - \frac{1}{2\tau} \mathbf{w}^{\mathsf{T}} \Sigma \mathbf{w}$$
$$\text{subject to } w_0 + \mathbf{w}^{\mathsf{T}} \mathbf{1} \le V_0.$$

This is the kind of quadratic optimization problem that we encountered earlier in Chap. 4. Its solution $(w_0, \mathbf{w})$ is given by $\mathbf{w} = \tau \Sigma^{-1}(\boldsymbol{\mu} - R_0 \mathbf{1})$ and $w_0 = V_0 - \mathbf{w}^{\mathsf{T}} \mathbf{1}$.

5.2   A Horse Race Example

Consider an investment situation where bets can be placed on the occurrence of one out of n mutually exclusive events that are assigned strictly positive probabilities. Simultaneous bets on the occurrence of different events are allowed. The situation corresponds to a horse race with n horses where bets can be placed on the winner and where simultaneous bets on different horses are allowed. Alternatively, the investment situation corresponds to the opportunity to buy digital options on the future value of an asset where only one digital option will give a nonzero payoff and where there is zero probability that none of them will. Let X_k be a random variable that takes the value 1 if the kth outcome occurs (the kth horse wins) and 0 otherwise, and let $q_k > 0$ be the price of the payoff X_k. Then $1/q_k$ can be interpreted as the odds for the kth horse winning the race or alternatively the nonzero outcome of the return of the kth digital option. Since $X_1 + \cdots + X_n = 1$ and precisely one X_k will take the value 1, investing the amount q_k on each payoff X_k gives the payoff 1 for sure. In particular, this investment corresponds to buying a synthetic zero-coupon bond with face value 1. In the horse-race setting, when the time value of money can be neglected, a fair bookmaker who does not demand profits for offering the game would set the odds so that $q_1 + \cdots + q_n = 1$. We assume that $q_1 + \cdots + q_n \ge 1$ in order to rule out opportunities to make a risk-free profit.

Here we consider an investor with an amount V_0 to invest in a combination of bets on the n outcomes in order to maximize the expected utility of the sum of the payoffs. Let $p_k \in (0, 1)$ be the subjective probability that the investor assigns to the event $X_k = 1$. An amount w_k invested in the payoff X_k corresponds to buying w_k/q_k number of contracts, which results in a time 1 value of $(w_k/q_k)X_k$. The investment problem we consider is how much the investor should invest in each X_k, i.e., the expected utility maximization problem

$$\text{maximize E}\left[u\left(w_1 q_1^{-1} X_1 + \cdots + w_n q_n^{-1} X_n\right)\right]$$
$$\text{subject to } w_1 + \cdots + w_n \le V_0 \tag{5.8}$$
$$\tau + \gamma w_k/q_k > 0, \quad k = 1, \dots, n.$$

The requirement $\tau + \gamma w_k/q_k > 0$ for all k ensures that the expected value exists finitely. The sufficient condition (5.2) for an optimal solution $\mathbf{w}$ here translates into the existence of a nonnegative λ such that

$$\lambda = p_k u'\left(w_k \frac{1}{q_k}\right)\frac{1}{q_k} \quad \text{for} \quad k = 1, \dots, n \quad \text{and} \quad \mathbf{w}^{\mathsf{T}} \mathbf{1} = V_0. \tag{5.9}$$

We write $1/q_k$ in (5.9) since $1/q_k$ is a dimensionless constant, which is the nonzero outcome of the return X_k/q_k. As a comparison, q_k^{-1} in (5.8) has the dimension money to the power -1.

We will now investigate the effect of different parameterizations of the HARA utility function on the optimal solution to (5.8).

Logarithmic utility. Let $u(x) = \log(\tau + x)$, the HARA utility function for $\gamma = 1$, and consider the problem

$$\text{maximize } E[\log(\tau + w_1 q_1^{-1} X_1 + \cdots + w_n q_n^{-1} X_n)]$$
$$\text{subject to } w_1 + \cdots + w_n \leq V_0$$
$$\tau + w_k/q_k > 0, \quad k = 1, \ldots, n.$$

The requirement $\tau + w_k/q_k > 0$ for all k ensures that the expectation is finite. Combining the first and second parts of (5.9) gives $1/\lambda = V_0 + \tau \sum_j q_j$ and

$$w_k = V_0 p_k + \tau \left(p_k \sum_{j=1}^{n} q_j - q_k \right) \quad \text{for} \quad k = 1, \ldots, n. \tag{5.10}$$

In particular, if $V_0 + \tau \sum_j q_j > 0$, then $\lambda > 0$ and $w_k = p_k/\lambda - \tau q_k > -\tau q_k$ for all k, so the optimal solution is given by (5.10). For $\tau = 0$ the solution is $w_k = V_0 p_k$, which means that for each k the invested amount in the kth possible outcome is proportional to the subjective probability p_k assigned to the event $\{X_k = 1\}$ and does not depend on the price q_k. The solution corresponds to a rather extreme view in the sense that it is irrelevant how the probability p_k compares to the price q_k.

Exponential utility. Let $u(x) = -\tau e^{-x/\tau}$. The investment problem (5.8) is solved by verifying that there is a nonnegative λ and a vector $\mathbf{w}$ that solve (5.9) with $u'(x) = e^{-x/\tau}$. Straightforward, but somewhat tedious, calculations show that $\lambda > 0$ and

$$w_k = V_0 \frac{q_k}{\sum_{j=1}^{n} q_j} + \tau \left(q_k \log \left(\frac{p_k}{q_k} \right) - \frac{q_k}{\sum_{j=1}^{n} q_j} \sum_{j=1}^{n} q_j \log \left(\frac{p_j}{q_j} \right) \right). \tag{5.11}$$

Power utility. Here we consider the HARA utility function

$$u(x) = \frac{1}{\gamma - 1} (\tau + \gamma x)^{1 - 1/\gamma}, \quad \text{for } \gamma \neq 0, 1 \text{ and } \tau + \gamma x > 0.$$

Investment problem (5.8) takes the form

$$\text{maximize } \frac{1}{\gamma - 1} E \left[\left(\tau + \gamma (w_1 q_1^{-1} X_1 + \cdots + w_n q_n^{-1} X_n) \right)^{1 - 1/\gamma} \right]$$
$$\text{subject to } w_1 + \cdots + w_n \leq V_0 \tag{5.12}$$
$$\tau + \gamma w_k/q_k > 0, \quad k = 1, \ldots, n,$$

and is solved by verifying that there is a nonnegative λ and a vector $\mathbf{w}$ with $\tau + \gamma w_k/q_k > 0$ that solve (5.9) with $u'(x) = (\tau + \gamma x)^{-1/\gamma}$. After a bit of algebra, one arrives at

$$\lambda = \left(\frac{\gamma V_0 + \tau \sum_{j=1}^{n} q_j}{\sum_{j=1}^{n} q_j (p_j/q_j)^\gamma} \right)^{-1/\gamma}$$

and

$$w_k = V_0 \frac{q_k (p_k/q_k)^\gamma}{\sum_{j=1}^{n} q_j (p_j/q_j)^\gamma} + \frac{\tau}{\gamma} \left(\frac{q_k (p_k/q_k)^\gamma}{\sum_{j=1}^{n} q_j (p_j/q_j)^\gamma} \sum_{j=1}^{n} q_j - q_k \right). \quad (5.13)$$

In particular, if $\gamma V_0 + \tau \sum_{j=1}^{n} q_j > 0$, then $\lambda > 0$,

$$\tau + \gamma w_k/q_k = \frac{q_k (p_k/q_k)^\gamma}{\sum_{j=1}^{n} q_j (p_j/q_j)^\gamma} \left(\gamma V_0 + \tau \sum_{j=1}^{n} q_j \right) > 0,$$

and the optimal solution to (5.12) is given by (5.13).

One would expect that if we let $\gamma \to 1$, then the solution w_k in (5.13) should converge to the solution (5.10) for the logarithmic utility function. Indeed, simply setting $\gamma = 1$ in expression (5.13) gives expression (5.10).

Similarly, one would expect that if we let $\gamma \to 0$, then the solution w_k in (5.13) should converge to the solution (5.11) for the exponential utility function. Applying l'Hôpital's rule and the relation $(d/dx)a^x = a^x \log(a)$ shows that the limit of (5.13) as $\gamma \to 0$ is the expression (5.11).

Suppose that $\tau = 0$, and consider the optimal solution (5.13) to (5.12). Notice that for each k, the money invested in the derivative paying one unit if the kth outcome occurs is the price q_k of the derivative times the number of derivatives bought, which is $(p_k/q_k)^\gamma$ times a normalizing constant that does not depend on k. In particular, the number of derivatives bought is large if the subjective probability p_k assigned to the kth outcome is large in comparison to the price q_k of this derivative, and if γ is large (corresponding to an investor who is not very risk averse).

For the HARA utility function with $\tau = 0$, $u(x) \to x$ as $\gamma \to \infty$. Therefore, one would expect that the optimal solution (5.13) to investment problem (5.12) as $\gamma \to \infty$ converges to the solution to

$$\text{maximize} \, \mathrm{E}[w_1 q_1^{-1} X_1 + \cdots + w_n q_n^{-1} X_n]$$
$$\text{subject to} \, w_1 + \cdots + w_n \le V_0 \quad (5.14)$$
$$w_k \ge 0, \quad k = 1, \ldots, n.$$

The solution to (5.14) is simply $w_k = V_0$ for the k that maximizes the ratio p_k/q_k, i.e., the index k of the bet with the highest expected payoff. On the assumption that there is an index k_* for which $p_{k_*}/q_{k_*} > p_k/q_k$ for all $k \neq k_*$, we find that

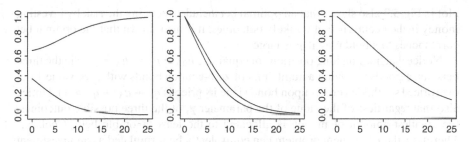

Fig. 5.3 *Left and middle plots*: optimal investments for investment problem in Example 5.6. *Left plot*: fractions of initial capital placed on bets on "Chelsea," "draw," and "Liverpool" as functions of γ. For a given value of γ, the value of the *lower curve* is the fraction of V_0 that is invested in the outcome "Chelsea," the difference in values between the *upper* and the *lower curve* is the fraction invested in the outcome "draw," and the remaining fraction is invested in the outcome "Liverpool." *Middle plot*: fractions of initial capital invested in synthetic risk-free asset, in "Liverpool," and in "draw, in that order, as functions of γ. The plot to the *right* refers to Example 5.7 and shows the fractions of the initial capital kept for later use (kept in the "pocket") and placed on the bet "draw"

$$\lim_{\gamma \to \infty} V_0 \frac{q_k(p_k/q_k)^\gamma}{\sum_{j=1}^n q_j(p_j/q_j)^\gamma} = \begin{cases} V_0 & \text{if } k = k_*, \\ 0 & \text{if } k \neq k_*, \end{cases}$$

i.e., the guess turned out to be correct.

Example 5.6 (Online sports betting II).   Consider Example 1.8, where a bookmaker offers odds on the outcome of a game between Chelsea and Liverpool. The odds correspond to the price $q_1 = 1/2.50$ for a digital option that pays 1 if the outcome of the game is "Chelsea," $q_X = 1/3.25$ for the outcome "draw," and $q_2 = 1/2.70$ for the outcome "Liverpool." We now consider optimal bets for an investor with a HARA utility function with $\tau = 0$ who assigns the probabilities $p_1 = p_X = p_2 = 1/3$ to the three possible outcomes of the game. From the investor's perspective, the bet "draw" is relatively cheap, whereas "Chelsea" and "Liverpool" are relatively expensive. The optimal bets are

$$w_k = V_0 \frac{q_k^{1-\gamma}}{q_1^{1-\gamma} + q_X^{1-\gamma} + q_2^{1-\gamma}} \quad \text{for } k = 1, X, 2.$$

The optimal bets as a function of γ are shown in the left plot in Fig. 5.3. For each γ the interval $[0, 1]$ is divided into three parts. For a given value of γ, the value of the lower curve is the fraction of V_0 that is invested in the first outcome, "Chelsea," the difference in values between the upper and the lower curve is the fraction invested in the second outcome, "draw," and the remaining fraction is invested in the third outcome, "Liverpool." We see that betting on "draw" is the most attractive bet from the investor's point of view (perceived as underpriced), whereas betting on the other possible outcomes are less attractive (perceived as overpriced). However, the left plot in Fig. 5.3 also shows that an optimal bet includes reducing the risk by investing money in the occurrence of unlikely outcomes. It is also shown that the optimal bet corresponds to a bold play if γ is large.

Notice that holding long positions of equal size $h_0 (= h_1 = h_X = h_2)$ in the three bets corresponds to having a number h_0 of zero-coupon bonds with face value 1 (an overpriced synthetic zero-coupon bond since its price is $q_1 + q_X + q_2 > 1$). Notice also that regardless of the value of the parameter γ, of the three possible outcomes, the smallest amount of money is allocated to the least favorable bet, "Chelsea." Therefore, the investment problem can equivalently be formulated as an investment problem with a risk-free asset with return $1/(q_1 + q_X + q_2)$ and with the bets on "Liverpool" and "draw" as risky assets. The middle plot in Fig. 5.3 shows the money (as fractions of the initial capital V_0) invested in the synthetic bond, in the outcome "Liverpool," and in the outcome "draw" as functions of γ. For a given value of γ, the value of the lower curve is the fraction of V_0 that is invested in the synthetic bond, the difference in values between the upper and the lower curve is the fraction invested in the outcome "Liverpool," and the remaining fraction is invested in the outcome "draw."

Example 5.7 (Online sports betting III). A more realistic version of the sports betting problem in Example 5.6 is obtained by including a risk-free asset with return $R_0 = 1$, corresponding to keeping some of the initial capital for use later (keeping it in the "pocket," say). With this modification the sufficient conditions (5.2) for an optimal solution (w_0, w_1, w_X, w_2) to the investment problem yields the condition $q_1 + q_X + q_2 = 1$, which is violated here. In particular, no optimal solution is obtained from (5.2) when all investment opportunities are considered. The left plot in Fig. 5.3 indicates that to find the optimal solution, we should set $w_1 = 0$ (omit bets on the outcome "Chelsea") and apply (5.2) to the modified problem. Since a position with $w_1, w_X, w_2 > 0$ can be interpreted as an investment in a synthetic risk-free asset with a return of less than one, it is hardly surprising that such a position can never be optimal here. The investment problem we consider here is as follows:

$$\text{maximize } \frac{\gamma^{1-1/\gamma}}{\gamma-1}\left(p_1 w_0^{1-1/\gamma} + p_X(w_0 + w_X q_X^{-1})^{1-1/\gamma} + p_2(w_0 + w_2 q_2^{-1})^{1-1/\gamma}\right)$$
$$\text{subject to } w_0 + w_X + w_2 \leq V_0$$
$$w_X, w_2 \geq 0.$$

If we temporarily ignore the nonnegativity condition $w_X, w_2 \geq 0$, then the sufficient conditions (5.2) translate into

$$\gamma^{-1/\gamma}\frac{p_X}{q_X}(w_0 + w_X q_X^{-1})^{-1/\gamma} = \lambda,$$

$$\gamma^{-1/\gamma}\frac{p_2}{q_2}(w_0 + w_2 q_2^{-1})^{-1/\gamma} = \lambda,$$

$$\gamma^{-1/\gamma}\left(p_1 w_0^{-1/\gamma} + p_X(w_0 + w_X q_X^{-1})^{-1/\gamma} + p_2(w_0 + w_2 q_2^{-1})^{-1/\gamma}\right) = \lambda,$$

$$w_0 + w_X + w_2 = V_0.$$

Solving for (w_0, w_X, w_2, λ) yields

$$\lambda = (\gamma V_0)^{-1/\gamma} \left(p_1^\gamma (1 - q_X - q_2)^{1-\gamma} + p_X^\gamma q_X^{1-\gamma} + p_2^\gamma q_2^{1-\gamma} \right)^{1/\gamma} > 0,$$

$$w_0 = \frac{1}{\gamma} \left(\frac{p_1}{\lambda (1 - q_X - q_2)} \right)^\gamma,$$

$$w_X = q_X \left(\frac{1}{\gamma} \left(\frac{p_X}{\lambda q_X} \right)^\gamma - w_0 \right),$$

$$w_2 = q_2 \left(\frac{1}{\gamma} \left(\frac{p_2}{\lambda q_2} \right)^\gamma - w_0 \right).$$

It remains to check whether or not the solution is feasible for the full problem with the condition $w_X, w_2 \geq 0$ included. Here $p_1 = p_X = p_2 = 1/3$ and $q_X = 1/3.25$, $q_2 = 1/2.75$, which leads to an infeasible solution because $w_2 < 0$ (other values of the parameters may lead to a feasible solution). We conclude that the optimal solution must be obtained at the boundary of the set of triplets (w_0, w_X, w_2) satisfying $w_X, w_2 \geq 0$. The least attractive bet is "Liverpool," so removing "Liverpool" as a possible bet (setting $w_2 = 0$) leaves us with the possibility of putting some initial money in the pocket and using the remaining money to bet on "draw." The maximization problem therefore reduces to

$$\text{maximize } \frac{\gamma^{1-1/\gamma}}{\gamma - 1} \left((p_1 + p_2) w_0^{1-1/\gamma} + p_X (w_0 + w_X q_X^{-1})^{1-1/\gamma} \right)$$
$$\text{subject to } w_0 + w_X \leq V_0$$
$$w_X \geq 0.$$

If we temporarily ignore the nonnegativity condition $w_X \geq 0$, then the sufficient conditions (5.2) translate into

$$\gamma^{-1/\gamma} \frac{p_X}{q_X} (w_0 + w_X q_X^{-1})^{-1/\gamma} = \lambda,$$

$$\gamma^{-1/\gamma} \left((p_1 + p_2) w_0^{-1/\gamma} + p_X (w_0 + w_X q_X^{-1})^{-1/\gamma} \right) = \lambda,$$

$$w_0 + w_X = V_0.$$

Solving for (w_0, w_X, λ) yields

$$\lambda = (\gamma V_0)^{-1/\gamma} \left((p_1 + p_2)^\gamma (1 - q_X)^{1-\gamma} + p_X^\gamma q_X^{1-\gamma} \right)^{1/\gamma} > 0,$$

$$w_0 = \frac{1}{\gamma} \left(\frac{p_1 + p_2}{\lambda (1 - q_X)} \right)^\gamma,$$

$$w_X = q_X \left(\frac{1}{\gamma} \left(\frac{p_X}{\lambda q_X} \right)^\gamma - w_0 \right) = \frac{q_X}{\gamma \lambda^\gamma} \left(\left(\frac{p_X}{q_X} \right)^\gamma - \left(\frac{p_1 + p_2}{1 - q_X} \right)^\gamma \right).$$

Since $\lambda > 0$ and $w_X \geq 0$, we have found the optimal solution. The optimal bets as a function of γ are shown in the right plot in Fig. 5.3. For each γ, the value of the curve is the fraction of V_0 that is kept "in the pocket." The remaining capital is used to bet on "draw."

5.3 The Optimal Derivative Position

In this section we consider a problem that can be viewed as the continuous limit of the horse-race problem, as the number of horses tends to infinity. We will consider the problem of how to design the optimal derivative on the future value X of some asset, e.g., a stock market index, given subjective views, risk profiles, budget constraints, and possibly the presence of liabilities.

We consider an investor who can formulate his subjective view on the random value X at time 1 in terms of a probability density $p(x)$. The investor's attitude toward risk and potential reward at time 1 is described by a strictly concave utility function u whose derivative u' can take any value in $(0, \infty)$. We assume that the investor can observe or ask for the price of an arbitrary derivative on X and that this price can be represented as a discounted expected value of the derivative payoff, where the expected value is computed as an integral with respect to a probability density $q(x)$. More precisely, the price of a derivative that pays $h(X)$ at time 1 is $B_0 \int h(x)q(x)dx$, where B_0 is the price of a zero-coupon bond maturing at time 1 with face value 1. The density functions p and q are assumed to satisfy the property $p(x)/q(x) \in (0, \infty)$ for all $x \in (0, \infty)$.

Our aim here is to determine a function h that solves the optimization problem

$$\begin{aligned} &\text{maximize } \mathrm{E}[u(h(X))] \\ &\text{subject to } \int h(x)q(x)dx \leq V_0/B_0. \end{aligned} \qquad (5.15)$$

We set $w(x) = B_0 h(x)q(x)$ and note that $\int_a^b w(x)dx$ is the amount of money that is invested in the occurrence of the event $\{X \in (a, b)\}$ by holding a derivative contract with payoff function h. Similarly, $\int w(x)dx$ is the total invested amount. Since $-u$ is convex, the optimization problem is equivalent to

$$\begin{aligned} &\text{minimize } -\mathrm{E}[u(h(X))] \\ &\text{subject to } \int h(x)q(x)dx \leq V_0/B_0. \end{aligned}$$

We may write $-\mathrm{E}[u(h(X))] = -\int u(h(x))p(x)dx$, and it is easily verified that this is a convex function of h. Proposition 2.4 gives sufficient conditions for an optimal solution h to our optimization problem (5.15).

The first condition in Proposition 2.4 here takes the form

$$\int \Big(u'(h(x))p(x) - \lambda q(x) \Big)(g(x) - h(x))dx = 0 \quad \text{for all } g.$$

This means that $u'(h(x))p(x) - \lambda q(x) = 0$ for all x, which implies that, for $p(x) > 0$,

$$h(x) = (u')^{-1}\left(\lambda\frac{q(x)}{p(x)}\right). \tag{5.16}$$

Note that u' is a strictly decreasing function that can take any value in $(0, \infty)$, and therefore $(u')^{-1}$ is well defined as a function on $(0, \infty)$, and in particular the right-hand side of (5.16) is well defined. The remaining conditions in Proposition 2.4 are $\int h(x)q(x)dx \leq V_0/B_0$, $\lambda(\int h(x)q(x)dx - V_0/B_0) = 0$, and $\lambda \geq 0$.

If there is a positive λ such that

$$\frac{V_0}{B_0} = \int (u')^{-1}\left(\lambda\frac{q(x)}{p(x)}\right)q(x)dx, \tag{5.17}$$

then (5.16) is the unique optimal solution to the optimization problem (5.15). If the right-hand side of (5.17) exists finitely for some positive λ_0, then it is a continuous and strictly decreasing function of λ on (λ_0, ∞).

Under the assumption of a positive λ solving (5.17), the amount of money invested in the outcome of X in the interval (a, b) is

$$\int_a^b w(x)dx = B_0 \int_a^b h(x)q(x)dx = B_0 \int_a^b (u')^{-1}\left(\lambda\frac{q(x)}{p(x)}\right)q(x)dx. \tag{5.18}$$

Before proceeding with some examples, let us comment on the expression for the optimal payoff function $h(x)$. Since the utility function u is concave, u' is decreasing, and therefore $(u')^{-1}$ is also decreasing. Thus, the investor seeks a higher payoff in states where $q(x)/p(x)$ is small, that is, in states where $p(x) > q(x)$, and a lower payoff in states where $p(x) < q(x)$. Since both $p(x)$ and $q(x)$ integrate to one, both cases will occur, unless $p = q$.

Example 5.8 (Risk-neutral beliefs). Suppose that $p = q$, i.e., that the investor's beliefs coincide with those reflected in the prices. From (5.16) we see that h is constant. Therefore, the budget constraint implies that $f = B_0/V_0$, which is the payoff function of a risk-free bond. The conclusion is that, in this case, we value all payoffs equally, and therefore it only makes sense to invest in the safest one.

Let u be a HARA utility function with $\gamma \neq 0$. If $\gamma = 1$, then $u(x) = \log(\tau + x)$ for $\tau + x > 0$, and otherwise

$$u(x) = \frac{1}{\gamma - 1}(\tau + \gamma x)^{1-1/\gamma} \quad \text{for } \tau + \gamma x > 0.$$

We find that $u'(x) = (\tau + \gamma x)^{-1/\gamma}$, which gives $(u')^{-1}(y) = (y^{-\gamma} - \tau)/\gamma$. Therefore,

$$w(x) = B_0 q(x) h(x) = B_0 q(x)(u')^{-1}\left(\lambda \frac{q(x)}{p(x)}\right)$$

$$= \frac{B_0}{\gamma} q(x)\left(\lambda^{-\gamma}\left(\frac{p(x)}{q(x)}\right)^{\gamma} - \tau\right). \tag{5.19}$$

Solving the equation $\int w(x) dx = V_0$ for $\lambda^{-\gamma}$ gives

$$\lambda^{-\gamma} = \left(\int q(x)\left(\frac{p(x)}{q(x)}\right)^{\gamma} dx\right)^{-1}\left(\frac{\gamma V_0}{B_0} + \tau\right),$$

which, inserted into (5.19), gives

$$h(x) = V_0 \frac{\left(\frac{p(x)}{q(x)}\right)^{\gamma}}{B_0 \int q(y)\left(\frac{p(y)}{q(y)}\right)^{\gamma} dy} + \frac{\tau}{\gamma}\left(\frac{\left(\frac{p(x)}{q(x)}\right)^{\gamma}}{\int q(y)\left(\frac{p(y)}{q(y)}\right)^{\gamma} dy} - 1\right) \tag{5.20}$$

and

$$w(x) = V_0 \frac{q(x)\left(\frac{p(x)}{q(x)}\right)^{\gamma}}{\int q(y)\left(\frac{p(y)}{q(y)}\right)^{\gamma} dy} + \frac{B_0 \tau}{\gamma}\left(\frac{q(x)\left(\frac{p(x)}{q(x)}\right)^{\gamma}}{\int q(y)\left(\frac{p(y)}{q(y)}\right)^{\gamma} dy} - q(x)\right). \tag{5.21}$$

Notice that (5.21) is the analog of expression (5.13) in the context considered here. In addition, if $\gamma = 1$, corresponding to logarithmic utility, then (5.21) takes the form

$$w(x) = V_0 p(x) + B_0 \tau(p(x) - q(x)),$$

which is the analog of (5.10).

Example 5.9 (Exponential utility). If $u(x) = -\tau e^{-x/\tau}$, then $u'(x) = e^{-x/\tau}$ and $(u')^{-1}(y) = -\tau \log(y)$ for $y > 0$. The optimal payoff function is given by

$$h(x) = \tau\left(\log p(x) - \log q(x) - \log \lambda\right).$$

Solving the equation $\int w(x) dx$, where $w(x) = B_0 h(x) q(x)$, for $\log \lambda$ gives

$$\log \lambda = -\frac{V_0}{B_0 \tau} + \int q(x) \log\left(\frac{p(x)}{q(x)}\right) dx,$$

and therefore

$$h(x) = \frac{V_0}{B_0} + \tau\left(\log\left(\frac{p(x)}{q(x)}\right) - \int q(y) \log\left(\frac{p(y)}{q(y)}\right) dy\right)$$

and

$$w(x) = V_0 q(x) + \tau B_0 q(x) \left(\log \left(\frac{p(x)}{q(x)} \right) - \int q(y) \log \left(\frac{p(y)}{q(y)} \right) dy \right).$$

5.3.1   Examples with Lognormal Distributions

To better understand the interplay between subjective views, prices, and degrees of risk aversion, we now turn to explicit examples on the assumption that both densities p and q are densities of lognormal probability distributions.

Suppose the subjective probability is that of a $LN(\mu_p, \sigma_p^2)$ distribution and that the probability derived from prices is that of a $LN(\mu_q, \sigma_q^2)$ distribution. That is,

$$p(x) = \frac{1}{x\sigma_p\sqrt{2\pi}} \exp\left(-\frac{1}{2\sigma_p^2}(\log x - \mu_p)^2 \right),$$

$$q(x) = \frac{1}{x\sigma_q\sqrt{2\pi}} \exp\left(-\frac{1}{2\sigma_q^2}(\log x - \mu_q)^2 \right),$$

$$\frac{p(x)}{q(x)} = \frac{\sigma_q}{\sigma_p} \exp\left(\frac{1}{2\sigma_q^2}(\log x - \mu_q)^2 - \frac{1}{2\sigma_p^2}(\log x - \mu_p)^2 \right).$$

Because of the lognormal densities, it is convenient to write $z = \log x$, which makes $L(z) = \log p(e^z) - \log q(e^z)$ a second-order polynomial in z and $p(x)/q(x) = e^{L(z)}$. Indeed,

$$L(z) = \log p(e^z) - \log q(e^z)$$

$$= \log\left(\frac{\sigma_q}{\sigma_p} \right) + \frac{1}{2\sigma_q^2}(z - \mu_q)^2 - \frac{1}{2\sigma_p^2}(z - \mu_p)^2$$

$$= \left(\log\left(\frac{\sigma_q}{\sigma_p} \right) + \frac{\mu_q^2}{2\sigma_q^2} - \frac{\mu_p^2}{2\sigma_p^2} \right) + \left(\frac{\mu_p}{\sigma_p^2} - \frac{\mu_q}{\sigma_q^2} \right)z + \left(\frac{1}{2\sigma_q^2} - \frac{1}{2\sigma_p^2} \right)z^2$$

$$= a_0 + a_1 z + a_2 z^2,$$

where

$$a_0 = \log\left(\frac{\sigma_q}{\sigma_p} \right) + \frac{\mu_q^2}{2\sigma_q^2} - \frac{\mu_p^2}{2\sigma_p^2},$$

$$a_1 = \frac{\mu_p}{\sigma_p^2} - \frac{\mu_q}{\sigma_q^2},$$

$$a_2 = \frac{1}{2\sigma_q^2} - \frac{1}{2\sigma_p^2}.$$

The optimal payoff function can be written as a function of z as

$$h(x) = (u')^{-1}\left(\lambda \frac{q(x)}{p(x)}\right) = (u')^{-1}(\lambda e^{-L(z)}) = \frac{1}{\gamma}(\lambda^{-\gamma}e^{\gamma L(z)} - \tau). \qquad (5.22)$$

To determine λ, we solve the equation $B_0 \int q(x)h(x)dx = V_0$ for λ.

$$\begin{aligned}
\frac{V_0}{B_0} &= \int_0^\infty q(x)h(x)dx \\
&= \int_{-\infty}^\infty q(e^z)h(e^z)e^z dz \\
&= -\frac{\tau}{\gamma} + \frac{\lambda^{-\gamma}}{\gamma}\int_{-\infty}^\infty \frac{1}{\sigma_q\sqrt{2\pi}}\exp\left\{-\frac{1}{2\sigma_q^2}(z-\mu_q)^2 + \gamma L(z)\right\}dz. \qquad (5.23)
\end{aligned}$$

The exponent in the integrand in (5.23) can be written as

$$\left(\gamma a_0 - \frac{\mu_q^2}{2\sigma_q^2}\right) + \left(\gamma a_1 + \frac{\mu_q}{\sigma_q^2}\right)z - \left(\frac{1}{2\sigma_q^2} - \gamma a_2\right)z^2 = b_0 + b_1 z - b_2 z^2,$$

and we notice that the integral in (5.23) exists finitely if

$$b_2 = \frac{1}{2\sigma_q^2} - \gamma a_2 = \frac{\gamma}{2\sigma_p^2} - \frac{\gamma-1}{2\sigma_q^2} > 0.$$

On the assumption that $b_2 > 0$, we may write

$$b_0 + b_1 z - b_2 z^2 = -\frac{1}{2(2b_2)^{-1}}\left(z - \frac{b_1}{2b_2}\right)^2 + \frac{b_1^2}{4b_2} + b_0.$$

Since the integral of any normal density over the entire real line equals one, the integral in (5.23) can be computed as

$$\begin{aligned}
&\frac{1}{\sigma_q\sqrt{2b_2}}\exp\left\{\frac{b_1^2}{4b_2} + b_0\right\} \\
&= \frac{1}{\sqrt{1-2\sigma_q^2\gamma a_2}}\exp\left\{\gamma a_0 - \frac{\mu_q^2}{2\sigma_q^2} + \left(\frac{(\sigma_q\gamma a_1)^2}{2} + \mu_q\gamma a_1 + \frac{\mu_q^2}{2\sigma_q^2}\right)\frac{1}{1-2\sigma_q^2\gamma a_2}\right\} \\
&= c(a_0, a_1, a_2, \gamma, \mu_q, \sigma_q).
\end{aligned}$$

Putting everything together yields

$$\lambda = \left(\frac{\gamma V_0}{B_0} + \tau\right)^{-1/\gamma} c(a_0, a_1, a_2, \gamma, \mu_q, \sigma_q)^{1/\gamma}. \qquad (5.24)$$

If $\sigma_p = \sigma_q$, then $a_2 = 0$ and (5.24) simplifies into

$$\lambda = \left(\frac{\gamma V_0}{B_0} + \tau\right)^{-1/\gamma} \exp\left\{a_0 + \mu_q a_1 + \gamma \frac{(\sigma_q a_1)^2}{2}\right\}. \tag{5.25}$$

Example 5.10 (Equal volatilities). Suppose that $\sigma_p = \sigma_q$, which gives $a_2 = 0$ and, therefore, from (5.22) that

$$h(x) = \frac{1}{\gamma}(\lambda^{-\gamma} e^{\gamma a_0} x^{\gamma a_1} - \tau),$$

where λ is given by (5.25).

If further $\mu_p > \mu_q$, then the investor agrees with the market view on the volatility but is overall more optimistic about the future value of the underlying asset. In this case, $a_1 > 0$, and the payoff function $h(x)$ is large for large values of x and small for small values of x. Two examples of such payoff functions are shown in Fig. 5.4. An investor who holds this derivative is betting on a market up-swing. This type of payoff may be achieved by, for instance, a long position in the underlying asset, possibly combined with long positions in call options and possibly also short positions in put options.

If $\mu_p < \mu_q$, then the investor agrees with the market view on the volatility but is overall more pessimistic about the future value of the underlying asset. In this case, $a_1 < 0$, and the payoff function $h(x)$ is large for small values of x and small for large values of x. An example of such a payoff function is shown in Fig. 5.4. An investor who holds this derivative is betting on a market down-swing.

If $\gamma a_1 > 1$, or equivalently $\mu_p > \mu_q + \sigma_p^2/\gamma$, then h is strictly convex and increasing, whereas if $\gamma a_1 \in (0, 1)$, or equivalently $\mu_q < \mu_p < \mu_q + \sigma_p^2/\gamma$, then h is strictly concave and increasing. If $\gamma a_1 = 1$, then h grows linearly.

Example 5.11 (Different volatilities). Suppose that the expected future value $E[S_1]$, from the investor's point of view, of the underlying asset equals the current forward price G_0:

$$e^{\mu_q + \sigma_q^2/2} = G_0 = E[S_1] = e^{\mu_p + \sigma_p^2/2}. \tag{5.26}$$

Suppose also that the investor believes that the volatility in the market will differ from the implied volatility σ_q.

In the case $\sigma_p > \sigma_q$, it follows from (5.26) that $\mu_p < \mu_q$, and therefore the optimal payoff function is given by

$$h(x) = \frac{1}{\gamma}(\lambda^{-\gamma} e^{\gamma(a_0 + a_1 \log x + a_2 (\log x)^2)} - \tau), \tag{5.27}$$

with $a_1 < 0$ and $a_2 > 0$ and λ given by (5.24). Here $h(x)$ is large for small and large values of x and small for intermediate values of x. An example of such a

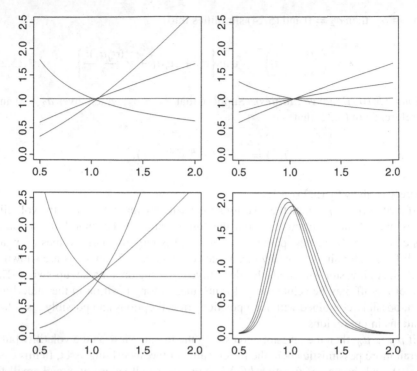

Fig. 5.4 Optimal derivative payoff functions for investment problem considered in Example 5.10. Here $V_0 = 1$, $\tau = 0$, $B_0 = e^{-0.05}$, and $\sigma_p = \sigma_q = 0.2$. In each plot the graphs of the payoff functions correspond to $\mu_p = 0.02 - \sigma_p^2/2$, $\mu_p = 0.05 - \sigma_p^2/2$, $\mu_p = 0.08 - \sigma_p^2/2$, and $\mu_p = 0.11 - \sigma_p^2/2$. The values of γ considered are $\gamma = 1$ (*upper left*), $\gamma = 1/2$ (*upper right*), and $\gamma = 2$ (*lower left*). *Lower-right plot*: four lognormal densities

payoff function is shown in Fig. 5.5. An investor holding this derivative is betting on large price fluctuations. This type of investment can be achieved by purchasing both at-the-money puts and at-the-money calls.

Similarly, if $\sigma_p < \sigma_q$, then it follows from (5.26) that $\mu_p > \mu_q$, and the optimal payoff function is given by (5.27) with $a_1 > 0$ and $a_2 < 0$. An example of such a payoff function is shown in Fig. 5.5. An investor holding this derivative is betting on small price fluctuations.

5.3.2   Investments in the Presence of Liabilities

Here we make the investment problem more complicated by assuming that the investor faces a random liability L at time 1. In this more general setting, the investor must take into account the distribution of L and its dependence with X when designing the optimal derivative payoff $h(X)$. More precisely, the investor

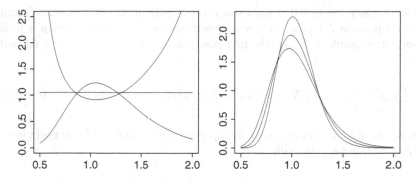

Fig. 5.5 The plot to the *left* shows optimal derivative payoff functions for the investment problem considered in Example 5.11. Here $V_0 = 1$, $\tau = 0$, $B_0 = e^{-0.05}$, $\sigma_q = 0.2$, and $\gamma = 1$. In each plot the graphs of the payoff functions correspond to $\mu_p = 0.05 - 0.17^2/2$, $\mu_p = 0.05 - 0.20^2/2 = \mu_q$, and $\mu_p = 0.05 - 0.23^2/2$. The plot to the *right* shows the three lognormal densities

wants to determine the payoff function h that does not cost more than the available capital V_0 and that maximizes the subjective expected utility of the future net worth $h(X) - L$.

The investor therefore searches for a function h that solves the optimization problem

$$\text{maximize } \mathrm{E}[u(h(X) - L)]$$
$$\text{subject to } \int h(x)q(x)dx \le V_0/B_0. \tag{5.28}$$

We set $w(x) = B_0 h(x) q(x)$ and note that $\int_a^b w(x)dx$ is the amount of money invested in the occurrence of the event $\{X \in (a, b)\}$ by holding a derivative contract with payoff function h. Similarly, $\int w(x)dx$ is the total invested amount. Since $v = -u$ is convex, the optimization problem is equivalent to

$$\text{minimize } -\mathrm{E}[u(h(X) - L)]$$
$$\text{subject to } \int h(x)q(x)dx \le V_0/B_0.$$

We assume that the conditional distribution of L, given an outcome x of X has a density function $p(l \mid x)$, and we write

$$-\mathrm{E}[u(h(X) - L)] = -\mathrm{E}[\mathrm{E}[u(h(X) - L) \mid X]]$$

$$= -\int \mathrm{E}[u(h(X) - L) \mid X = x]p(x)dx$$

$$= \int \left(-\int u(h(x) - l)p(l \mid x)dl \right) p(x)dx,$$

and it is easily verified that the expression in parentheses is a convex function
of h. Proposition 2.4 gives sufficient conditions for an optimal solution h to our
optimization problem (5.28). The first condition in Proposition 2.4 here takes the
form

$$\int \Big(\mathrm{E}[u'(h(X)-L) \mid X=x]p(x)-\lambda q(x)\Big)(g(x)-h(x))dx = 0 \quad \text{for all } g.$$

The remaining conditions are $\int h(x)q(x)dx \leq V_0/B_0$, $\lambda(\int h(x)q(x)dx - V_0/B_0)=0$, and $\lambda \geq 0$. With

$$\psi_x(y) = \mathrm{E}[u'(y-L) \mid X=x],$$

the first condition in Proposition 2.4 can be formulated as

$$\int \Big(\psi_x(h(x))p(x)-\lambda q(x)\Big)(g(x)-h(x))dx = 0 \quad \text{for all } g.$$

This means that $\psi_x(h(x))p(x)-\lambda q(x)=0$ for all x, which implies that, for
$p(x)>0$,

$$h(x) = \psi_x^{-1}\Big(\lambda \frac{q(x)}{p(x)}\Big). \tag{5.29}$$

Note that ψ_x is a strictly decreasing function that can take any value in $(0,\infty)$, and
therefore ψ_x^{-1} is well defined as a function on $(0,\infty)$, and in particular the right-
hand side of (5.29) is well defined.

If there is a positive λ such that

$$\frac{V_0}{B_0} = \int \psi_x^{-1}\Big(\lambda \frac{q(x)}{p(x)}\Big)q(x)dx, \tag{5.30}$$

then (5.29) is the unique optimal solution to optimization problem (5.28). Note
that if the right-hand side of (5.30) exists finitely for some positive λ_0, then it is
a continuous and strictly decreasing function of λ on (λ_0,∞).

Under the assumption of a positive λ solving (5.30), the amount of money
invested in the outcome of X in the interval (a,b) is

$$\int_a^b w(x)dx = B_0 \int_a^b h(x)q(x)dx = B_0 \int_a^b \psi_x^{-1}\Big(\lambda \frac{q(x)}{p(x)}\Big)q(x)dx. \tag{5.31}$$

Example 5.12 (Perfect hedge). Suppose $L=g(X)$ for some function g. Then it is
possible to hedge the liability perfectly. In this case, $\psi_x = u'(h(x)-g(x))$, which
implies that

$$h(x) = g(x) + (u')^{-1}\Big(\lambda \frac{q(x)}{p(x)}\Big). \qquad (5.32)$$

If there is a positive λ such that

$$V_0/B_0 = \int g(x)q(x)dx + \int (u')^{-1}\Big(\lambda \frac{q(x)}{p(x)}\Big)q(x)dx,$$

then (5.32) with this λ is the optimal solution.

If you can afford to buy the hedge, the optimal payoff function h is such that the investor first buys the perfect hedge $g(x)$ for the price $B_0 \int g(x)q(x)dx$ and then solves the optimal investment problem for the remaining capital $V_0 - B_0 \int g(x)q(x)dx$.

Example 5.13 (Qualitative interpretations). Consider a first-order approximation of $\psi_x(z)$ of the form

$$\psi_x(z) = E[u'(z - E[L \mid X = x] + E[L \mid X = x] - L) \mid X = x]$$

$$\approx u'(z - E[L \mid X = x])$$

$$+ E[u''(z - E[L \mid X = x])(E[L \mid X = x] - L) \mid X = x]$$

$$= u'(z - E[L \mid X = x]).$$

Using this approximation of ψ_x in place of ψ_x, we find that

$$h(x) \approx E[L \mid X = x] + (u')^{-1}\Big(\lambda \frac{q(x)}{p(x)}\Big).$$

We see that the optimal position is approximately to buy the quadratic hedge $E[L \mid X = x]$ for the liability and invest the remaining capital according to the optimal payoff function without the liability.

Example 5.14 (Exponential utility). Consider the utility function $u(x) = -\tau e^{-x/\tau}$ and a liability of the form $L = g(X)Y$, where X and Y are independent. Then

$$\psi_x(h(x)) = E[u'(h(x) - g(x)Y)] = e^{-h(x)/\tau} E[e^{g(x)Y/\tau}] = e^{-h(x)/\tau} M_Y(g(x)/\tau),$$

where $M_Y(s) = E[e^{sY}]$ is the moment-generating function of Y. Since $\psi_x(h(x))p(x) - \lambda q(x) = 0$, we find that

$$h(x) = \tau\Big(\log p(x) - \log q(x) - \log \lambda + \log M_Y(g(x)/\tau)\Big).$$

We may compare this expression for h to that in Example 5.9,

$$h(x) = \tau\Big(\log p(x) - \log q(x) - \log \lambda \Big),$$

in the case of no liability. In the presence of the liability $g(X)Y$, the investor first buys the derivative with payoff function $\tau \log M_Y(g(x)/\tau)$, which can be viewed as protection against uncertainty in the value of the liability, and invests the remaining capital as if there were no liability.

Solving the equation $\int w(x)dx$, where $w(x) = B_0 h(x)q(x)$, for $\log \lambda$ gives

$$\log \lambda = -\frac{V_0}{B_0 \tau} + \int q(x) \log\Big(\frac{p(x)}{q(x)}\Big)dx + \int q(x) \log M_Y(g(x)/\tau)dx,$$

and therefore

$$h(x) = \frac{V_0}{B_0} + \tau\Big(\log\Big(\frac{p(x)}{q(x)}\Big) - \int q(y) \log\Big(\frac{p(y)}{q(y)}\Big)dy \Big)$$
$$+ \tau\Big(\log M_Y(g(x)/\tau) - \int q(y) \log M_Y(g(y)/\tau)dy \Big).$$

5.4 Notes and Comments

The theory of decision under uncertainty is important to any structured approach to decision making for individuals and organizations. Financial decision making is just one out of many possible applications of this theory, and the presentation here is incomplete, to say the least, for readers who seek a good overview of the ideas and mathematical results that have been developed by influential researchers in this area. The work of Leonard Savage is a cornerstone in the theory of decision under uncertainty, and very readable presentations of his and related work are the books [43] by Savage and [19] by Itzhak Gilboa. Both books combine deep mathematics with interesting nontechnical discussions. Expected utility theory has played a prominent role in the development of the modern theory of economics and finance.

Our brief presentation of risk premia and coefficients of risk aversion consists of selected topics from the work of John Pratt in [36] and the work of Kenneth Arrow, see, e.g., [3].

Many different approaches to asset allocation problems for nonnormally distributed returns, including expected utility maximization, are presented in Attilio Meucci's book [33].

The material in Sect. 5.3 is based on the work of Peter Carr and Dilip Madan in [8]. We only considered the investment problem from an individual investor's perspective, whereas Carr and Madan also considered an economy with multiple investors that simultaneously optimize their positions and studied the effects of the heterogeneity of preferences and probability beliefs on prices and derivative positions held at equilibrium.

5.5 Exercises

In the exercises below it is assumed, wherever applicable, that you can take positions corresponding to fractions of assets.

Exercise 5.1 (Credit default swap). Consider an investor who is an expected-utility maximizer with a utility function $u(x) = \sqrt{x}$, $x > 0$. The investor has $100 to invest in long positions in a defaultable bond, a credit default swap on this bond, and in a risk-free government bond. One defaultable bond costs $96 today and pays $100 6 months from today if the issuer does not default and 0 in case the issuer defaults. The credit default swap costs $2 today and pays $100 6 months from today if the bond issuer defaults, and nothing otherwise. The risk-free bond costs $99 today and pays $100 6 months from now.

(a) The investor believes that the default probability is 0.02. How much of the $100 does the investor invest in the defaultable bond, in the risk-free bond, and in the credit default swap?

(b) Another investor is an expected-utility maximizer with a utility function $u(x) = x^\beta$ for β in $(0, 1)$. Also, this investor believes that the default probability is 0.02 and decides not to buy the bond. What can be said about β?

Exercise 5.2 (Bets on the credit rating). A credit rating agency gives a credit rating to every large company and country. Suppose the following credit ratings: Excellent, Good, Poor, and Default. Consider a bond issued by the Belgian government with a current credit rating of Good. In 6 months, Belgium will receive an updated credit rating. You have access to a market where you can buy the following contracts on the credit rating of Belgium:

1. A contract that costs $1,150 today that pays $10,000 in 6 months if the credit rating at that time is Excellent.
2. A contract that costs $8,100 today that pays $10,000 in 6 months if the credit rating at that time is Good.
3. A contract that costs $700 today that pays $10,000 in 6 months if the credit rating at that time is Poor.
4. A contract that costs $50 today that pays $10,000 in 6 months if the credit rating at that time is Default.

Your subjective probabilities of the credit rating in 6 months are

$$P(\text{Excellent}) = 0.11, \quad P(\text{Good}) = 0.80,$$
$$P(\text{Poor}) = 0.08, \quad P(\text{Default}) = 0.01.$$

Determine how to optimally invest the initial $10,000 capital in the four contracts if the objective is to maximize expected utility of your capital in 6 months; use a HARA utility function with $\tau = 0$ and $\gamma = 2.5$.

Exercise 5.3 (Hedging with electricity futures). Consider a company whose assets consist of a power-intensive aluminum smelter and processing facilities. The value of the company's assets 1 year from today and the income from sales during the year is modeled as the random variable X_1. The value of the company's liabilities and the production costs during the year is modeled as the random variable X_2. The company decides to take a position in electricity futures contracts to hedge its production costs during the year but also to capitalize on its knowledge of the electricity market. The profit (possibly negative) from a long position in an electricity futures strategy of a size corresponding to the predicted energy usage in production is modeled as the random variable X_3. Assume that $\mathbf{X} = (X_1, X_2, X_3)^{\mathsf{T}}$ is normally distributed with mean μ and covariance matrix Σ, where

$$\mu = \begin{pmatrix} 20 \\ 10 \\ -1 \end{pmatrix} \quad \text{and} \quad \Sigma = \begin{pmatrix} 2^2 & -3 & -2 \\ -3 & 3^2 & 5 \\ -2 & 5 & 2^2 \end{pmatrix}.$$

Determine the maximizer h of the expected utility $\mathrm{E}[u(X_1 - X_2 + hX_3)]$, where $u(x) = -\tau e^{-x/\tau}$.

Exercise 5.4 (Optimal payoff function). A model q for the implied forward density has been determined from option prices with a common maturity. An agent considers a model for the subjective probability density p of the form, for some real number θ,

$$\frac{p(x)}{q(x)} = e^{\theta x - \Lambda(\theta)}, \quad \text{where } \Lambda(\theta) = \log \int e^{\theta x} q(x) dx.$$

Determine the optimal payoff function h for an agent who is maximizing expected utility using a HARA utility function.

Project 5 (Subjective volatility smile). Consider prices of a number of liquidly traded call/put options written on a common future value of some asset. Determine the implied volatility smile $\sigma_I(K)$ by fitting a second-degree polynomial as in Example 1.9, and derive the resulting implied forward density q.

A market participant's subjective view of the distribution of the asset value on which the options are written can be expressed in terms of a subjective volatility smile $\sigma_S(K)$. The subjective probability density p is obtained by the same procedure as in Example 1.9, with the exception that the market smile σ_I is replaced by σ_S.

Investigate how different subjective volatility smiles affect the optimal derivative payoff function h when using a HARA utility function.

(a) Determine the optimal derivative payoff function h when the volatility smile σ_S is a parallel shift, up or down, of the implied volatility smile σ_I.

(b) Determine the optimal derivative payoff function h when the volatility smile σ_S has a higher/lower slope compared to the implied volatility smile σ_I.

(c) Determine the optimal derivative payoff function h when the volatility smile σ_S has more/less curvature than the implied volatility smile σ_I.

In (a)–(c), be aware that a careless choice of the function σ_S may lead to a function p that is not a probability density.

Chapter 6
Risk Measurement Principles

In this chapter, we take a close look at the principles of risk measurement. We argue that it is natural to quantify the riskiness of a position in monetary units so that the measurement of the risk of a position can be interpreted as the size of buffer capital that should be added to the position to provide a sufficient protection against undesirable outcomes. In the investment problems in Chap. 4, variance was used to quantify the riskiness of a portfolio. However, variance, being just the expected squared deviation from the mean value, does not differentiate between good positive deviations and bad negative deviations and cannot easily be translated into meaningful monetary values unless the future value we consider is close to normally distributed. The risk premium considered in Chap. 5 is more natural than the variance as a summary of the riskiness and potential reward of a position. However, the risk premium is difficult to use effectively to control the risk taking of a financial institution or to determine whether the aggregate position of a company or business unit is acceptable from a risk perspective. In this chapter, we will present measures of risk, including the widely used value-at-risk and expected shortfall, analyze their properties, and evaluate their performance in a large number of examples.

6.1 Risk Measurement

We now turn to the topic of how to measure risk. Consider two times, time 0, which is now, and a future time $\Delta t > 0$. We may choose to measure time in units of Δt and therefore take the future time to be 1.

Let V_1 represent the random value at time 1 of a portfolio. The precise meaning of portfolio is left unspecified but may include assets, liabilities, and any kind of contract that can be assigned a monetary value. To measure the risk of the portfolio, we analyze the probability distribution of V_1. The probability distributions assigned to V_1 are likely to vary among a group of individuals or organizations for which the

H. Hult et al., *Risk and Portfolio Analysis: Principles and Methods*, Springer Series
in Operations Research and Financial Engineering, DOI 10.1007/978-1-4614-4103-8_6,
© Springer Science+Business Media New York 2012

future portfolio value is of relevance. Moreover, the way the probability distribution of V_1 is transformed into a measurement of the riskiness of the portfolio may depend on the context.

An asset manager, whose main objective is to generate profits while controlling the risk of and size of losses, needs to consider the whole range of possible outcomes of V_1 together with possible externally imposed risk constraints and profitability requirements. A risk controller analyzes the part of the distribution of V_1 corresponding to unfavorable outcomes. In particular, the portfolio may be considered acceptable by the risk controller but not by the asset manager if it is not likely to produce a good return. Similarly, a portfolio that has good potential of producing high returns may be unacceptable to the risk controller who finds that the probabilities of large losses are too high and have been overlooked (or ignored) by the asset manager.

A regulator of a finance or insurance market wants to impose rules on risk taking that on the one hand prevents banks or insurance companies from taking too much risk, and thereby threatening financial stability, but on the other hand allows companies to be profitable. The rules must enable the supervisory authority to classify the overall position of a company as either acceptable or unacceptable. Moreover, the supervisor must be able to inform a company with an unacceptable position of suitable actions to obtain an acceptable position, for instance, the minimum additional capital that the company must raise and invest prudently in order to be allowed to continue its business.

Many properties of a portfolio can be understood in terms of the probability distribution (e.g., the density function or distribution function) of its future value V_1. However, probability distributions are difficult objects to compare. Therefore, it is tractable to come up with a good way to summarize, from a risk measurement perspective, the entire probability distribution in a single number. We now discuss how this can be done.

Suppose there is a reference instrument with percentage return R_0 from time 0 to 1. The precise meaning of the reference instrument may depend on the context in which we are quantifying risk. For simplicity, here we take it be risk-free zero-coupon bonds maturing at time 1. If B_0 is the current spot price of the bond with face value 1 at time 1, then $R_0 = 1/B_0$ is the percentage return on the risk-free zero-coupon bond.

Consider a linear vector space $\mathbb{X}$ of random variables X representing the values at time 1 of portfolios. We denote by ρ a function that assigns a real number (or $+\infty$) to each X in $\mathbb{X}$, representing a measurement of the risk of X. The number $\rho(X)$ is interpreted as the minimum capital that needs to be added to the portfolio at time 0 and invested in the reference instrument in order to make the position acceptable. If $\rho(X) \leq 0$, then X is the value at time 1 of an acceptable portfolio; no capital needs to be added. In principle, a risk measure ρ could assign different values to two equally distributed future portfolio values X_1 and X_2. Throughout the book, we will only consider risk measures ρ for which $\rho(X)$ depends on X only through its probability distribution.

Next we list and comment upon some properties that have been proposed as natural requirements for good risk measures.

Translation invariance. $\rho(X + cR_0) = \rho(X) - c$ for all real numbers c.

This property says that adding a certain amount c of cash (and buying zero-coupon bonds for this amount) will reduce risk by the same amount. In particular, for an unacceptable portfolio X, adding the amount $\rho(X)$ makes the position acceptable: $\rho(X + \rho(X)R_0) = \rho(X) - \rho(X) = 0$.

Monotonicity. If $X_2 \leq X_1$, then $\rho(X_1) \leq \rho(X_2)$.

This property says that if the first position has a greater value than the second position at time 1 for sure, then the first position must be considered less risky. A risk measure satisfying the properties translation invariance and monotonicity is called a monetary measure of risk.

It is often suggested that a risk measure should reward diversification. Loosely speaking, it is wise not to put all your eggs in the same basket. The following property describes how diversification should be rewarded.

Convexity. $\rho(\lambda X_1 + (1 - \lambda)X_2) \leq \lambda\rho(X_1) + (1 - \lambda)\rho(X_2)$ for all real numbers λ in $[0, 1]$.

In particular, if $\rho(X_1) \leq \rho(X_2)$ and ρ has the convexity property, then

$$\rho(\lambda X_1 + (1 - \lambda)X_2) \leq \lambda\rho(X_1) + (1 - \lambda)\rho(X_2) \leq \rho(X_2).$$

For example, investing a fraction of the initial capital in one stock and the remaining capital in another stock, rather than everything in the more risky stock, reduces the overall risk. A risk measure satisfying the properties translation invariance, monotonicity, and convexity is called a convex measure of risk.

Normalization. $\rho(0) = 0$.

The normalization property says that it is acceptable not to take any position at all. Note that convexity and normalization imply that for λ in $[0, 1]$

$$\rho(\lambda X) = \rho(\lambda X + (1 - \lambda)0) \leq \lambda\rho(X),$$

which in turn implies that for $\lambda \geq 1$

$$\lambda\rho(X) = \lambda\rho\left(\frac{1}{\lambda}\lambda X\right) \leq \lambda\frac{1}{\lambda}\rho(\lambda X) = \rho(\lambda X).$$

We conclude that the risk increases at least linearly in the size of the position. A strict inequality for large λ would reflect the well-known difficulty of selling off a large position within a short amount of time without affecting the price too much.

Positive homogeneity. $\rho(\lambda X) = \lambda\rho(X)$ for all $\lambda \geq 0$.

This property means that if we double the size of the position, then we double the risk. Moreover, taking $\lambda = 0$ we find that $\rho(0) = 0$, i.e., the positive homogeneity property implies the normalization property.

Subadditivity. $\rho(X_1 + X_2) \leq \rho(X_1) + \rho(X_2)$.

This property says that diversification should be rewarded. A bank consisting of two units should be required to put aside less buffer capital than the sum of the buffer capital for the two units considered as separate entities. In particular, if the regulator enforces the use of a subadditive risk measure, then it does not encourage companies to break up into parts in order to reduce the buffer capital requirement. Note that convexity together with positive homogeneity implies subadditivity.

A risk measure ρ satisfying the properties of translation invariance, monotonicity, positive homogeneity, and subadditivity is called a coherent measure of risk. Whereas a coherent risk measure is also a convex risk measure, a convex risk measure need not be coherent.

It may seem unintuitive at first to define the risk measure ρ on the value of a portfolio at time 1. Suppose $R_0 = 1$ and consider a fund manager who at time 0 invests $V_0 = \$10$ million in a giveaway portfolio with a value of $V_1 = \$1$ million .at time 1 (\$9 million is given away). If the risk measure ρ satisfies the translation invariance and normalization properties, then the risk measure applied to the future value of the portfolio yields $\rho(\$1 \text{ million}) = -\$1$ million, which corresponds to an acceptable investment. How can giving away money be an acceptable investment? The explanation lies in the interpretation of the future value X. Let us consider two stylized cases. In the first case, the fund manager is managing his own money, $X = V_1$, and there is nothing unacceptable about letting the fund manager give away some or all of his capital. In the second case, the money of the fund belongs to the fund's investors. In this case, the initial capital should be viewed as a liability to the fund's investors and $X = V_1 - V_0$. Therefore, $\rho(X) = \$9$ million, which corresponds to an unacceptable investment.

Example 6.1 (Solvency capital requirement). In the Solvency II framework, which is a regulatory framework for the insurance industry, a company is considered solvent if $\rho(A_1 - L_1) \leq 0$, where A_1 and L_1 are the values of its assets and liabilities 1 year from now and ρ a monetary (translation invariant and monotone) risk measure. It is quite common to illustrate the solvency graphically in terms of a picture of the balance sheet of the insurance company with the current value of assets to the left and the current value of liabilities to the right, and with the insurer being solvent if the height of the left column exceeds that of the right column (Fig. 6.1).

Let A_0 be the current market value of the assets, and let L_0 be the current market value (or best estimate) of the liabilities. Since ρ is translation invariant, we may write

$$\rho(A_1 - L_1) = \rho([A_0 - L_0]R_0 + [A_1 - A_0R_0] - [L_1 - L_0R_0])$$
$$= L_0 - A_0 + \rho(\underbrace{[A_1 - A_0R_0] - [L_1 - L_0R_0]}_{\Delta}).$$

Fig. 6.1 Balance sheet. *Left*: present value of assets (*gray*). *Right*: present value of liabilities (*gray*). The solvency capital requirement is illustrated in *white*. The company is solvent if the present value of the assets is greater than the present value of the liabilities plus the solvency capital requirement

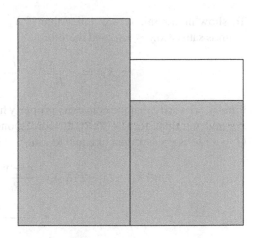

The quantity $\rho(\Delta)$ is called the solvency capital requirement and is denoted by SCR. A portfolio with a future value $A_1 - L_1$ is acceptable if $\rho(A_1 - L_1) \leq 0$, which is equivalent to $A_0 \geq L_0 + \text{SCR}$. The latter says that the current value of the assets exceeds the current value of the liabilities plus the solvency capital requirement. The balance sheet illustration of solvency may give the false impression, if not correctly interpreted, that solvency is about current asset and liability values, whereas solvency is really about future asset and liability values.

Example 6.2 (An absolute lower bound). Suppose that acceptable portfolios are those that are certain not to be below a fixed number c. This gives the risk measure

$$\rho(X) = \min\{m : mR_0 + X \geq c\}.$$

Define x_0 to be the smallest value that X can take (if no such value exists, then take x_0 to be the largest value smaller than all the values that X can take), and notice that

$$\rho(X) = \min\{m : mR_0 + X \geq c\} = \frac{c - x_0}{R_0}, \tag{6.1}$$

i.e., the discounted difference between the required capital c at time 1 and the worst possible outcome for the value of the portfolio at time 1. In particular, we note that if the portfolio contains short positions in some asset with an unbounded value at time 1 so that $x_0 = -\infty$, then $\rho(X) = +\infty$.

We claim that the risk measure ρ given by (6.1) is a convex measure of risk. To verify this claim, we need to show the translation invariance, monotonicity, and convexity. Translation invariance is shown by noticing that for all real numbers a,

$$\rho(X + aR_0) = \frac{c - (x_0 + aR_0)}{R_0} = \rho(X) - a.$$

To show monotonicity, we notice that if $X_2 \leq X_1$, then the corresponding lower bounds satisfy $x_{02} \leq x_{01}$, and therefore

$$\rho(X_1) = \frac{c - x_{01}}{R_0} \leq \frac{c - x_{02}}{R_0} = \rho(X_2).$$

Finally, we verify that the convexity property holds. If X_1 and X_2 have lower bounds x_{01} and x_{02}, then, for $\lambda \in [0, 1]$, the corresponding lower bound y_0 for $Y = \lambda X_1 + (1 - \lambda)X_2$ is greater than or equal to $\lambda x_{01} - (1 - \lambda)x_{02}$. Therefore,

$$
\begin{aligned}
\rho(\lambda X_1 + (1 - \lambda)X_2) &= \frac{c - y_0}{R_0} \\
&\leq \frac{c - \lambda x_{01} - (1 - \lambda)x_{02}}{R_0} \\
&= \lambda \frac{c - x_{01}}{R_0} + (1 - \lambda)\frac{c - x_{02}}{R_0} \\
&= \lambda \rho(X_1) + (1 - \lambda)\rho(X_2).
\end{aligned}
$$

Example 6.3 (Mean–variance risk measures). Consider portfolios whose future values X have finite variances and a risk measure of the form

$$\rho(X) = -\,\mathrm{E}[X/R_0] + c\sqrt{\mathrm{Var}(X/R_0)}, \quad c > 0. \tag{6.2}$$

By standard properties of the expected value and variance, it follows that ρ is translation invariant and positively homogeneous. Moreover, ρ is subadditive. This follows from the fact that

$$
\begin{aligned}
\mathrm{Var}(X_1 + X_2) &= \mathrm{Var}(X_1) + \mathrm{Var}(X_2) + 2\,\mathrm{Cor}(X_1, X_2)\sqrt{\mathrm{Var}(X_1)\,\mathrm{Var}(X_2)} \\
&\leq \left(\sqrt{\mathrm{Var}(X_1)} + \sqrt{\mathrm{Var}(X_2)}\right)^2,
\end{aligned}
$$

i.e., the standard deviation of the sum is less than or equal to the sum of the standard deviations for the two terms. Since ρ is both positively homogeneous and subadditive, it is also convex. However, the monotonicity condition is in general not satisfied, so ρ in (6.2) is not a convex measure of risk. The following example illustrates the lack of monotonicity. Let $X_1 = -R_0$ with probability one and let X_2 be a random variable that may take the values R_0 and $-R_0$, each with probability $1/2$. Then $X_1 \leq X_2$, and if $c > 1$ in (6.2), then

$$\rho(X_2) = c\sqrt{\mathrm{Var}(X_2/R_0)} = c > 1 = -\,\mathrm{E}[X_1/R_0] = \rho(X_1).$$

The lack of monotonicity is a serious flaw and limits the use of the mean–variance risk measure. However, for normally distributed random variables the mean–variance risk measure is canonical. If X is normally distributed, then we may write $X \stackrel{d}{=} E[X] + \sqrt{\text{Var}(X)}Z$, where Z is a standard normally distributed random variable. For any translation-invariant, positively homogeneous risk measure ρ we find that

$$\rho(X) = \rho(E[X] + \sqrt{\text{Var}(X)}Z)$$
$$= -E[X/R_0] + \sqrt{\text{Var}(X/R_0)}R_0\rho(Z).$$

We conclude that as long as X is normally distributed, any translation-invariant, positively homogeneous risk measure satisfies the defining property (6.2) of mean–variance risk measures.

6.2 Value-at-Risk

The value-at-risk (VaR) at level $p \in (0, 1)$ of a portfolio with value X at time 1 is

$$\text{VaR}_p(X) = \min\{m : P(mR_0 + X < 0) \le p\}, \tag{6.3}$$

where R_0 is the percentage return of a risk-free asset. In words, the VaR of a position with value X at time 1 is the smallest amount of money that if added to the position now and invested in the risk-free asset ensures that the probability of a strictly negative value at time 1 is not greater than p.

From (6.3) we see that $X \ge 0$ implies that $\text{VaR}_p(X) \le 0$. In order for VaR_p to be a sensible choice of risk measure for typical asset portfolios with mainly long positions, it is common to take the following view: at the current time 0 one starts from scratch and takes a risk-free loan of size V_0 (which is the current portfolio value), uses the capital to purchase the asset portfolio, and ends up with the net value $X = V_1 - V_0R_0$ at time 1. Therefore, the portfolio is classified as acceptable if the difference between the actual future portfolio value and the value that would be obtained by instead investing the current portfolio value in a risk-free asset is VaR_p-acceptable.

Before investigating the properties of VaR we first need to make sure that the minimum in (6.3) is attained so the definition really makes sense. To this end, note that

$$\{m : P(mR_0 + X < 0) \le p\}$$
$$= \{m : P(-X/R_0 > m) \le p\}$$
$$= \{m : 1 - P(-X/R_0 \le m) \le p\}$$
$$= \{m : P(-X/R_0 \le m) \ge 1 - p\}. \tag{6.4}$$

Since a distribution function F is right continuous ($F(x) \downarrow F(x_0)$ as $x \downarrow x_0$) and increasing, $\{m : F(m) \ge 1 - p\} = [m_0, \infty)$ for some m_0, and therefore there exists a smallest element.

Set $L = -X/R_0$. If $X = V_1 - V_0 R_0$ is the net gain from the investment, where the current portfolio value V_0 is viewed as a liability, then $L = -X/R_0 = V_0 - V_1/R_0$ has a natural interpretation as the discounted loss. The identities in (6.4) give an alternative (equivalent) formulation of $\mathrm{VaR}_p(X)$ in terms of L:

$$\mathrm{VaR}_p(X) = \min\{m : \mathrm{P}(L \leq m) \geq 1 - p\}. \tag{6.5}$$

We may interpret $\mathrm{VaR}_p(X)$ as the smallest value m such that the probability of the discounted portfolio loss $L = -X/R_0$ being at most m is at least $1 - p$. Expressed differently, $\mathrm{VaR}_p(X)$ is the smallest amount of money that, if put aside and invested in a risk-free asset at time 0, will be sufficient to cover a potential loss at time 1 with a probability of at least $1 - p$. Commonly encountered values for p are 5%, 1%, and 0.5%, which shows that $\mathrm{VaR}_p(X)$ describes (to some extent) the right tail of the probability distribution of the discounted loss L. The length in physical time of the time period over which the discounted loss is modeled is often taken to reflect the time it may take to move out of an unfavorable position in the face of adverse price movements. In market risk measurement (e.g., stocks, bond and financial derivatives), the length of the time period is typically 1 day or 10 days, whereas 1 year is typical for credit and insurance risk measurement (e.g., retail or corporate loans or the aggregate value of assets and liabilities of an insurance company).

In statistical terms, $\mathrm{VaR}_p(X)$ is the $(1 - p)$-quantile of L. The u-quantile of a random variable L with distribution function F_L is defined as

$$F_L^{-1}(u) = \min\{m : F_L(m) \geq u\},$$

and F_L^{-1} is just the ordinary inverse if F_L is strictly increasing. If F_L is both continuous and strictly increasing, then $F_L^{-1}(u)$ is the unique value m such that $F_L(m) = u$. For a general F_L, the quantile value $F_L^{-1}(u)$ is obtained by plotting the graph of F_L and setting $F_L^{-1}(u)$ to be the smallest value m for which $F_L(m) \geq u$. With this notation it follows that

$$\mathrm{VaR}_p(X) = F_L^{-1}(1 - p). \tag{6.6}$$

To better understand the properties of the risk measure VaR_p, we first study the quantile function in more detail. We denote the uniform distribution on the interval $(0, 1)$ by $U(0, 1)$, i.e., the probability distribution of a random variable U satisfying $\mathrm{P}(U \leq u) = u$ for u in $(0, 1)$.

Proposition 6.1. *Let F be a distribution function on $\mathbb{R}$. Then:*

(i) *$u \leq F(x)$ if and only if $F^{-1}(u) \leq x$.*
(ii) *If F is continuous, then $F(F^{-1}(u)) = u$.*
(iii) *(Quantile transform) If U is $U(0, 1)$-distributed, then $\mathrm{P}(F^{-1}(U) \leq x) = F(x)$.*
(iv) *(Probability transform) If X has distribution function F, then $F(X)$ is $U(0, 1)$-distributed if and only if F is continuous.*

Proof. (i): Suppose $F^{-1}(u) \leq x$. By definition, $F(F^{-1}(u)) = F(\min\{y : F(y) \geq u\}) \geq u$. Since F is nondecreasing, $u \leq F(F^{-1}(u)) \leq F(x)$. Suppose now that $u \leq F(x)$. Since F is nondecreasing, $F^{-1}(F(x)) = \min\{y : F(y) \geq F(x)\} \leq x$. Since F^{-1} also is nondecreasing, $F^{-1}(u) \leq F^{-1}(F(x)) \leq x$.

(ii): As in (i) we have $u \leq F(F^{-1}(u))$. Take $y < F^{-1}(u)$ and note that by (i) this is equivalent to $F(y) < u$. Now, if $F(y) - P(X \leq y) < u$ for all $y < F^{-1}(u)$, then $P(X < F^{-1}(u)) \leq u$. Then

$$u \leq F(F^{-1}(u)) = P(X < F^{-1}(u)) + P(X = F^{-1}(u))$$

$$\leq u + P(X = F^{-1}(u)).$$

The continuity of F implies that $P(X = F^{-1}(u)) = 0$. We conclude that $u = F(F^{-1}(u))$.

(iii): $U \leq F(x)$ if and only if $F^{-1}(U) \leq x$ by (i). Hence, $P(F^{-1}(U) \leq x) = P(U \leq F(x)) = F(x)$.

(iv): Suppose F is continuous. By the quantile transform and (ii),

$$P(F(X) = u) = P(F(F^{-1}(U)) = u) = P(U = u) = 0.$$

Hence, by (i)

$$P(F(X) \leq u) = P(F(X) < u) + P(F(X) = u)$$

$$= P(F(X) < u)$$

$$= 1 - P(F(X) \geq u)$$

$$= 1 - P(X \geq F^{-1}(u))$$

$$= P(X < F^{-1}(u))$$

$$= F(F^{-1}(u)).$$

It now follows from (ii) that $F(X)$ is $U(0, 1)$-distributed.

To show the converse we show the equivalent statement, that if F is not continuous, then $F(X)$ is not $U(0, 1)$-distributed. If F is discontinuous at x, then $0 < P(X = x) \leq P(F(X) = F(x))$. Hence, $F(X)$ has a point mass and therefore cannot be $U(0, 1)$-distributed. □

It is not difficult to see that the quantile function, and therefore also VaR, is translation invariant and positive homogeneous. For constants c_1, c_2 with $c_2 > 0$,

$$F_{c_1+c_2L}^{-1}(p) = \min\{m : P(c_1 + c_2L \leq m) \geq p\}$$

$$= \min\{m : F_L((m - c_1)/c_2) \geq p\}$$

$$= \{\text{put } m' = (m - c_1)/c_2\}$$
$$= \min\{c_1 + c_2 m' : F_L(m') \geq p\}$$
$$= c_1 + c_2 \min\{m' : F_L(m') \geq p\}$$
$$= c_1 + c_2 F_L^{-1}(p). \tag{6.7}$$

Moreover, the quantile function, and therefore also VaR, satisfies the monotonicity condition. This follows from the fact that $L_2 \leq L_1$ implies $F_{L_1}(m) \leq F_{L_2}(m)$ and therefore

$$F_{L_1}^{-1}(p) = \min\{m : F_{L_1}(m) \geq p\}$$
$$\geq \min\{m : F_{L_2}(m) \geq p\} = F_{L_2}^{-1}(p).$$

Here we summarize the properties for VaR that have been established up to this point.

Proposition 6.2. *The properties translation invariance, monotonicity, and positive homogeneity hold for* VaR$_p$.

Examples of the lack of subadditivity of VaR$_p$ can be found even for sums of independent and identically distributed random variables. One such example is obtained by combining Examples 6.9 and 6.10.

Example 6.4 (A crude upper bound). Sometimes we need to estimate the quantile $F_L^{-1}(p)$ for $p \in (0, 1)$ close to one, although the distribution of L is far from being well understood. Suppose, for instance, that only the mean $E[L]$ and the variance $\mathrm{Var}(L)$ are available to us. Cantelli's inequality, the one-sided version of Chebyshev's inequality, says that

$$P(L - E[L] \geq y) \leq \frac{\mathrm{Var}(L)}{y^2 + \mathrm{Var}(L)}$$

or equivalently that

$$P(L \geq y) \leq \frac{\mathrm{Var}(L)}{(y - E[L])^2 + \mathrm{Var}(L)}.$$

Now we can turn this upper bound for the tail probability into an upper bound for the p-quantile of the distribution of L:

$$F_L^{-1}(p) \leq \min\left\{y : \frac{\mathrm{Var}(L)}{(y - E[L])^2 + \mathrm{Var}(L)} \leq 1 - p\right\} = E[L] + \left(\frac{\mathrm{Var}(L)p}{1 - p}\right)^{1/2}.$$

The upper bound on the quantile is not necessarily a good estimate, but it is the smallest upper bound (best conservative estimate) if no information about the distribution of L is available besides the mean and the variance.

Example 6.5 (Lognormal distribution). Consider a stock with spot price S_0 today and random spot price S_1 tomorrow, and assume that the 1-day interest rate is zero. We want to compute $\mathrm{VaR}_p(S_1 - S_0)$ under the assumption that the log return $\log(S_1/S_0)$ is normally distributed. Note that $\mathrm{VaR}_p(S_1 - S_0) = F^{-1}_{S_0 - S_1}(1 - p)$ and that

$$S_0 - S_1 = -S_0(e^{\log(S_1/S_0)} - 1) \stackrel{\mathrm{d}}{=} -S_0(e^{\mu + \sigma Z} - 1),$$

where Z is standard normally distributed. Write $L = S_0 - S_1$ and notice that $L = -g(Z)$, where g is a continuous and strictly increasing function. To compute $\mathrm{VaR}_p(S_1 - S_0) = F^{-1}_L(1 - p)$, we will combine the two relations

$$F^{-1}_{-g(Z)}(1 - p) = -F^{-1}_{g(Z)}(p), \tag{6.8}$$

$$F^{-1}_{g(Z)}(p) = g(F^{-1}_Z(p)), \tag{6.9}$$

to obtain

$$\mathrm{VaR}_p(S_1 - S_0) = -g(F^{-1}_Z(p)) = S_0(1 - e^{\mu + \sigma \Phi^{-1}(p)}).$$

Let us first show relation (6.8). Since $\mathrm{P}(g(Z) = x) = 0$ for every x, it holds that

$$F_{-g(Z)}(x) = \mathrm{P}(-g(Z) \le x) = \mathrm{P}(g(Z) \ge -x) = 1 - F_{g(Z)}(-x),$$

and therefore solving $F_{-g(Z)}(x) = 1 - p$ for x is equivalent to solving $F_{g(Z)}(-x) = p$, which in turn is equivalent to $x = -F^{-1}_{g(Z)}(p)$. Let us now show relation (6.9). We notice that

$$F_{g(Z)}(x) = \mathrm{P}(g(Z) \le x) = \mathrm{P}(Z \le g^{-1}(x)) = F_Z(g^{-1}(x)),$$

and therefore solving $F_{g(Z)}(x) = p$ for x is equivalent to solving $g^{-1}(x) = F^{-1}_Z(p)$, which in turn is equivalent to $x = g(F^{-1}_Z(p))$.

As the previous example illustrates, a common situation is when we want to compute $\mathrm{VaR}_p(X)$ when $X = g(Z)$ for a continuous and monotone function g and a random variable Z. In the preceding example, g was continuous and strictly increasing and Z had a normal distribution. The payoff function of a call option, $(S_1 - K)_+$, is nondecreasing but not strictly increasing, so the preceding calculation does not apply. The following two results show that also the more general situation can be handled without too much difficulty.

Proposition 6.3. *If $g : \mathbb{R} \to \mathbb{R}$ is nondecreasing and left continuous, then for any random variable Z it holds that $F_{g(Z)}^{-1}(p) = g(F_Z^{-1}(p))$ for all $p \in (0,1)$.*

Proof. First we show that, with $X = g(Z)$, $F_X^{-1}(p) \leq g(F_Z^{-1}(p))$. To see this, first observe that since g is nondecreasing, $Z \leq F_Z^{-1}(p)$ implies that $g(Z) \leq g(F_Z^{-1}(p))$. Moreover, since F_Z is right continuous, it holds that $F_Z(F_Z^{-1}(p)) \geq p$. Therefore,

$$P(X \leq g(F_Z^{-1}(p))) = P(g(Z) \leq g(F_Z^{-1}(p))) \geq P(Z \leq F_Z^{-1}(p)) \geq p.$$

Since $F_X^{-1}(p)$ is the smallest number m such that $P(X \leq m) \geq p$, we have shown that $F_X^{-1}(p) \leq g(F_Z^{-1}(p))$.

To show the reverse inequality $F_X^{-1}(p) \geq g(F_Z^{-1}(p))$, we use the left continuity of g. Since g is nondecreasing and left continuous, there exists for each $y \in \mathbb{R}$ and $\varepsilon > 0$ a $\delta > 0$ such that

$$\{z : z \in (y - \delta, y]\} \subset \{z : g(z) \in (g(y) - \varepsilon, g(y)]\}.$$

Moreover, since g is nondecreasing, we have

$$\{z : g(z) \leq g(y)\} = \{z : z \leq y\} \cup \{z : g(z) = g(y), z > y\},$$

$$\{z : g(z) \leq g(y) - \varepsilon\} \subset \{z : z \leq y\}.$$

Combining the preceding three set relations yields

$$\{z : g(z) \leq g(y) - \varepsilon\} \subset \{z : z \leq y - \delta\}.$$

Therefore,

$$P(X \leq g(F_Z^{-1}(p)) - \varepsilon) = P(g(Z) \leq g(F_Z^{-1}(p)) - \varepsilon)$$
$$\leq P(Z \leq F_Z^{-1}(p) - \delta)$$
$$< p,$$

where in the last step we used the fact that the right continuity of F_Z implies that for every $\delta > 0$ we have $F_Z(F_Z^{-1}(p) - \delta) < p$. It follows that $g(F_Z^{-1}(p)) - \varepsilon < F_X^{-1}(p)$. Since $\varepsilon > 0$ was arbitrary, we conclude that $g(F_Z^{-1}(p)) \leq F_X^{-1}(p)$. This completes the proof. $\square$

The following proposition, combined with Proposition 6.3, enables efficient computations of $\text{VaR}_p(X)$ in a wide range of applications. We denote by

$$F_X^{-1}((1-p)+) = \lim_{\varepsilon \downarrow 0} F_X^{-1}(1 - p + \varepsilon)$$

the limit from the right of the quantile function of X, F_X^{-1}, at the point $1 - p$.

Proposition 6.4. *For any random variable X, $F_{-X}^{-1}(p) = -F_X^{-1}((1-p)+)$ for all $p \in (0,1)$. In particular, if F_X is continuous and strictly increasing, then $F_{-X}^{-1}(p) = -F_X^{-1}(1-p)$.*

The best way to verify the equality $F_{-X}^{-1}(p) = -F_X^{-1}((1-p)+)$ is by selecting a random variable X whose distribution function F_X has both a flat part and a jump and to draw and inspect the graphs of F_X, F_{-X}, F_X^{-1}, and F_{-X}^{-1}. Without loss of generality we may choose a random variable with distribution function $x \mapsto F_X(x)$ whose graph is shown in the upper left plot in Fig. 6.2. To draw the graph of $p \mapsto -F_X^{-1}((1-p)+)$, we proceed as follows. The graph of $p \mapsto F_X^{-1}(p)$ (lower left plot in Fig. 6.2) is obtained by reflecting the graph of F_X in the line $y = x$. Note that the quantile function F_X^{-1} is left continuous. Finally, we draw the graph of $p \mapsto -F_X^{-1}((1-p)+)$ (lower right plot in Fig. 6.2) by first reflecting the graph of F_X^{-1} in x-axis, then reflecting the resulting graph in the line $p = 1/2$, and finally taking limits from the left of the resulting function of p (which corresponds to taking limits from the right if the function is viewed as a function of $1 - p$). To draw the graph of $p \mapsto F_{-X}^{-1}(p)$, we proceed as follows. First draw the graph of $x \mapsto F_{-X}(x)$ (upper right plot in Fig. 6.2). Then draw the graph of $p \mapsto F_{-X}^{-1}(p)$ (lower right plot in Fig. 6.2) by reflecting the previous graph in the line $y = x$.

A formal proof of Proposition 6.4 goes as follows.

Proof. First note that

$$F_{-X}^{-1}(p) = \min\{m : P(-X \leq m) \geq p\}$$
$$= \min\{m : P(X \geq -m) \geq p\}$$
$$= \min\{m : P(X < -m) \leq 1 - p\}$$
$$= -\max\{m : P(X < m) \leq 1 - p\}.$$

It remains to show that $\max\{m : P(X < m) \leq 1 - p\} = \lim_{\varepsilon \downarrow 0} F_X^{-1}(1 - p + \varepsilon)$. Let $m_{1-p} = \max\{m : P(X < m) \leq 1 - p\}$, and note that it follows from the definition of the quantile F_X^{-1} that $F_X(F_X^{-1}(u)) = F_X(\min\{m : F_X(m) \geq u\}) \geq u$. Therefore, for $\varepsilon \in (0, p)$,

$$P(X < m_{1-p}) \leq 1 - p < 1 - p + \varepsilon \leq P(X \leq F_X^{-1}(1 - p + \varepsilon)),$$

from which it follows that $m_{1-p} \leq F_X^{-1}(1 - p + \varepsilon)$. Since the inequalities hold for any $\varepsilon \in (0, p)$, we may take the limit as $\varepsilon \downarrow 0$ and therefore conclude that

$$m_{1-p} \leq \lim_{\varepsilon \downarrow 0} F_X^{-1}(1 - p + \varepsilon).$$

To show the reverse inequality, we first take an arbitrary $\delta > 0$ and note that the definition of m_{1-p} implies that $P(X < m_{1-p} + \delta) > 1 - p$, which in turn implies that

$$P(X < m_{1-p} + \delta) > 1 - p + \varepsilon$$

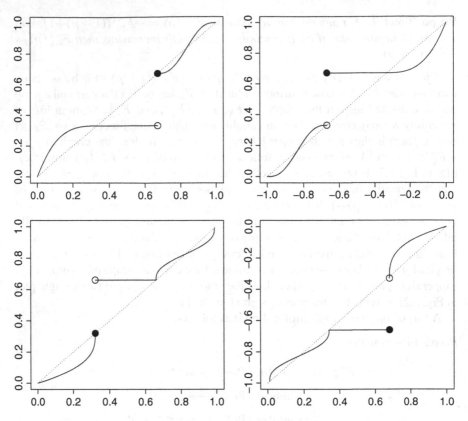

Fig. 6.2 *Upper left plot*: distribution functions $F_X(x)$; *upper right plot*: distribution function $F_{-X}(x)$. *Lower left plot*: quantile function $F_X^{-1}(p)$; *lower right plot*: function $-F_X^{-1}((1-p)+)$

for all sufficiently small $\varepsilon > 0$. Therefore, $m_{1-p} + \delta \geq F_X^{-1}(1-p+\varepsilon)$ for all sufficiently small $\varepsilon > 0$, and taking the limit as $\varepsilon \downarrow 0$ gives

$$m_{1-p} + \delta \geq \lim_{\varepsilon \downarrow 0} F_X^{-1}(1-p+\varepsilon).$$

Since $\delta > 0$ was arbitrary, it follows that

$$m_{1-p} \geq \lim_{\varepsilon \downarrow 0} F_X^{-1}(1-p+\varepsilon).$$

This completes the proof.   □

Example 6.6 (Put spread). Consider a portfolio with a value at time 1 given by

$$(K_2 - S_1)_+ - (K_1 - S_1)_+ \quad \text{for } K_1 < K_2.$$

This is the value at maturity of a put spread; a long position in a put option with strike K_2 and price C_2 and a short position of the same size in a put option with a lower strike K_1 and price C_1. The net value X at time 1, considering the cost of the put spread as a risk-free loan to be paid at time 1, is

$$X = (K_2 - S_1)_+ - (K_1 - S_1)_+ - (C_2 - C_1)R_0.$$

We write $L = -X/R_0 = g(S_1)$, where

$$g(y) = \frac{1}{R_0}\left((K_1 - y)_+ - (K_2 - y)_+\right) + C_2 - C_1$$

and note that the function g is continuous and nondecreasing. By Proposition 6.3 it follows that

$$\mathrm{VaR}_p(X) = F_L^{-1}(1 - p) = F_{g(S_1)}^{-1}(1 - p) = g(F_{S_1}^{-1}(1 - p))$$

$$= \frac{1}{R_0}\left((K_1 - F_{S_1}^{-1}(1 - p))_+ - (K_2 - F_{S_1}^{-1}(1 - p))_+\right) + C_2 - C_1.$$

Example 6.7 (Structured product). Financial contracts that are combinations of a bond, giving the buyer a guaranteed return on investment, and a derivative contract, giving the buyer the possibility of a high return, are often called structured products. The simplest form of a structured product with maturity in 1 year is a portfolio consisting of a long position of size h_0 in a risk-free bond that pays 1 to its holder 1 year from now and a long position of size h_1 in a European call option on the value S_1 of a stock index 1 year from now with a strike price K. Suppose that S_1 is lognormally distributed, $\mathrm{LN}(\mu, \sigma^2)$. If the current spot prices of the bond and option are B_0 and C_0, respectively, then the current value of the portfolio is $V_0 = h_0 B_0 + h_1 C_0$.

To evaluate the riskiness of this portfolio, we want to compute $\mathrm{VaR}_p(X)$, where $X = V_1 - V_0/B_0$ and $V_1 = h_0 + h_1(S_1 - K)_+$ is the value of the portfolio at maturity. Write

$$\mathrm{VaR}_p(X) = F_{-B_0 V_1 + V_0}^{-1}(1 - p) = F_{-g(Z)}^{-1}(1 - p),$$

where Z is standard normally distributed and g is given by

$$g(Z) = B_0(h_0 + h_1(e^{\mu + \sigma Z} - K)_+) - V_0.$$

Note that g is continuous and nondecreasing. Applying first Proposition 6.4 and then Proposition 6.3 gives

$$F_{-g(Z)}^{-1}(1 - p) = -F_{g(Z)}^{-1}(p+) = -g(F_Z^{-1}(p+)).$$

Since Z is standard normally distributed with strictly increasing distribution function Φ, we find that $F_Z^{-1}(p+) = \Phi^{-1}(p)$ and conclude that

$$\mathrm{VaR}_p(X) = -B_0(h_0 + h_1(e^{\mu+\sigma\Phi^{-1}(p)} - K)_+) + V_0$$

$$= h_1\Big(C_0 - B_0(e^{\mu+\sigma\Phi^{-1}(p)} - K)_+)\Big).$$

Suppose value X at time 1 of a portfolio can be expressed as $X = f(Z)$ for a smooth nonlinear function f and Z having a standard distribution. If f is not monotone, then it is difficult (or impossible) to express the quantiles of X in terms of the quantiles of Z. One way to overcome this difficulty is by approximating f by a first-order Taylor expansion and then approximate X by

$$X \approx f(\mathrm{E}[Z]) + \frac{df}{dz}(\mathrm{E}[Z])(Z - \mathrm{E}[Z]).$$

This approach is referred to as linearization. The following example gives an illustration of linearization in a simple example where explicit calculations are possible and linearization is not really needed. The example also shows that linearization (like any other approximation) must be used wisely; careless use may result in serious errors.

Example 6.8 (Linearization). You hold a portfolio consisting of a long position of 5 shares of stock A. The stock price today is $S_0 = 100$, and we assume a zero interest rate. The daily log returns

$$Y_1 = \log(S_1/S_0), Y_2 = \log(S_2/S_1), \ldots$$

of stock A are assumed to have a normal distribution with zero mean and standard deviation $\sigma = 0.01$. Let V_0 be the current value of the portfolio, and let $V_1 = S_0 e^{Y_1} = S_0 e^{0.01Z}$, where $Z = Y_1/0.01$ is standard normally distributed, be the value of the portfolio tomorrow.

We first consider the effect of linearization over a 1-day horizon. We start by explicitly computing $\mathrm{VaR}_{0.01}(V_1 - V_0)$ and then compute the approximation obtained by replacing V_1 by its first-order Taylor approximation with respect to Z. Notice that $\mathrm{VaR}_{0.01}(V_1 - V_0) = F_{V_0-V_1}^{-1}(0.99)$ and $V_0 - V_1 = -500(e^{Y_1} - 1) = -500(e^{0.01Z} - 1)$. Therefore, as in Example 6.5,

$$\mathrm{VaR}_{0.01}(V_1 - V_0) = 500(1 - e^{0.01\Phi^{-1}(0.01)}) = 11.5.$$

The first-order Taylor approximation of V_1 is $V_1 = 500e^{0.01Z} \approx 500(0.01Z + 1)$, which gives

$$\mathrm{VaR}_{0.01}(V_1 - V_0) \approx \mathrm{VaR}_{0.01}(5Z) = 5\varPhi^{-1}(0.99) \approx 11.6.$$

The relative error of the VaR approximation is 1.2%, which is rather small.

We now consider the effect of linearization over a longer time horizon and illustrate that the error due to linearization may be substantial. We make the simplifying assumption that log returns over nonoverlapping time periods are independent. We consider the effect of holding the aforementioned portfolio for 100 (trading) days and let V_{100} be the value of the portfolio 100 days from now. We start by explicitly computing $\mathrm{VaR}_{0.01}(V_{100} - V_0)$ and then compute the approximation obtained by replacing V_{100} seen as a function of a standard normal variable Z by its first-order Taylor approximation with respect to Z. As previously, we ignore interest rates. We may write $\mathrm{VaR}_{0.01}(V_{100} - V_0) = F_{V_0 - V_{100}}^{-1}(0.99)$, where $V_0 - V_{100} = -500(e^{Y_{100}} - 1)$ with Y_{100} denoting the 100-day log return. Note that

$$Y_{100} = \log S_{100}/S_0 = \log S_1/S_0 + \cdots + \log S_{100}/S_{99},$$

which shows that Y_{100} is a sum of 100 independent $N(0, 0.01^2)$-distributed random variables. Therefore, $Y_{100} \overset{d}{=} 0.1Z$, where Z is standard normally distributed. In particular, $V_0 - V_{100} \overset{d}{=} -500(e^{0.1Z} - 1)$ and

$$\mathrm{VaR}_{0.01}(V_{100} - V_0) = 500(1 - e^{0.1\varPhi^{-1}(0.01)}) \approx 103.8.$$

Using a first-order Taylor approximation gives $V_{100} \overset{d}{=} 500e^{0.1Z} \approx 500(0.1Z + 1)$, which gives the approximation

$$\mathrm{VaR}_{0.01}(V_{100} - V_0) \approx \mathrm{VaR}_{0.01}(50Z) = 50\varPhi^{-1}(0.99) \approx 116.3.$$

The relative error of the VaR approximation is 12.1%.

Next follows the first two in a series of four examples on credit default swaps (CDSs). The examples treat portfolios containing defaultable bonds and CDSs. There are two general messages communicated by these examples. The first message is that VaR at level p does not provide any information about the worst-case outcomes corresponding to an event whose probability is less than p. Using VaR, therefore, enables investors to hide risk in the right tail of the distribution of L. The second message is that, besides not investing at all, there are essentially two ways to reduce risk. One way is to hedge a risk by buying protection against undesired events. Hedging may be viewed as buying insurance. In the following examples, hedging amounts to buying credit default swaps. Another way to reduce risk is by diversification. A well-diversified position has a future value that depends on many independent sources of randomness such that the exposure to each one of them is small. Diversification is the key principle for an insurer and the opposite of buying insurance.

A problem with VaR is that it does not necessarily reward diversification. In Example 6.10 it is shown that a diversified portfolio may have higher risk, measured by VaR, than a comparable nondiversified portfolio. The example also shows that VaR is not subadditive in general.

Example 6.9 (Credit default swap I). Consider an investor with \$100 who has the opportunity to take long positions in a defaultable bond and a credit default swap (CDS) on this bond. One bond costs \$97 now and pays \$100 6 months from now if the issuer does not default and 0 if the issuer defaults. The CDS costs \$4 and pays \$100 6 months from now if the bond issuer defaults and nothing otherwise. For simplicity we assume that a risk-free bond with maturity in 6 months has zero interest rate, so $B_0 = 1$. The investor believes that the default probability is 0.02 and wants to maximize the expected value of V_1, the value in dollars of the investor's position at the maturity of the bond, subject to the risk constraint $\text{VaR}_{0.05}(V_1 - 100) \leq 10$ and a budget constraint. It is assumed throughout that the investor can only take long positions. Otherwise, with the prices given previously, there would be an arbitrage opportunity. Why? How much of the \$100 does the investor invest in the bond? How much in the CDS?

Let w_1 and w_2 be the amounts invested in bonds and CDSs in the portfolio, respectively. Let $c_1 = 97$ and $c_2 = 4$ be the prices of the bond and the CDS, respectively. Then the value at time 1 (after 6 months) is $V_1 = w_1 c_1^{-1} 100(1 - I) + w_2 c_2^{-1} 100 I$, where I is the default indicator, $I = 1$ if the issuer defaults, and $I = 0$ otherwise, with $P(I = 1) = 0.02$. Then

$$E[V_1] = 98 w_1 c_1^{-1} + 2 w_2 c_2^{-1} = \frac{98}{97} w_1 + \frac{1}{2} w_2,$$

from which it is clear that the investor wants to invest as much as possible in the bond without violating the constraints. Moreover,

$$\text{VaR}_p(V_1 - 100) = 100 + \text{VaR}_p\left(100 w_1 c_1^{-1} + 100\left(w_2 c_2^{-1} - w_1 c_1^{-1}\right) I\right),$$

which gives

$$\text{VaR}_p(V_1 - 100) = 100 - 100 w_1 c_1^{-1}$$
$$+ \begin{cases} 100(w_2 c_2^{-1} - w_1 c_1^{-1}) \, \text{VaR}_p(I) & \text{if } w_2 c_2^{-1} \geq w_1 c_1^{-1}, \\ 100(w_1 c_1^{-1} - w_2 c_2^{-1}) \, \text{VaR}_p(-I) & \text{if } w_2 c_2^{-1} < w_1 c_1^{-1}. \end{cases}$$

By (6.6) we have $\text{VaR}_p(I) = F_{-I}^{-1}(1 - p)$ and $\text{VaR}_p(-I) = F_I^{-1}(1 - p)$ where, by Proposition 6.4,

$$F_{-I}^{-1}(1 - p) = -F_I^{-1}(p+) = \begin{cases} -1 & \text{if } p \in [0.98, 1], \\ 0 & \text{if } p \in [0, 0.98), \end{cases}$$

and

$$F_I^{-1}(1-p) = \begin{cases} 0 \text{ if } p \in [0.02, 1], \\ 1 \text{ if } p \in [0, 0.02). \end{cases}$$

This implies that

$$\mathrm{VaR}_p(V_1 - 100) = 100 - \begin{cases} 100\max(w_1c_1^{-1}, w_2c_2^{-1}) \text{ if } p \in [0.98, 1], \\ 100w_1c_1^{-1} \quad\quad\quad\quad \text{ if } p \in [0.02, 0.98), \\ 100\min(w_1c_1^{-1}, w_2c_2^{-1}) \text{ if } p \in [0, 0.02). \end{cases}$$

In particular, $\mathrm{VaR}_{0.05}(V_1 - 100) = 100 - 100w_1c_1^{-1}$, and therefore $w_2 = 100 - w_1$, together with $\mathrm{VaR}_{0.05}(V_1 - 100) \leq 10$, is equivalent to $w_1 \geq 87.3$. Since a dollar invested in the bond gives a much better expected return than a dollar invested in the CDS, the investor wants to maximize w_1 subject to the constraints. Therefore, the solution to the optimization problem with the VaR constraint is $(w_1, w_2) = (100, 0)$. That is, buy defaultable bonds only. The catch here is that VaR at level 0.05 does not take into account the possibility of default, which occurs with probability 0.02. This enables the investor to hide the default risk in the tail.

Example 6.10 (Credit default swap II). Let us look a bit closer at the optimal solution $(w_1, w_2) = (100, 0)$ to the investment problem in Example 6.9. The optimal weights give the optimal portfolio value $V_1 = (100^2/97)(1 - I)$ at maturity. Moreover, we have seen that

$$\mathrm{VaR}_{0.05}(V_1 - 100) = \mathrm{VaR}_{0.05}\left(\frac{100^2}{97}(1 - I) - 100\right)$$

$$= 100 - \frac{100^2}{97} + \frac{100^2}{97}\mathrm{VaR}_{0.05}(-I)$$

$$= 100\left(1 - \frac{100}{97}\right) < 0. \tag{6.10}$$

The negative value highlights the fact that at the 5% level VaR does not pick up the default risk. In particular, it treats the defaultable bond as a risk-free bond. Suppose, in contrast, that we have 100 identical bonds whose default events are independent and that the investor invests one dollar in each of them (which gives the same expected portfolio value as for the optimal solution in Example 6.9). The risk of the new portfolio, in terms of $\mathrm{VaR}_{0.05}$, is

$$\mathrm{VaR}_{0.05}(V_1 - 100) = \mathrm{VaR}_{0.05}\left(\frac{100}{97}\sum_{k=1}^{100}(1 - I_k) - 100\right)$$

$$= 100 - \frac{100^2}{97} + \frac{100}{97}\mathrm{VaR}_{0.05}\left(-\sum_{k=1}^{100} I_k\right).$$

Since $Z = \sum_{k=1}^{100} I_k$ is $\text{Bin}(100, 0.02)$-distributed and $\text{VaR}_{0.05}(-Z) = F_Z^{-1}(0.95)$, it follows that

$$\text{VaR}_{0.05}(V_1 - 100) = 100 - \frac{100^2}{97} + \frac{100}{97} F_Z^{-1}(0.95).$$

We can compute $P(Z \leq 4) \approx 0.949$ and $P(Z \leq 5) \approx 0.985$. Therefore, $F_Z^{-1}(0.95) = \min\{m : P(Z \leq m) \geq 0.95\} = 5$, which implies that

$$\text{VaR}_{0.05}(V_1 - 100) = 100 + \frac{100}{97}(-100 + 5)$$

$$= 100\left(1 - \frac{95}{97}\right) > 0. \qquad (6.11)$$

That is, in this example diversification increases the risk! The reason is that diversification here makes $\text{VaR}_{0.05}$ take into account the default risk that for the nondiversified investment was hidden in the tail. In particular, we conclude that VaR is not subadditive since (6.10) and (6.11) imply

$$\text{VaR}_{0.05}\left(\sum_{k=1}^{100}(1 - I_k)\right) > \sum_{k=1}^{100} \text{VaR}_{0.05}(1 - I_k).$$

6.3 Expected Shortfall

Although VaR is probably the most commonly used risk measure for risk control in the financial industry, it has several limitations. Its biggest weakness is that it ignores the left tail (beyond level p) of the distribution of X. (The fact that it is just a quantile value means that it ignores most of the distribution of X.) In particular, it allows a careless/dishonest risk manager to miss/hide unlikely but catastrophic risks in the left tail.

A natural remedy for not considering catastrophic loss events with small probabilities would be to consider the average VaR values below the level p. This average of VaR values gives the risk measure expected shortfall (ES) at level p, which is defined as

$$\text{ES}_p(X) = \frac{1}{p} \int_0^p \text{VaR}_u(X) du.$$

With minor technical modifications, ES is also called Average VaR (AVaR), Conditional VaR (CVaR), Tail VaR (TVaR), or Tail Conditional Expectation (TCE).

ES is often proposed as a superior alternative to VaR because it considers all of the left tail of the probability distribution of X and because it is a coherent measure of risk. The coherence of ES, Proposition 6.6, implies that it is also convex, and the latter property is essential for ensuring that investment problems with ES constraints are convex optimization problems. To show the coherence of ES and also to use it effectively in optimization problems, we first present useful alternative representations of ES.

Proposition 6.5. *(i) ES has the following representations:*

$$\mathrm{ES}_p(X) = \frac{1}{p} \int_{1-p}^{1} F_L^{-1}(u)du, \quad L = -X/R_0, \tag{6.12}$$

$$\mathrm{ES}_p(X) = -\frac{1}{p} \int_0^p F_{X/R_0}^{-1}(u)du, \tag{6.13}$$

$$\mathrm{ES}_p(X) = -\frac{1}{p} \mathrm{E}[X/R_0 I\{X/R_0 \le F_{X/R_0}^{-1}(p)]$$

$$- F_{X/R_0}^{-1}(p)\left(1 - \frac{F_{X/R_0}(F_{X/R_0}^{-1}(p))}{p}\right), \tag{6.14}$$

$$\mathrm{ES}_p(X) = \min_c -c + \frac{1}{p} \mathrm{E}[(c - X/R_0)_+]. \tag{6.15}$$

(ii) If X has a continuous distribution function, then, with $L = -X/R_0$,

$$\mathrm{ES}_p(X) = \mathrm{E}[L \mid L \ge \mathrm{VaR}_p(X)] = \mathrm{E}[L \mid L \ge F_L^{-1}(1 - p)]. \tag{6.16}$$

The right-hand side of (6.15) is often called CVaR. This representation is useful in portfolio optimization problems. From (6.16) we find that if X has a continuous distribution function, then ES is the average loss conditional on the loss being larger than or equal to the VaR at the level p. This expression motivates the name ES.

Proof. (i) From the definition we see that ES is simply an average of quantile values of L:

$$\mathrm{ES}_p(X) = \frac{1}{p} \int_0^p F_L^{-1}(1 - u)du = \frac{1}{p} \int_{1-p}^{1} F_L^{-1}(u)du.$$

This proves the first representation (6.12). To prove the second representation, recall from Proposition 6.4 that $F_{-X/R_0}^{-1}(1 - u) = -F_{X/R_0}^{-1}(u+)$. But $F_{X/R_0}^{-1}(u+)$ is not equal to $F_{X/R_0}^{-1}(u)$ in general. However, we do have equality for almost all u in the sense that if we draw U uniformly on $(0, 1)$, then $F_{X/R_0}^{-1}(U+) = F_{X/R_0}^{-1}(U)$ with probability one. In particular, when U has a uniform distribution on $(0, p)$, it holds that

$$\mathrm{ES}_p(X) = \frac{1}{p} \int_0^p F_L^{-1}(1 - u)du$$

$$= E[F_L^{-1}(1 - U)]$$

$$= -E[F_{X/R_0}^{-1}(U+)]$$

$$= -E[F_{X/R_0}^{-1}(U)]$$

$$= -\frac{1}{p} \int_0^p F_{X/R_0}^{-1}(u) du,$$

which proves (6.13). Let us prove (6.14). The only difficulty is when F_{X/R_0} has a jump at $F_{X/R_0}^{-1}(p)$ and $F_{X/R_0}(F_{X/R_0}^{-1}(p)) > p$. Using statements (i) and (iii) of Proposition 6.1 shows that

$$E[X/R_0 I\{X/R_0 \le F_{X/R_0}^{-1}(p)\}]$$

$$= E[F_{X/R_0}^{-1}(U) I\{F_{X/R_0}^{-1}(U) \le F_{X/R_0}^{-1}(p)\}]$$

$$= E[F_{X/R_0}^{-1}(U) I\{U \le F_{X/R_0}(F_{X/R_0}^{-1}(p))\}]$$

$$= E[F_{X/R_0}^{-1}(U) I\{U \le p\}] + E[F_{X/R_0}^{-1}(U) I\{p < U \le F_{X/R_0}(F_{X/R_0}^{-1}(p))\}]$$

$$= \int_0^p F_{X/R_0}^{-1}(u) du + F_{X/R_0}^{-1}(p)\Big(F_{X/R_0}(F_{X/R_0}^{-1}(p)) - p\Big).$$

Therefore,

$$-\frac{1}{p} E[X/R_0 I\{X/R_0 \le F_{X/R_0}^{-1}(p)\}] - F_{X/R_0}^{-1}(p)\left(1 - \frac{F_{X/R_0}(F_{X/R_0}^{-1}(p))}{p}\right)$$

$$= -\frac{1}{p} \int_0^p F_{X/R_0}^{-1}(u) du,$$

from which the conclusion follows from (6.13). To prove (6.15) we consider the function

$$G(c) = -c + \frac{1}{p} E[(c - X/R_0)_+] = -c + \frac{1}{p} \int_{-\infty}^c F_{X/R_0}(x) dx$$

and note that G is convex. Since

$$E[(c - X/R_0)_+] = E[(c - X/R_0) I\{c - X/R_0 > 0\}]$$

$$= \int_0^\infty P(c - X/R_0 > t) dt$$

$$= \int_{-\infty}^c P(X/R_0 < u) du,$$

we find that G is differentiable except at the points where F_{X/R_0} has jumps and, except for those points,

$$G'(c) = -1 + \frac{1}{p} F_{X/R_0}(c).$$

It follows that $G'(c) \leq 0$ for c such that $F_{X/R_0}(c) \leq p$ and that $G'(c) \geq 0$ for c such that $F_{X/R_0}(c) \geq p$. Therefore, G has a (not necessarily unique) minimum at $\min\{c : F_{X/R_0}(c) \geq p\} = F_{X/R_0}^{-1}(p)$. Evaluating G at this point gives

$$G(F_{X/R_0}^{-1}(p)) = -F_{X/R_0}^{-1}(p) + \frac{1}{p} \mathrm{E}[(F_{X/R_0}^{-1}(p) - X/R_0)_+]$$

$$= -F_{X/R_0}^{-1}(p) + \frac{1}{p} \mathrm{E}\left[\left(F_{X/R_0}^{-1}(p) - X/R_0\right) I\left\{X/R_0 \leq F_{X/R_0}^{-1}(p)\right\}\right]$$

$$= -\frac{1}{p} \mathrm{E}\left[X/R_0 I\left\{X/R_0 \leq F_{X/R_0}^{-1}(p)\right\}\right]$$

$$- F_{X/R_0}^{-1}(p)\left(1 - \frac{F_{X/R_0}(F_{X/R_0}^{-1}(p))}{p}\right),$$

from which the conclusion follows from (6.14).

(ii) Suppose that X has a continuous distribution function. Recall from point (iii) of Proposition 6.1 that if U is uniformly distributed on $(0, 1)$, then $F_L^{-1}(U)$ has distribution function F_L. In particular, the random variables L, $F_L^{-1}(U)$, and $F_L^{-1}(1 - U)$ all have the same distribution function F_L. Moreover, if F_L is continuous, then $F_L(F_L^{-1}(u)) = u$ by Proposition 6.1(ii). Therefore,

$$\mathrm{E}[L \mid L \geq F_L^{-1}(1 - p)] = \frac{\mathrm{E}[L\, I\{L \geq F_L^{-1}(1 - p)\}]}{\mathrm{P}(L \geq F_L^{-1}(1 - p))}$$

$$= \frac{1}{p} \mathrm{E}[F_L^{-1}(1 - U)\, I\{F_L^{-1}(1 - U) \geq F_L^{-1}(1 - p)\}]$$

$$= \frac{1}{p} \mathrm{E}[F_L^{-1}(1 - U)\, I\{1 - U \geq 1 - p\}]$$

$$= \frac{1}{p} \mathrm{E}[F_L^{-1}(1 - U)\, I\{U \leq p\}]$$

$$= \frac{1}{p} \int_0^p \mathrm{VaR}_u(X)\,du.$$

$\square$

We are now well equipped to prove that ES is a coherent measure of risk.

Proposition 6.6. *ES is a coherent measure of risk.*

Proof. It follows immediately from the definition that ES inherits the properties translation invariance, monotonicity, and positive homogeneity from VaR. It only remains to prove subadditivity. Consider two future portfolio values X_1 and X_2 and write $Y_k = X_k/R_0$ for $k = 1, 2$. We will use representation (6.15) of ES to prove subadditivity, i.e., that $\mathrm{ES}_p(X_1 + X_2) \leq \mathrm{ES}_p(X_1) + \mathrm{ES}_p(X_2)$. For $k = 1, 2$ let c_k^* be a minimizer of

$$-c + \frac{1}{p} \mathrm{E}[(c - Y_k)_+].$$

Note that

$$\mathrm{ES}_p(X_1 + X_2) = \min_c -c + \frac{1}{p} \mathrm{E}[(c - Y_1 - Y_2)_+]$$

$$\leq -(c_1^* + c_2^*) + \frac{1}{p} \mathrm{E}\left[(c_1^* + c_2^* - Y_1 - Y_2)_+\right].$$

The proof is complete if we show the nonnegativity of the difference

$$\mathrm{ES}_p(X_1) + \mathrm{ES}_p(X_2) - \mathrm{ES}_p(X_1 + X_2)$$

$$\geq -c_1^* + \frac{1}{p} \mathrm{E}[(c_1^* - Y_1)_+] - c_2^* + \frac{1}{p} \mathrm{E}[(c_2^* - Y_2)_+]$$

$$+ (c_1^* + c_2^*) - \frac{1}{p} \mathrm{E}[(c_1^* + c_2^* - Y_1 - Y_2)_+]$$

$$= \frac{1}{p} \mathrm{E}[(c_1^* - Y_1)(I\{Y_1 \leq c_1^*\} - I\{Y_1 + Y_2 \leq c_1^* + c_2^*\})]$$

$$+ \frac{1}{p} \mathrm{E}[(c_2^* - Y_2)(I\{Y_2 \leq c_2^*\} - I\{Y_1 + Y_2 \leq c_1^* + c_2^*\})].$$

We claim that the last two terms above are nonnegative. Indeed,

$$\mathrm{E}[(c_1^* - Y_1)(I\{Y_1 \leq c_1^*\} - I\{Y_1 + Y_2 \leq c_1^* + c_2^*\})]$$

$$= \mathrm{E}[(c_1^* - Y_1)(I\{Y_1 \leq c_1^*\} - I\{Y_1 + Y_2 \leq c_1^* + c_2^*\})I\{Y_1 \leq c_1^*\}]$$

$$+ \mathrm{E}[(c_1^* - Y_1)(I\{Y_1 \leq c_1^*\} - I\{Y_1 + Y_2 \leq c_1^* + c_2^*\})I\{Y_1 > c_1^*\}]$$

$$\geq \mathrm{E}[(c_1^* - Y_1)I\{Y_1 \leq c_1^*\}] - \mathrm{E}[(c_1^* - Y_1)I\{Y_1 > c_1^*\}]$$

$$\geq 0,$$

which shows the nonnegativity of the first term. An identical argument shows that the second term is nonnegative too. The proof is complete.   □

Next we continue the sequence of examples on defaultable bonds and CDSs. Here the risk measure VaR is replaced by ES, and this changes the portfolio selection problem substantially.

Example 6.11 (Credit default swap III). Consider the investor and the investment opportunities in Example 6.9. Here the risk constraint $VaR_{0.05}(V_1 - 100) \leq 10$ is replaced by $ES_{0.05}(V_1 - 100) \leq 10$.

Recall that $VaR_p(V_1 - 100)$ was computed in Example 6.9:

$$VaR_p(V_1 - 100) = 100 - \begin{cases} 100\max(w_1 c_1^{-1}, w_2 c_2^{-1}) & \text{if } p \in [0.98, 1], \\ 100 w_1 c_1^{-1} & \text{if } p \in [0.02, 0.98), \\ 100\min(w_1 c_1^{-1}, w_2 c_2^{-1}) & \text{if } p \in [0, 0.02). \end{cases}$$

Then $ES_{0.05}(V_1 - 100)$ can be computed as

$$ES_{0.05}(V_1 - 100) = \frac{1}{0.05}\int_0^{0.05} VaR_p(V_1 - 100)dp$$

$$= \begin{cases} 100 - 100 w_1 c_1^{-1} & \text{if } w_1 c_1^{-1} < w_2 c_2^{-1}, \\ 100 - 100\frac{3}{5}w_1 c_1^{-1} - 100\frac{2}{5}w_2 c_2^{-1} & \text{if } w_1 c_1^{-1} \geq w_2 c_2^{-1}. \end{cases}$$

Recall that $c_1 = 97$ and $c_2 = 4$. With $w_2 = 100 - w_1$ we find that $w_1 c_1^{-1} < w_2 c_2^{-1}$ is equivalent to $w_1 < 96.0396$. We want to take w_1 as large as possible and therefore consider the case $w_1 \geq 96.0396$. In this case, $ES_{0.05}(V_1 - 100) \leq 10$, together with $w_2 = 100 - w_1$, is equivalent to $w_1 \leq 97$. Since a dollar invested in the bond gives a much better expected return than a dollar invested in the CDS, the investor wants to maximize w_1 subject to the constraints. Therefore, the solution to the optimization problem with the ES constraint is $(w_1, w_2) = (97, 3)$. Since ES takes into account the entire tail, there is no way to hide the default risk in the tail. This is reflected in the optimal portfolio.

Example 6.12 (Credit default swap IV). Consider an investor who has $100 and may invest the capital in long positions in 100 bonds and CDSs that are identical to those in Example 6.10. It is assumed that the corresponding indicator variables I_k (I_k takes the value 1 if the kth bond issuer defaults) are independent. The value of the investor's portfolio at the maturity of the bonds is

$$V_1 = \sum_{k=1}^{100} \frac{100}{97} w_k (1 - I_k) + \sum_{k=1}^{100} \frac{100}{4} w_{100+k} I_k,$$

where $w_1, \ldots, w_{100}$ is the capital invested in the bonds and $w_{101}, \ldots, w_{200}$ is the capital invested in the CDSs. The investor wants to maximize the expected value

$$E[V_1] = \sum_{k=1}^{100} \frac{98}{97} w_k + \sum_{k=1}^{100} \frac{1}{2} w_{100+k},$$

from which it is seen that the investor wants to invest as much as possible in the bonds. The risk constraint is given by $\text{ES}_{0.05}(V_1 - 100) \leq 10$. In Example 6.11, we saw that with only one bond and one CDS the optimal solution was $(w_1, w_2) = (97, 3)$. Here it seems plausible that a diversified position in the bonds leads to lower risk and therefore that it will be possible to invest less capital in the CDSs with the low expected returns. We now verify that this is indeed the case. Just as in Example 6.10 we have

$$\text{VaR}_p(V_1 - 100) = 100 - \frac{100^2}{97} + \frac{100}{97} F_Z^{-1}(1 - p),$$

where $Z = \sum_{k=1}^{100} I_k$ is $\text{Bin}(100, 0.02)$-distributed. This gives

$$\text{ES}_p(V_1 - 100) = 100 - \frac{100^2}{97} + \frac{100}{97} \left(\frac{1}{0.05} \int_0^{0.05} F_Z^{-1}(1 - p)dp \right),$$

where

$$\frac{1}{0.05} \int_0^{0.05} F_Z^{-1}(1 - p)dp = 20\Big((\text{P}(Z \leq 5) - 0.95)5$$

$$+ \sum_{k=6}^{100} k(\text{P}(Z \leq k) - \text{P}(Z \leq k - 1))\Big)$$

$$\approx 5.41416,$$

and therefore $\text{ES}_{0.05}(V_1 - 100) \approx 2.488825 < 10$. We conclude that the investor may invest the entire capital in the bonds without violating the risk constraint. Thus, an optimal portfolio is $w_1 = \cdots = w_{100} = 1$, $w_{101} = \cdots = w_{200} = 0$.

Next we study some standard models for log returns of asset prices where ES can be explicitly computed.

Example 6.13 (Normal and Student's t distribution). Consider a 1-day investment in a risky asset. Suppose the influence of interest rates for such a short time period can be neglected. Let $X = V_1 - V_0 = \mu + \sigma Z$, where Z is a standard normally distributed random variable, and let Φ and ϕ denote the distribution and density function of Z, respectively. Then $\text{VaR}_p(X) = -\mu + \sigma \Phi^{-1}(1 - p)$ and

$$\text{ES}_p(X) = -\mu + \frac{\sigma}{p} \int_{1-p}^1 \Phi^{-1}(u)du$$

$$= \{\text{set } l = \Phi^{-1}(u)\}$$

$$= -\mu + \frac{\sigma}{p} \int_{\Phi^{-1}(1-p)}^\infty l\phi(l)dl$$

$$= -\mu + \frac{\sigma}{p} \int_{\Phi^{-1}(1-p)}^{\infty} l \frac{1}{\sqrt{2\pi}} e^{-l^2/2} dl$$

$$= -\mu + \frac{\sigma}{p} \left[-\frac{1}{\sqrt{2\pi}} e^{-l^2/2} \right]_{\Phi^{-1}(1-p)}^{\infty}$$

$$= \mu + \sigma \frac{\phi(\Phi^{-1}(1-p))}{p}.$$

Now let Z have a standard Student's t distribution with $\nu > 0$ degrees of freedom. Then Z has a density

$$g_\nu(x) = C \left(1 + \frac{x^2}{\nu} \right)^{-(\nu+1)/2}, \quad \text{where } C = \frac{\Gamma((\nu+1)/2)}{\sqrt{\nu\pi}\,\Gamma(\nu/2)}.$$

If t_ν is the distribution function of Z, then $\text{VaR}_p(X) = -\mu + \sigma t_\nu^{-1}(1-p)$ and, if $\nu > 1$, then

$$\text{ES}_p(X) = -\mu + \frac{\sigma}{p} \int_{t_\nu^{-1}(1-p)}^{\infty} l g_\nu(l) dl$$

$$= -\mu + \frac{\sigma}{p} \left[\frac{C\nu/2}{-(\nu+1)/2+1} \left(1 + \frac{l^2}{\nu} \right)^{-(\nu+1)/2+1} \right]_{t_\nu^{-1}(1-p)}^{\infty}$$

$$= -\mu + \sigma \frac{g_\nu(t_\nu^{-1}(1-p))}{p} \left(\frac{\nu + (t_\nu^{-1}(p))^2}{\nu-1} \right).$$

Example 6.14 (Normal and Student's t: a comparison). The normal distribution and the Student's t distribution are simple and popular distributions for modeling log returns of asset prices. An important difference between the two is that Student's t distributions have heavier tails, i.e., they place more mass far away from the mean. This can be observed directly from the density function. The standard Student's t distribution with ν degrees of freedom has a density that decays roughly as $|x|^{-\nu}$ for large $|x|$ (called polynomial decay), whereas the standard normal density decays much faster, as $e^{-x^2/2}$. This implication of heavy tails is that there is a higher probability of extreme outcomes. Let us compare the risk measures VaR and ES for the two distributions.

First we compare $\text{VaR}_p(X)$ and $\text{ES}_p(X)$ as a function of p (left plot in Fig. 6.3). The plot shows the ratio $\text{ES}_p(X)/\text{VaR}_p(X)$ as a function of p for the standard normal distribution (lower graph) and the standard Student's t distribution with 3, 2, and 1.1 degrees of freedom (second lowest to upper graph). For the normal distribution the ratio is slightly above one, indicating that for small values of p most of the remaining probability mass in the tail to the left of $\Phi^{-1}(p)$ is concentrated very close to $\Phi^{-1}(p)$. Note that for heavier tails, i.e., smaller degree of freedom

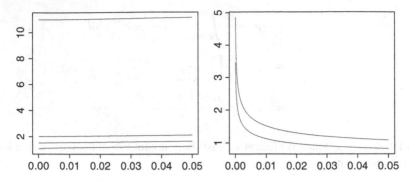

Fig. 6.3 *Left plot*: graphs of $\mathrm{ES}_p(X)/\mathrm{VaR}_p(X)$ as a function of p for X standard normal distribution (*lowest graph*) and Student's t-distribution for $\nu = 3, 2, 1.1$. *Right plot*: graphs of $\mathrm{VaR}_p(X)/\mathrm{VaR}_p(Y)$ (*lower graph*) and $\mathrm{ES}_p(X)/\mathrm{ES}_p(Y)$ (*upper graph*) as functions of p, where X is t-distributed with $\nu = 3$ and variance 1, and Y is standard normally distributed

parameters ν, the ratio is higher, indicating that the probability mass to the left of $t_\nu^{-1}(p)$ is spread out to the left of this value and spread out more the smaller the value of ν is.

In the right plot in Fig. 6.3, we compare VaR for a t_3-distribution and a normal distribution with unit variance, and similarly for ES by plotting the ratios $\mathrm{VaR}_p(X)/\mathrm{VaR}_p(Y)$ and $\mathrm{ES}_p(X)/\mathrm{ES}_p(Y)$ as functions of p, where X is t-distributed with $\nu = 3$ and variance 1, and Y is standard normally distributed. If Z has a standard t_ν-distribution, then its variance is $\nu(\nu - 2)^{-1}$, so in this example, $X = Z/\sqrt{3}$, which implies that X and Y both have unit variance. We observe that for small p the ratios are greater than one. This is a result of the heavier tails of the t_3-distribution.

Example 6.15 (Lognormal distribution). Consider the current and future values V_0 and V_1 of an asset. By borrowing the amount V_0 to finance the long position in the asset, the future net value of the position is $X = V_1 - V_0 R_0$, where $V_0 R_0$ is the future value of the debt. If $Z_1 = \log(V_1/V_0)$ is the log return of the asset, then $X = V_0(\exp\{Z_1\} - R_0)$.

We will analyze $\mathrm{ES}_p(X)$ under the assumption that Z_1 has either a normal distribution or a Student's t distribution. Applying Proposition 6.3, with $g(z) = V_0(e^z - R_0)$, and Proposition 6.4 gives

$$\mathrm{VaR}_u(X) = F^{-1}_{-g(Z_1)/R_0}(1 - u) = -g\left(F_{Z_1}^{-1}(u)\right) = V_0\left(1 - \frac{1}{R_0}e^{F_{Z_1}^{-1}(u)}\right).$$

If Z_1 is $\mathrm{N}(\mu, \sigma^2)$-distributed, then $F_{Z_1}^{-1}(u) = \mu + \sigma\Phi^{-1}(u)$

$$\mathrm{ES}_p(X) = \frac{1}{p}\int_0^p \mathrm{VaR}_u(X)du = V_0\left(1 - \frac{1}{pR_0}\int_0^p e^{\mu + \sigma\Phi^{-1}(u)}du\right).$$

With $q(u) = \Phi^{-1}(u)$ we have $dq(u)/du = 1/\phi(\Phi^{-1}(u))$, and the integral to the right above can be written as

$$
\int_0^p e^{\mu + \sigma \Phi^{-1}(u)} du = \int_{-\infty}^{\Phi^{-1}(p)} \frac{1}{\sqrt{2\pi}} e^{\mu + \sigma q - q^2/2} dq
$$

$$
= e^{\mu + \sigma^2/2} \int_{-\infty}^{\Phi^{-1}(p)} \frac{1}{\sqrt{2\pi}} e^{-(q-\sigma)^2/2} dq
$$

$$
= \Phi(\Phi^{-1}(p) - \sigma) e^{\mu + \sigma^2/2}.
$$

We have found that if Z_1 is $N(\mu, \sigma^2)$-distributed, then

$$
ES_p(X) = V_0 \left(1 - \frac{\Phi(\Phi^{-1}(p) - \sigma) e^{\mu + \sigma^2/2}}{p R_0} \right).
$$

Similarly, if Z_1 is distributed as $\mu + \sigma Y$, where Y has a standard Student's t distribution with ν degrees of freedom, then $F_{Z_1}^{-1}(u) = \mu + \sigma t_\nu^{-1}(u)$ and

$$
ES_p(X) = V_0 \left(1 - \frac{1}{p R_0} \int_0^p e^{\mu + \sigma t_\nu^{-1}(u)} du \right).
$$

The integral expression can be evaluated by numerical integration.

6.4 Risk Measures Based on Utility Functions

Consider a concave and strictly increasing function u, that is, a utility function. Suppose that we consider a portfolio with value X at time 1 acceptable if it satisfies $E[u(X)] \geq u(C)$ for a predetermined number C, i.e., if its certainty equivalent is at least C. Let

$$
\rho_u(X) = \min\{m : E[u(mR_0 + X)] \geq u(C)\}, \tag{6.17}
$$

and note that $\rho_u(X)$ is the smallest amount of money that needs to be added and invested in a risk-free asset to make the corresponding position acceptable. In fact, $\rho_u(X)$ is the unique number m satisfying $E[u(mR_0 + X)] = u(C)$. Let us prove this claim. Since u is strictly increasing, the function $m \mapsto E[u(mR_0 + X)]$ is also strictly increasing, so there is at most one such number m. Since $m \mapsto u(mR_0 + x)$ is concave, then $m \mapsto E[u(mR_0 + X)]$ is also concave and, therefore, also continuous. Therefore, there is at least one such number m.

Proposition 6.7. *The risk measure ρ_u in (6.17) is a convex measure of risk.*

Before proving the proposition we remark that ρ_u is in general not a coherent measure of risk. If $C = 0$, then the normalization property holds, but ρ_u will typically not be positively homogeneous.

Proof. We have

$$\rho_u(X + yR_0) = \min\{m : \mathrm{E}[u((m + y)R_0 + X)] \geq u(C)\}$$
$$= \min\{k : \mathrm{E}[u(kR_0 + X)] \geq u(C)\} - y$$
$$= \rho_u(X) - y,$$

which shows that ρ_u is translation invariant. Since u is increasing, $X_2 \leq X_1$ implies that $\mathrm{E}[u(mR_0 + X_2)] \leq \mathrm{E}[u(mR_0 + X_1)]$, and therefore

$$\rho_u(X_2) = \min\{m : \mathrm{E}[u(mR_0 + X_2)] \geq u(C)\}$$
$$\geq \min\{m : \mathrm{E}[u(mR_0 + X_1)] \geq u(C)\}$$
$$= \rho_u(X_1),$$

which proves the monotonicity of ρ_u. By the definition of ρ_u, it holds that

$$\rho_u(\lambda X_1 + (1 - \lambda)X_2) = \min\{m : \mathrm{E}[u(mR_0 + \lambda X_1 + (1 - \lambda)X_2)] \geq u(C)]\}.$$

Therefore, the convexity of ρ_u follows if we show that $m_0 = \lambda\rho_u(X_1) + (1 - \lambda)$ $\rho_u(X_2)$ satisfies $\mathrm{E}[u(m_0 R_0 + \lambda X_1 + (1 - \lambda)X_2)] \geq u(C)$. Indeed,

$$\mathrm{E}[u([\lambda\rho_u(X_1) + (1 - \lambda)\rho_u(X_2)]R_0 + \lambda X_1 + (1 - \lambda)X_2)]$$
$$\geq \lambda\,\mathrm{E}[u(\rho_u(X_1)R_0 + X_1)] + (1 - \lambda)\,\mathrm{E}[u(\rho_u(X_2)R_0 + X_2)]$$
$$= \lambda u(C) + (1 - \lambda)u(C)$$
$$= u(C),$$

where the first inequality holds because u is concave and where the second to last equality holds because $\mathrm{E}[u(\rho_u(X_k)R_0 + X_k)] = u(C)$ by definition of ρ_u. The proof is complete.   □

6.5 Spectral Risk Measures

Consider a random variable X representing the value at time 1 of a portfolio. Let R_0 be the return of a zero-coupon bond maturing at time 1, and let F_{X/R_0} be the distribution function of X/R_0, i.e., the discounted future portfolio value. A natural set of risk measures consists of risk measures that can be written as

-1 times a weighted average of the quantile values $F_{X/R_0}^{-1}(p)$. We have seen that $\text{VaR}_p(X) = -F_{X/R_0}^{-1}(p)$ for those x where $F_{X/R_0}(x)$ is neither flat nor has a jump and that

$$\text{ES}_p(X) = -\frac{1}{p}\int_0^p F_{X/R_0}^{-1}(u)du.$$

In particular, ES_p puts equal weight on all the quantiles $F_{X/R_0}^{-1}(u)$ for $u < p$. It is not at all evident that this is the most natural choice. Consider a nonnegative function ϕ on $(0, 1)$ that is decreasing and integrates to 1, and define

$$\rho_\phi(X) = -\int_0^1 \phi(u)F_{X/R_0}^{-1}(u)du. \tag{6.18}$$

A risk measure ρ_ϕ with this representation is called a spectral risk measure, and the function ϕ is called the risk aversion function. A tractable property of spectral risk measures is that, like risk measures based on utility functions, all quantile values of the probability distribution of the considered portfolio value can be taken into account—not just those corresponding to the left tail. We see that ES_p is a spectral risk measure with risk aversion function $p^{-1}I_{(0,p)}$. This risk aversion function says that the worst fractions p of quantile values are weighted equally as they enter only through their mean value. In particular, extreme losses are not considered worse (receive higher weights) than less extreme losses. In general, the risk aversion function lets you specify your attitude toward risk. In spirit, it is similar to a utility function. The difference is that the utility function relates how much you value x units of cash over y units of cash, whereas the risk aversion function relates how highly you penalize the quantile at level p over the quantile at level q. Two examples of risk aversion functions are the polynomial and exponential risk aversion functions given by

$$\phi_{\text{pol},\beta}(p) = \frac{1}{\beta}(1 - p)^{\beta-1}, \quad \beta \geq 1,$$

$$\phi_{\text{exp},\gamma}(p) = \frac{\gamma \exp\{-\gamma p\}}{1 - \exp\{-\gamma\}}, \quad \gamma > 0.$$

Note that both functions are decreasing and integrate to 1. For the most part we will in the sequel assume that the risk aversion function ϕ is differentiable. This assumption is made purely for convenience. The results presented below hold also without this assumption.

We begin with two useful representations of a spectral risk measure. The first one shows, using integration by parts, that ρ_ϕ can be viewed as a weighted average of ES. The second representation is similar to representation (6.15) for ES but requires the more general convex optimization from Sect. 2.2.

Proposition 6.8. *If ϕ is differentiable, then ρ_ϕ in (6.18) satisfies*

$$\rho_\phi(X) = -\int_0^1 \frac{d\phi}{du}(u)u\,\mathrm{ES}_u(X)du - \phi(1)\,\mathrm{E}[X/R_0],, \tag{6.19}$$

$$\rho_\phi(X) = \min_f \int_0^1 \frac{d\phi}{du}(u)\{uf(u) - \mathrm{E}[(f(u) - X/R_0)_+]du - \phi(1)\,\mathrm{E}[X/R_0], \tag{6.20}$$

where the minimum is taken over all functions f.

Proof. First observe that

$$\int_0^1 F_{X/R_0}^{-1}(u)du = \mathrm{E}[X/R_0].$$

This follows, for instance, from (6.14) with $p = 1$. Then, upon changing the order of integration in the third equality below, we find that

$$
\begin{aligned}
\rho_\phi(X) &= -\int_0^1 \phi(v) F_{X/R_0}^{-1}(v)dv \\
&= \int_0^1 \left[\int_v^1 \frac{d\phi}{du}(u)du - \phi(1)\right] F_{X/R_0}^{-1}(v)dv \\
&= \int_0^1 \frac{d\phi}{du}(u) \left[\int_0^u F_{X/R_0}^{-1}(v)dv\right] du - \phi(1)\int_0^1 F_{X/R_0}^{-1}(v)dv \\
&= -\int_0^1 \frac{d\phi}{du}(u)u \left[\left(-\frac{1}{u}\right)\int_0^u F_{X/R_0}^{-1}(v)dv\right] du - \phi(1)\,\mathrm{E}[X/R_0] \\
&= -\int_0^1 \frac{d\phi}{du}(u)u\,\mathrm{ES}_u(X)du - \phi(1)\,\mathrm{E}[X/R_0].
\end{aligned}
$$

This proves (6.19). Informally, the second representation (6.20) follows from the representation of $\mathrm{ES}_u(X)$ in (6.15). Write

$$\mathrm{ES}_u(X) = \min_{f(u)} -f(u) + \frac{1}{u}\,\mathrm{E}[(f(u) - X/R_0)_+],$$

insert this expression into (6.19), and finally move the coordinatewise minimum inside the integral out of the integral to get (6.20). Now we consider a more formal argument, in the context of Sect. 2.2. Let

$$F(f) = \int_0^1 \frac{d\phi}{du}(u)\{uf(u) - \mathrm{E}[(f(u) - X/R_0)_+]du - \phi(1)\,\mathrm{E}[X/R_0].$$

Since there are no constraints on f, here we have

$$H(f, g) = \int_0^1 \frac{d\phi}{du}(u)\{u - F_{X/R_0}(f(u))\}(g(u) - f(u))du.$$

If $H(f, g) = 0$ for each g, then f must satisfy $F_{X/R_0}(f(u)) = u$ for each u. However, for such a function f it follows, as in the proof of (6.15), that

$$-f(u) + \frac{1}{u}E[(f(u) - X/R_0)_+] = \mathrm{ES}_u(X),$$

and therefore it follows from (6.19) that the minimum of $F(f)$ is given by

$$-\int_0^1 \frac{d\phi}{du}(u)u\,\mathrm{ES}_u(X)du - \phi(1)\,\mathrm{E}[X/R_0] = \rho_\phi(X).$$

The proof is complete. $\qquad\qquad\qquad\qquad\qquad\qquad\qquad\qquad\qquad\qquad\square$

From representation (6.19) we observe that many properties of spectral risk measures follow from properties of ES. In particular, spectral risk measures are coherent.

Proposition 6.9. *The spectral risk measure ρ_ϕ in (6.18) is a coherent measure of risk.*

Proof. Since ϕ is nonnegative and integrates to 1, the properties of the quantile function imply that ρ_ϕ is translation invariant, monotone, and positively homogeneous. To prove subadditivity, we make the additional assumption that the risk aversion function ϕ is differentiable. Then the subadditivity of ρ_ϕ follows from the subadditivity of ES. Indeed, for two future portfolio values X_1 and X_2 we have

$$\rho_\phi(X_1 + X_2) = -\int_0^1 \frac{d\phi}{du}(u)u\,\mathrm{ES}_u(X_1 + X_2)du - \phi(1)\,\mathrm{E}[(X_1 + X_2)/R_0]$$

$$\le -\int_0^1 \frac{d\phi}{du}(u)u\Big(\mathrm{ES}_u(X_1) + \mathrm{ES}_u(X_2)\Big)du - \phi(1)\,\mathrm{E}[(X_1 + X_2)/R_0]$$

$$= \rho_\phi(X_1) + \rho_\phi(X_2). \qquad\qquad\qquad\qquad\qquad\qquad\qquad\square$$

6.6 Notes and Comments

An extensive account of VaR for financial risk management is given in the book [24] by Philippe Jorion. The concept of coherent measures of risk was proposed by Philippe Artzner, Freddy Delbaen, Jean-Marc Eber and David Heath [4]. For an extensive account of convex and coherent measures of risk see the book [17] by Hans Föllmer and Alexander Schied. The coherence of ES was proved by Carlo Acerbi and Dirk Tasche in [2]. An introduction to and properties of spectral risk measures can be found in Acerbi's work [1]. Portfolio optimization with ES constraints was considered by Tyrrell Rockafellar and Stan Uryasev in [38] and extended to so-called generalized deviations, which are closely related to spectral risk measures, in works by Rockafellar, Uryasev, and Michael Zabarankin [39–41].

6.7 Exercises

In the exercises below, it is assumed, wherever applicable, that you can take positions corresponding to fractions of assets.

Exercise 6.1 (Convexity and subadditivity). Show that a positively homogeneous risk measure is convex if and only if it is subadditive.

Exercise 6.2 (Stop-loss reinsurance). Suppose that the total claim amount S in 1 year for an insurance company has a standard exponential distribution. The insurance company can buy so-called stop-loss reinsurance so that a claim amount exceeding $F_S^{-1}(0.95)$ is paid by the reinsurer. In this case, the insurance company has to pay $L = \min(S, F_S^{-1}(0.95)) + p$, where p is the premium paid for the stop-loss reinsurance. Determine the premium p for which $F_S^{-1}(0.99) = F_L^{-1}(0.99)$.

Exercise 6.3 (Quantile bound). Let Z denote the daily log return of an asset. Empirical studies suggest that Z has zero mean, standard deviation 0.01, and a symmetric density function. Someone claims that $F_Z^{-1}(0.99) = 0.1$. Use Chebyshev's inequality $P(|Z - E[Z]| > x) \leq x^{-2} \operatorname{Var}(Z)$, for $x > 0$, to show that this claim is false.

Exercise 6.4 (Tail conditional median). The tail conditional median $\mathrm{TCM}_p(X) = \mathrm{median}[L \mid L \geq \mathrm{VaR}_p(X)]$, where $L = -X/R_0$, has been proposed as a more robust alternative to $\mathrm{ES}_p(X)$ since $\mathrm{TCM}_p(X)$ is not as sensitive as $\mathrm{ES}_p(X)$ to the behavior of the left tail of the distribution of X.

Let Y have a standard Student's t distribution with ν degrees of freedom, and set $X = e^{0.01Y} - 1$. Compute and plot the graphs of $\mathrm{ES}_{0.01}(X)$ and $\mathrm{TCM}_{0.01}(X)$ as functions of $\nu \in [1, 15]$.

Exercise 6.5 (Production planning). Consider a company that has the option to start production of a volume $t \geq 0$ of a certain good during the next year. The company has capital of \$10,000 to use for the production. Any capital not spent on production is deposited in a bank account that does not pay interest. The cost for producing a volume $t > 0$ of the good is t thousand dollars plus a startup cost of \$5,000. The income from selling a volume t of the good is $5t$ thousand dollars. The unknown demand for the good (the maximum volume the company can sell) is modeled as a random variable with distribution function $1 - x^{-2}$, $x \geq 1$.

(a) How much should the company produce to maximize $E[V_1(t)]$, where $V_1(t)$ is the income from sales plus money in the bank account at the end of next year when producing volume t of the good?

(b) Compute $\text{VaR}_p(V_1(t) - 10{,}000)$ with $V_1(t)$ as in (a) and where t is the maximizer of $E[V_1(t)]$.

Exercise 6.6 (Risky bonds). Consider a market with an asset with a risk-free 1-year return of $R_0 = 1.05$. There are also two defaultable bonds on the market whose issuers can be assumed to default independently of each other. Both bonds have maturity in 1 year and a face value of $\$100{,}000$, which is paid in the case of no default before the end of the year. For each bond a default event makes the bond worthless. Both bonds have the same price of $100{,}000(1 - q)/R_0$ dollars today, where $q = 0.025$ can be interpreted as the market's implied default probability. You believe that the market is overestimating the default probability, which you believe is $p = 0.024$. You have $V_0 = \$1$ million to invest in the risky bonds and in the risk-free asset.

(a) Determine the portfolio that maximizes your expected return given that the standard deviation of your portfolio does not exceed $\$25{,}000$. You are not allowed to take short positions in the risky bonds or in the risk-free asset.

(b) Determine the expected value and the standard deviation of the value at the end of the year of the optimal portfolio in (a).

(c) Compute $\text{VaR}_{0.05}(V_1 - V_0 R_0)$ and $\text{ES}_{0.05}(V_1 - V_0 R_0)$, where V_1 is the value of the optimal portfolio in (a) at the end of the year.

(d) Shortly after you buy the portfolio, a financial crisis breaks out and you realize that one of the issuers is in serious financial distress. You update the default probability to 0.91 for one of the bonds. The other bond is unaffected by the crisis, and its default probability remains 0.024. You can assume that the default events are independent. Compute $\text{VaR}_{0.05}(V_1 - V_0 R_0)$ and $\text{ES}_{0.05}(V_1 - V_0 R_0)$, where V_1 is the value of the optimal portfolio in (a) at the end of the year.

Exercise 6.7 (Leverage and margin calls). Consider the portfolio in Exercise 3.3(c).

(a) Compute $\text{VaR}_p(V_2)$ for $p \leq 0.05$, where V_2 is the value in 2 months of the portfolio in Exercise 3.3 (c) that maximizes the expected payoff in 2 months.

(b) Compute $\text{ES}_p(V_2)$ for $p \leq 0.05$, where V_2 is as in (a).

Exercise 6.8 (Risk and diversification). Consider the setup in Example 6.10 with 100 identical bonds whose default events are independent. Consider an investor with initial capital of $V_0 = \$1$ million who invests this capital in long positions of equal size in $n \leq 100$ of the bonds. The value of the bond portfolio at maturity of the bonds is denoted by $V_1(n)$.

(a) Plot $\text{VaR}_{0.05}(V_1(n) - V_0)$ as a function of n, where n ranges from 1 to 100.

(b) Plot $\text{ES}_{0.05}(V_1(n) - V_0)$ as a function of n, where n ranges from 1 to 100.

Project 6 (Collar options). A private investor owns a large quantity of shares of a single stock and is worried about the position being too risky in the near future. A bank offers the investor the opportunity to implement a collar option as protection against falling share prices. The collar option considered here is a long position in a European put option on the future share price with a strike price below the current share price and a short position of the same size in a European call option with strike price above the current share price with the same time to maturity as for the put option.

Suppose that the investor holds 1,000 shares and the current share price is $100. Suppose further that the strike prices of the put and call options are $95 and $105, respectively, and both options expire in 2 months. Suppose that the stock pays no dividends within the next 2 months, that all interest rates are zero, and that the put and call prices correspond to implied volatilities of 0.25 and 0.2, respectively, per year if the Black–Scholes formulas for European put and call options are used.

Suppose that the log return of the share price from today until half a month from today is $0.04X$, where X has a standard Student's t distribution with 4 degrees of freedom, and that the implied volatilities in half a month from today are the same as today.

(a) The investor decides to take a collar option position corresponding to 1,000 puts and calls. The investor's collar option position is financed by a zero-interest-rate loan if the initial value is positive. If the value is negative, then the investor receives cash that is deposited in an account that pays no interest. Express V_1, the value in half a month from today of the shares and the collar option position minus the current value of the collar option position, as a function of the log return $0.04X$.

(b) Consider the same situation as in (a) and compute $\text{VaR}_{0.05}(V_1 - V_0)$, where V_0 is the current value of the shares. Compare the result to the corresponding result in a situation where the investor decides not to take a collar option position (only shares).

(c) The investor decides to take a collar option position corresponding to $h \in [0, 1,000]$ puts and calls. Vary h and study the effect on the density function of V_1, where V_1 is the value in half a month from today of the shares and the collar option position minus the current value of the collar option position.

Part II
Methods

Chapter 7
Empirical Methods

In this chapter we consider a modeling approach that uses a set of historical data, such as bond prices, share prices, claim sizes, or exchange rates, to model the value at a future time $T > 0$ of portfolios whose values depend on a given set of assets and possibly also liabilities. Here we want the data to speak for themselves in the sense that the model for the future values should only be based on information available in the given historical data samples. The assumption we make is therefore that the information in the samples is representative of future values and that no additional probability beliefs of the modeler are relevant.

Historical share prices $S_{-n}, \dots, S_0$ of a stock over the last $n + 1$ time periods are not necessarily good representatives of possible values for the future share price S_1. But the sample of historical returns $R_{-k} = S_{-k+1}/S_{-k}$, for $k = n - 1, \dots, 1$, may be assumed to be a good representative of possible values for the future return $R_1 = S_1/S_0$ over the next time period. Similarly, the historical zero rates $r_{-n}, \dots, r_0$, corresponding to a given time to maturity, may be transformed into zero rate changes $r_{-k+1} - r_{-k}$, for $k = n - 1, \dots, 1$, that can be viewed as good representatives of the possible zero rate change $r_1 - r_0$ over the next time period. If we believe in this approach, then appropriate transformations of the historical samples produce samples of the random values, e.g., returns, that determine the future portfolio values. If the generated sample of returns or value changes can be viewed as samples from independent and identically distributed random variables, then standard statistical techniques can be used to investigate the probability distribution of future portfolio values, expressed as known functions of future returns or value changes.

In this chapter, we will investigate this approach to modeling the future. This is a subjective approach just as any other approach (such as assigning a parametric probability distribution to the future portfolio value). However, it is fully nonparametric and is a reasonable approach if we believe that the mechanism that produced the returns in the past is the same as the mechanism that will produce returns in the future, even if the mechanism is unknown to us.

H. Hult et al., *Risk and Portfolio Analysis: Principles and Methods*, Springer Series
in Operations Research and Financial Engineering, DOI 10.1007/978-1-4614-4103-8_7,
© Springer Science+Business Media New York 2012

The first topic of this chapter is how to turn historical prices into a sample from the distribution of the future portfolio value under the assumption that returns over the next time period will be similar to those returns. This material is presented in Sect. 7.1. In Sect. 7.2 we consider the empirical distribution, which is the probability distribution derived from a data sample. The quantile function of the empirical distribution is the empirical quantile studied in Sect. 7.3. The main objective in these two sections is to investigate how the accuracy of empirical probabilities and quantiles, relative to the true unknown quantities they are measuring, varies with the sample size and characteristics of the unknown distribution from which the sample of observations is generated. Empirical distributions and quantiles provide natural estimators of value-at-risk (VaR) and expected shortfall (ES), which are presented in Sect. 7.4. Point estimates of risk measures are not particularly useful unless they are accompanied by estimates of their accuracy. Therefore, we analyze in detail methods for constructing confidence intervals for the quantities estimated by empirical estimators. In Sect. 7.5, we present a method for constructing exact confidence intervals for quantiles and a method for constructing approximative confidence intervals using the nonparametric bootstrap procedure. The latter method is further studied in Sect. 7.6, which deals with the uncertainty in estimates for solvency capital requirements for a nonlife insurer.

7.1 Sample Preparation

Denote the current time by 0, and consider a future time that we call time 1. Let V_1 be the random value of some portfolio at time 1 that we can express as a function of the vector $\mathbf{S}_1$ of asset prices at time 1. For the sake of clarity of presentation we take $\mathbf{S}_1$ to be the share prices of some stocks. It is assumed that we have access to a sample $\{\mathbf{S}_{-n}, \mathbf{S}_{-n+1}, \ldots, \mathbf{S}_0\}$ of vectors of historical prices from the n previous equally spaced points in time (e.g., days, weeks) and from the current time. It is clear that the sample points may be strongly dependent (the share price on any given day is strongly dependent on the previous day's price). Moreover, it is likely that the asset prices $\mathbf{S}_{-k}$, from k time periods ago, are quite different from what can be anticipated for $\mathbf{S}_1$, at least if k is large. The sample of historical asset prices may, however, be transformed into a sample of vectors of returns $\mathbf{R}_{-n+1}, \ldots, \mathbf{R}_0$, where

$$\mathbf{R}_{-k} = \left(R_{-k}^1, \ldots, R_{-k}^d\right)^{\mathrm{T}} \text{ with } R_{-k}^l = S_{-k}^l / S_{-k-1}^l$$

for $k = 0, \ldots, n-1$, and $l = 1, \ldots, d$. It is often reasonable to assume, supported by statistical analysis, that the points of the sample $\{\mathbf{R}_{-n+1}, \ldots, \mathbf{R}_0\}$ are weakly dependent and close to identically distributed and have distributional characteristics that are representative also for $\mathbf{R}_1$, the vector of percentage returns for the next time period. The portfolio value at time 1 is $V_1 = f(\mathbf{R}_1)$ for some function f that depends on information available at time 0 such as the current asset prices $\mathbf{S}_0$. Then the sample $\{\mathbf{R}_{-n+1}, \ldots, \mathbf{R}_0\}$ of return vectors can be transformed into the sample $\{f(\mathbf{R}_{-n+1}), \ldots, f(\mathbf{R}_0)\}$ from the probability distribution of $V_1 = f(\mathbf{R}_1)$. If the vectors in the former sample are approximately independent copies of $\mathbf{R}_1$, then the vectors in the latter sample are approximately independent copies of V_1.

The transformation of historical prices into historical returns is not essential for the sample preparation scheme to work. Returns could, for instance, be replaced by something else, such as price differences. The essential point is that the original sample $\{\mathbf{S}_{-n}, \mathbf{S}_{-n+1}, \ldots, \mathbf{S}_0\}$ is transformed into a sample $\{\mathbf{Z}_{-n+1}, \ldots, \mathbf{Z}_0\}$, which in turn could be transformed into a sample $\{f(\mathbf{Z}_{-n+1}), \ldots, f(\mathbf{Z}_0)\}$ whose points may be viewed as independent copies of the future portfolio value V_1. This situation is the desired starting point for statistical analysis. From a sample of independent and identically distributed random variables drawn from the unknown probability distribution of V_1, statistical methods can be applied to investigate the probability distribution of the future portfolio value V_1.

The approach presented for generating a sample from the probability distribution of the future portfolio value is based on the assumption that changes in values in the past contain relevant information for assessing the probability distribution of changes in value from now until the future time we are considering. Determining the extent to which this assumption is reasonable requires some serious thinking. Big changes in the legal or political environment, monetary policies of governments or central banks, or other events may make it hard to justify this assumption.

Throughout the book we write $\{\mathbf{S}_{-n}, \mathbf{S}_{-n+1}, \ldots, \mathbf{S}_0\}$ for the random vectors of historical prices (and similarly for the sample of returns) and $\mathbf{s}_{-n}, \mathbf{s}_{-n+1}, \ldots, \mathbf{s}_0$ for the actual observations of the historical prices. The following example illustrates the sample preparation approach.

Example 7.1 (Sample preparation). Consider a portfolio consisting of long positions in two different assets, one unit of the first asset and two units of the second asset. The daily prices per unit of the two assets over the last 20 days are given by S_t^1 and S_t^2 for $t = -20, \ldots, 0$. Suppose the corresponding pairs of returns

$$\mathbf{R}_t = (R_t^1, R_t^2) = (S_t^1/S_{t-1}^1, S_t^2/S_{t-1}^2), \quad t = -19, \ldots, 0,$$

are independent and identically distributed. If V_1 is the value of the portfolio at time 1, then

$$V_1 = S_1^1 + 2S_1^2 = S_0^1 \frac{S_1^1}{S_0^1} + 2S_0^2 \frac{S_1^2}{S_0^2} = S_0^1 R_1^1 + 2S_0^2 R_1^2 = f(\mathbf{R}_1),$$

where $f(x, y) = S_0^1 x + 2S_0^2 y$. The random variables $\{f(\mathbf{R}_{-20+1}), \ldots, f(\mathbf{R}_0)\}$ can be viewed as a sample of independent copies of V_1.

It may happen that we have access to daily historical prices and want to use the data to investigate the probability distribution of the value of a portfolio a week (month or year) from now. Then there are different options available. Consider the sample $\{\mathbf{S}_{-n}, \mathbf{S}_{-n+1}, \dots, \mathbf{S}_0\}$ of vectors of historical prices and suppose we want to investigate the distribution of V_T, where $T > 1$. We assume that the original sample can be transformed into a sample $\{\mathbf{R}_{-n+1}, \dots, \mathbf{R}_0\}$ of vectors of returns such that the vectors are approximately independent copies of $\mathbf{R}_1$ and that $V_T = f(\mathbf{R}_1 \cdots \mathbf{R}_T)$, where $\mathbf{R}_1 \cdots \mathbf{R}_T$ is interpreted as componentwise multiplication and $\mathbf{R}_1 \cdots \mathbf{R}_T$ is the vector of returns over the next period of length T.

Example 7.2 (Thinning of the sample). One way of obtaining a sample of vectors of returns over time periods of length T would be to start with the sample $\{\mathbf{S}_{-T[n/T]}, \dots, \mathbf{S}_{-T}, \mathbf{S}_0\}$ and set

$$\mathbf{R}_{-k}^{(T)} = ((R^{(T)})_{-k}^1, \dots, (R^{(T)})_{-k}^d)^{\mathsf{T}} \text{ with } (R^{(T)})_{-k}^l = S_{-Tk}^l / S_{-T(k+1)}^l$$

for $k = 0, \dots, [n/T] - 1$, and $l = 1, \dots, d$. Here $[y]$ denotes the largest integer smaller than or equal to y, i.e., $[y] = \max\{k \in \mathbb{N} : k \le y\}$. The sample $\{\mathbf{R}_{-[n/T]+1}^{(T)}, \dots, \mathbf{R}_0^{(T)}\}$ is a sample of vectors of returns over nonoverlapping time periods of length T. If these return vectors are independent copies of $\mathbf{R}_1 \cdots \mathbf{R}_T$, then $f(\mathbf{R}_{-[n/T]+1}^{(T)}), \dots, f(\mathbf{R}_0^{(T)})$ are independent copies of V_T. The problem with this approach is that much of the possibly relevant information in the original sample $\{\mathbf{S}_{-n}, \mathbf{S}_{-n+1}, \dots, \mathbf{S}_0\}$ is ignored and the sample size is reduced from n to $[n/T]$.

Example 7.3 (Historical simulation). An approach that, unlike the approach in Example 7.2, uses the entire original sample is to draw with replacement T vectors from the sample $\{\mathbf{R}_{-n+1}, \dots, \mathbf{R}_0\}$ and form the componentwise product of these vectors, denoted by $\mathbf{R}_1^{*(T)}$. Repeat the procedure m times to obtain the sample $\{\mathbf{R}_1^{*(T)}, \dots, \mathbf{R}_m^{*(T)}\}$ of fictive return vectors over time periods of length T. If the original return vectors $\mathbf{R}_{-n+1}, \dots, \mathbf{R}_0$ are independent and identically distributed, then the vectors $\mathbf{R}_1^{*(T)}, \dots, \mathbf{R}_m^{*(T)}$ are identically distributed but not independent since some of the random indices may take the same index value, but they are conditionally independent given $\mathbf{R}_{-n+1}, \dots, \mathbf{R}_0$.

The sample $\{f(\mathbf{R}_1^{*(T)}), \dots, f(\mathbf{R}_m^{*(T)})\}$ is a sample of size m, where the sample points are approximately distributed as V_T. This approach to generating a sample from the distribution of V_T is called a historical simulation. On the one hand, all the original sample points are used and the sample size m can be chosen arbitrarily large. On the other hand, the original return vectors appear as factors in more than one of the fictive return vectors $\mathbf{R}_k^{*(T)}$, so there may be substantial redundancy in the constructed sample of return vectors over periods of length T.

7.2 Empirical Distributions

Consider observations $\mathbf{x}_1, \dots, \mathbf{x}_n$ of independent and identically distributed d-dimensional random vectors $\mathbf{X}_1, \dots, \mathbf{X}_n$ with a common unknown distribution function $F(\mathbf{x}) = \mathrm{P}(\mathbf{X} \le \mathbf{x})$, where $\mathbf{X}$ is an independent copy of $\mathbf{X}_k$ and $\mathbf{X} \le \mathbf{x}$ is interpreted as an inequality for all the components; $\mathbf{X} \leq \mathbf{x}$ if and only if $X_j \leq x_j$ for $j = 1, \ldots, d$. Suppose that we want to compute some quantity $\theta = \theta(F)$ that depends on F, for instance, the mean, the variance, a quantile, or a risk measure. It is impossible to compute θ since F is unknown, but the observations $\mathbf{x}_1, \ldots, \mathbf{x}_n$ allow us to approximate the unknown distribution by that obtained from assigning a probability weight $1/n$ to each of the $\mathbf{x}_k$. That is, approximating the unknown $F(\mathbf{x})$ by the fraction $F_n(\mathbf{x})$ of the $\mathbf{x}_k$ that are smaller than or equal to $\mathbf{x}$,

$$F_n(\mathbf{x}) = \frac{1}{n} \sum_{k=1}^{n} I\{\mathbf{x}_k \leq \mathbf{x}\}.$$

The distribution function F_n is called the empirical distribution function of $\mathbf{x}_1, \ldots, \mathbf{x}_n$. The random counterpart, which is the empirical distribution associated with the random sample $\{\mathbf{X}_1, \ldots, \mathbf{X}_n\}$, is given by

$$F_{n,\mathbf{X}}(\mathbf{x}) = \frac{1}{n} \sum_{k=1}^{n} I\{\mathbf{X}_k \leq \mathbf{x}\}.$$

Note that $F_{n,\mathbf{X}}$ is a random object whose outcome F_n is a distribution function.

The (strong) law of large numbers says that if $Z_1, Z_2, \ldots$ is a sequence of independent copies of a random variable Z for which the expected value $\mathrm{E}[Z]$ exists finitely, then

$$\frac{1}{n} \sum_{k=1}^{n} Z_k \to \mathrm{E}[Z] \quad \text{with probability 1 as } n \to \infty.$$

If we choose $Z_k = I\{\mathbf{X}_k \leq \mathbf{x}\}$, then $\mathrm{E}[Z_k] = \mathrm{P}(\mathbf{X}_k \leq \mathbf{x}) = F(\mathbf{x})$ and the law of large numbers implies that, with probability one, $\lim_{n\to\infty} F_{n,\mathbf{X}}(\mathbf{x}) = F(\mathbf{x})$. In particular, the empirical distribution function $F_{n,\mathbf{X}}$ is a good approximation of the unknown distribution function F as long as the sample size n is sufficiently large. Similarly, if we choose $Z_k = h(\mathbf{X}_k)$, then $\mathrm{E}[Z_k] = \mathrm{E}[h(\mathbf{X}_k)] = \mathrm{E}[h(\mathbf{X})]$ and the law of large numbers implies that, with probability one,

$$\int h(\mathbf{x}) dF_{n,\mathbf{X}}(\mathbf{x}) = \frac{1}{n} \sum_{k=1}^{n} h(\mathbf{X}_k) \to \mathrm{E}[h(\mathbf{X})] = \int h(\mathbf{x}) dF(\mathbf{x}) \quad \text{as } n \to \infty. \quad (7.1)$$

In particular, the expression on the left-hand side of (7.1) is a good approximation of the expression on the right-hand side as long as the sample size n is sufficiently large.

Example 7.4 (Sample mean and variance). Consider a sample $\{x_1, \ldots, x_n\}$ and the corresponding empirical distribution function F_n. The sample mean

$\overline{x} = (x_1 + \cdots + x_n)/n$ is simply the expected value of a random variable with the distribution function F_n:

$$\int x \, dF_n(x) = \frac{1}{n} \sum_{k=1}^{n} x_k = \overline{x}.$$

We know from (7.1) that $\overline{X} = (X_1 + \cdots + X_n)/n \to E[X]$ with probability one as $n \to \infty$, and it is easy to see that $E[\overline{X}] = E[X]$. The variance of F_n is

$$\int x^2 \, dF_n(x) - \left(\int x \, dF_n(x)\right)^2 = \frac{1}{n} \sum_{k=1}^{n} x_k^2 - \frac{1}{n^2} \left(\sum_{k=1}^{n} x_k\right)^2$$

$$= \frac{1}{n} \left(\sum_{k=1}^{n} x_k^2 - n\overline{x}^2\right)$$

$$= \frac{1}{n} \left(\sum_{k=1}^{n} x_k^2 - 2 \sum_{k=1}^{n} \overline{x} x_k + \sum_{k=1}^{n} \overline{x}^2\right)$$

$$= \frac{1}{n} \sum_{k=1}^{n} (x_k - \overline{x})^2.$$

We know from (7.1) that, with probability one,

$$\lim_{n \to \infty} \frac{1}{n} \sum_{k=1}^{n} (X_k - \overline{X})^2 = \text{Var}(X).$$

However, the expected value of the variance estimator is not equal to $\text{Var}(X)$. Therefore, the variance is typically estimated by the sample variance

$$S^2 = \frac{1}{n-1} \sum_{k=1}^{n} (X_k - \overline{X})^2,$$

which satisfies $E[S^2] = \text{Var}(X)$.

Consider a subset B of $\mathbb{R}^d$ and suppose that we want to estimate the probability $P(B) = P(\mathbf{X} \in B)$. Similarly to the empirical distribution function, we form the empirical estimator

$$P_{n,\mathbf{X}}(B) = \frac{1}{n} \sum_{k=1}^{n} I\{\mathbf{X}_k \in B\}. \tag{7.2}$$

Notice that $P_{n,\mathbf{x}}(B) = F_{n,\mathbf{x}}(\mathbf{x})$ if $B = \{\mathbf{y} : \mathbf{y} \leq \mathbf{x}\}$ and that the sum in (7.2) is $\mathrm{Bin}(n, P(B))$-distributed. In particular, from the expected value $nP(B)$ and variance $nP(B)(1 - P(B))$ of the binomially distributed sum in (7.2) we find that

$$E[P_{n,\mathbf{x}}(B)] = P(B) \quad \text{and} \quad \mathrm{Var}(P_{n,\mathbf{x}}(B)) = \frac{1}{n}P(B)(1 - P(B)).$$

Moreover, it follows from the law of large numbers that $\lim_{n \to \infty} P_{n,\mathbf{x}}(B) = P(B)$ with probability one.

Example 7.5 (Estimation of small probabilities). In this example, we investigate the sample size needed for accurate empirical estimation of a small probability $P(B)$. A common measure of the accuracy of an estimator is the relative error—the standard deviation of the estimator divided by the estimated quantity. In this context, the relative error is given by

$$\frac{\mathrm{Var}(P_{n,\mathbf{x}}(B))^{1/2}}{P(B)} = n^{-1/2}\Big(\frac{1}{P(B)} - 1\Big)^{1/2}.$$

It is natural to require that the standard deviation of the estimator must be at least no greater than the probability to be estimated. Under this requirement, since $P(B)$ is assumed to be small, we find that $n \approx 1/P(B)$, which corresponds to a very large required sample size if $P(B)$ is small.

The accuracy of the estimator can be investigated by considering the probability

$$\mathrm{P}\Big(\Big|\frac{P_{n,\mathbf{x}}(B) - P(B)}{P(B)}\Big| < \varepsilon\Big) = \mathrm{P}\Big(1 - \varepsilon < \frac{P_{n,\mathbf{x}}(B)}{P(B)} < 1 + \varepsilon\Big).$$

Since the sum in (7.2) is $\mathrm{Bin}(n, P(B))$-distributed, we find that

$$\mathrm{P}\Big(\frac{P_{n,\mathbf{x}}(B)}{P(B)} < 1 + \varepsilon\Big) = \sum_{k=0}^{[n(1+\varepsilon)P(B)]} \binom{n}{k} P(B)^k (1 - P(B))^{n-k},$$

and similarly with $1 - \varepsilon$ instead of $1 + \varepsilon$.

Another approach to investigating the accuracy of the estimator when the sample size n is large is to apply the central limit theorem. If $Z_1, Z_2, \ldots$ is a sequence of independent copies of a random variable Z with finite expected value μ and standard deviation σ, then

$$\lim_{n \to \infty} \mathrm{P}\Big(\frac{Z_1 + \cdots + Z_n - n\mu}{n^{1/2}\sigma} \leq x\Big) = \Phi(x) \quad \text{for all } x,$$

where Φ denotes the standard normal distribution function. Taking $Z_k = I\{\mathbf{X}_k \in B\}/n$ we find that

$$\lim_{n\to\infty} P\left(\left(\frac{n}{P(B)(1-P(B))}\right)^{1/2}(P_{n,\mathbf{X}}(B) - P(B)) \le x\right) = \Phi(x) \quad \text{for all } x.$$

In particular, $P_{n,\mathbf{X}}(B)$ is approximately $N(P(B), P(B)(1 - P(B))/n))$-distributed if n is large.

7.3 Empirical Quantiles

Here we consider observations $x_1, \ldots, x_n$ from independent and identically distributed random variables $X_1, \ldots, X_n$ with a common unknown distribution function F defined on the real line $\mathbb{R}$. The empirical quantile function F_n^{-1} is the quantile function of the empirical distribution function F_n and therefore given by

$$F_n^{-1}(p) = \min\{x : F_n(x) \ge p\}.$$

Similarly, the empirical quantile function $F_{n,X}^{-1}$ is the quantile function of $F_{n,X}$. We will now show that the empirical quantile $F_{n,X}^{-1}(p)$ is the kth largest of the sample points $X_1, \ldots, X_n$ (and therefore the same holds for F_n^{-1} in terms of the sample points $x_1, \ldots, x_n$), where $k = k(n, p)$ depends on n and p. It turns out to be useful to order the sample $\{X_1, \ldots, X_n\}$ such that $X_{1,n} \ge \cdots \ge X_{n,n}$ (if F is continuous, then with probability one there are no $j \ne k$ such that $X_j = X_k$, i.e., no ties). Note that

$$\min\{x : F_{n,X}(x) \ge p\} = \min\left\{x : \sum_{k=1}^{n} I\{X_{k,n} \le x\} \ge np\right\}. \tag{7.3}$$

Since the sum $\sum_{k=1}^{n} I\{X_{k,n} \le x\}$ can only take integer values, we see that the right-hand side of (7.3) is equal to $X_{j,n}$ for some j. Which j? Note that for any j in the set $\{1, \ldots, n\}$,

$$\sum_{k=1}^{n} I\{X_{k,n} \le X_{j,n}\} = \sum_{k=j}^{n} I\{X_{k,n} \le X_{j,n}\} = n - j + 1,$$

and we must look for the largest j such that the last expression is greater than or equal to np. If we take $j = [n(1 - p)] + 1$, then

$$n - j + 1 = n - [n(1 - p)] \ge n - n(1 - p) = np,$$

with equality if and only if np is an integer. In particular, every $j \ge [n(1 - p)] + 2$ gives $n - j + 1 < np$. We conclude that the empirical quantile function is given by

$$F_{n,X}^{-1}(p) = X_{[n(1-p)]+1,n}, \quad p \in (0, 1),$$

a piecewise constant function on $(0, 1)$ with

$$X_{[n(1-p)]+1,n} = X_{k,n} \text{ if } p \in (1 - k/n, 1 - (k-1)/n]. \tag{7.4}$$

It can be shown that if F is strictly increasing, then $P(\lim_{n \to \infty} F_{n,X}^{-1}(p) = F^{-1}(p)) = 1$ for all $p \in (0, 1)$. Therefore, the empirical quantile is an arbitrary good approximation of the true but unknown quantile if the sample size n is sufficiently large. We prove the following slightly weaker statement.

Proposition 7.1. *Let $X_1, X_2, \ldots$ be a sequence of independent and identically distributed random variables with common distribution function F, and let $F_{n,X}$ be the empirical distribution function of the first n elements of the sequence. If F is strictly increasing in a neighborhood of $F^{-1}(p)$, then $\lim_{n \to \infty} P(|F_{n,X}^{-1}(p) - F^{-1}(p)| > \varepsilon) = 0$ for every $\varepsilon > 0$.*

Proof. From the quantile transform, Proposition 6.1, we know that $F^{-1}(U)$ has distribution function F if U is uniformly distributed on $(0, 1)$. Therefore, we may consider a sequence of independent random variables $U_1, U_2, \ldots$ uniformly distributed on $(0, 1)$ and represent $X_1, \ldots, X_n$ as $F^{-1}(U_1), \ldots, F^{-1}(U_n)$. Write $U_{1,n} \geq \cdots \geq U_{n,n}$ for the ordered U_k. Note that $F_{n,X}^{-1}(p) = F^{-1}(U_{[n(1-p)]+1,n})$, and since F is strictly increasing in a neighborhood of $F^{-1}(p)$, it follows that F^{-1} is continuous at p. Note also that

$$\{u : |F^{-1}(u) - F^{-1}(p)| > \varepsilon\} = \{u : |F^{-1}(u) - F^{-1}(p)| > \varepsilon, |u - p| \geq \delta\}$$

$$\cup \{u : |F^{-1}(u) - F^{-1}(p)| > \varepsilon, |u - p| < \delta\}$$

$$\subset \{u : |u - p| \geq \delta\}$$

$$\cup \{u : |F^{-1}(u) - F^{-1}(p)| > \varepsilon, |u - p| < \delta\},$$

and the continuity at p implies that

$$\lim_{\delta \to 0} \{u : |F^{-1}(u) - F^{-1}(p)| > \varepsilon, |u - p| < \delta\}$$

is the empty set. Therefore, for all $\delta > 0$,

$$P(|F_{n,X}^{-1}(p) - F^{-1}(p)| > \varepsilon) \leq P(|U_{[n(1-p)]+1,n} - p| \geq \delta) + C_\delta,$$

and since $\lim_{\delta \to 0} C_\delta = 0$, to complete the proof it only remains to show that $\lim_{n \to \infty} P(|U_{[n(1-p)]+1,n} - p| \geq \delta) = 0$ for every $\delta > 0$.

We claim that $U_{k,n}$ is Beta$(n - k + 1, k)$-distributed. To verify this claim, we first recall that the Beta(a, b) distribution is a probability distribution on $(0, 1)$ with density function

$$f(x) = \frac{\Gamma(a+b)}{\Gamma(a)\Gamma(b)} x^{a-1}(1-x)^{b-1},$$

where $\Gamma(n) = (n-1)!$. The Beta(a, b) distribution has mean $a/(a+b)$ and variance $ab(a+b)^{-2}(a+b+1)^{-1}$. The density function $f_{U_{k,n}}$ of $U_{k,n}$ can be expressed as

$$f_{U_{k,n}}(x) = \frac{d}{dx} P(U_{k,n} \le x) = \lim_{\Delta \to 0} \frac{P(U_{k,n} \in [x, x+\Delta])}{\Delta}.$$

We want to compute the limit on the right-hand side above. To this end, we introduce the notation

$$A_x = \{n - k \text{ of the } U_j \text{ are in } (0, x) \text{ and}$$
$$1 \text{ of the } U_j \text{ is in } [x, x+\Delta] \text{ and}$$
$$k - 1 \text{ of the } U_j \text{ are in } (x+\Delta, 1)\}$$

and notice that

$$P(U_{k,n} \in [x, x+\Delta]) = P(A_x) + o(\Delta)$$

$$= \frac{n!}{(n-k)!(k-1)!} x^{n-k} \Delta^1 (1-x-\Delta)^{k-1} + o(\Delta),$$

where $o(\Delta)$ is the probability of the event $\{U_{k,n} \in [x, x+\Delta]\}$ when two or more of the U_j are in $[x, x+\Delta]$. Letting $\Delta \to 0$ gives

$$f_{U_{k,n}}(x) = \frac{n!}{(n-k)!(k-1)!} x^{n-k}(1-x)^{k-1}$$

$$= \frac{\Gamma(n+1)}{\Gamma(n-k+1)\Gamma(k)} x^{n-k+1-1}(1-x)^{k-1},$$

which confirms the claim that $U_{k,n}$ is Beta$(n - k + 1, k)$-distributed. In particular,

$$E[U_{k,n}] = \frac{n-k+1}{n+1} = 1 - \frac{k}{n+1}, \qquad (7.5)$$

$$E[U_{k,n}^2] = \frac{(n-k+1)(n-k+2)}{(n+1)(n+2)}. \qquad (7.6)$$

Finally, take $p \in (0, 1)$ and $k(n) = [n(1-p)] + 1$. Then from (7.5) and (7.6) we find that

$$E[U_{k(n),n}] = 1 - \frac{[n(1-p)]+1}{n+1} \to 1 - (1-p) = p \quad \text{as } n \to \infty$$

and further that

$$n\,\mathrm{E}[(U_{k(n),n} - p)^2] \to p(1 - p) \quad \text{as } n \to \infty. \tag{7.7}$$

The conclusion now follows from an application of Markov's inequality together with (7.7): for every $\delta > 0$

$$\mathrm{P}(|U_{k(n),n} - p| \geq \delta) \leq \delta^{-2}\,\mathrm{E}[(U_{k(n),n} - p)^2] \to 0 \quad \text{as } n \to \infty. \qquad \square$$

But how good is the empirical quantile as an approximation of the true quantile for finite sample sizes? It turns out that this question can be answered, at least in the sense that for a given distribution function F we can express the distribution of the empirical quantile in terms of F.

Let Y_x be the number of sample points exceeding x, i.e., the number of indices k for which $X_k > x$. It follows immediately that Y_x is $\mathrm{Bin}(n, q)$-distributed, where $q = \mathrm{P}(X_k > x) = 1 - F(x)$. We have

$$\mathrm{P}(X_{1,n} \leq x) = \mathrm{P}(Y_x = 0),$$

$$\mathrm{P}(X_{2,n} \leq x) = \mathrm{P}(Y_x \leq 1),$$

$$\vdots$$

$$\mathrm{P}(X_{j,n} \leq x) = \mathrm{P}(Y_x \leq j - 1).$$

Since $F_{n,X}^{-1}(p) = X_{[n(1-p)]+1,n}$, we have found that

$$\mathrm{P}(F_{n,X}^{-1}(p) \leq x) = \mathrm{P}(Y_x \leq [n(1-p)]) = \sum_{k=0}^{[n(1-p)]} \binom{n}{k}(1 - F(x))^k F(x)^{n-k}.$$

For a given F these probabilities are easily evaluated on a computer. In particular, we can compute probabilities of the kind

$$\mathrm{P}\left(\left|\frac{F_{n,X}^{-1}(0.95) - F^{-1}(0.95)}{F^{-1}(0.95)}\right| < \varepsilon\right) = \mathrm{P}\left(1 - \varepsilon < \frac{F_{n,X}^{-1}(0.95)}{F^{-1}(0.95)} < 1 + \varepsilon\right),$$

i.e., the probability that the relative error is at most ε. Graphs showing these probabilities for different sample sizes and distributions can be found in Fig. 7.1.

The graph of the density function gives information about the concentration of the probability mass that is more easily interpreted than the graph of the distribution function. Differentiating the distribution function of the empirical quantile

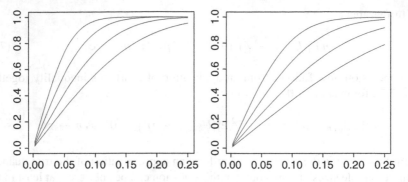

Fig. 7.1 Probabilities $P(1 - \varepsilon < F_{n,X}^{-1}(0.95)/F^{-1}(0.95) < 1 + \varepsilon)$ for ε in $(0, 0.25)$ for sample sizes $n = 100, 200, 400, 800$ (*lower* to *upper curve*), where F is the standard normal distribution function in the *left plot* and standard lognormal in the *right plot*

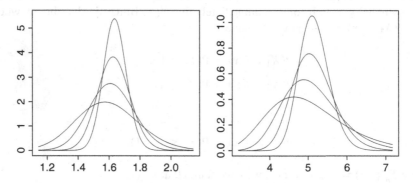

Fig. 7.2 Density functions of empirical quantile estimators $F_{n,X}^{-1}(0.95)$ for sample sizes $n = 100, 200, 400, 800$, where F is the standard normal distribution function in the *left plot* and standard lognormal in the *right plot*

$$P(F_{n,X}^{-1}(p) \le x) = \sum_{k=0}^{[n(1-p)]} \binom{n}{k} (1 - F(x))^k F(x)^{n-k},$$

i.e., the right-hand expression above, gives the density function of the empirical quantile (assuming that F has a density f). It is given by

$$\frac{n!}{[n(1-p)]!(n - [n(1-p)] - 1)!} (1 - F(x))^{[n(1-p)]} F(x)^{n-[n(1-p)]-1} f(x). \quad (7.8)$$

The graphs of density function (7.8) of the empirical quantile for different sample sizes and distributions are shown in Fig. 7.2.

Now we consider another approach to investigating the accuracy of the empirical quantile estimator $F_{n,X}^{-1}(p) = X_{[n(1-p)]+1,n}$ based on the sample $\{X_1, \ldots, X_n\}$ of independent copies of X with distribution function F. The approach considered here is appropriate for rather large sample sizes n. Here we want to determine the sample size n required for bounding the root mean square error (RMSE) of the empirical quantile by $10^{-d} F^{-1}(p)$ for some $d \geq 1$. More precisely, we want to determine the smallest integer n such that

$$E[(X_{[n(1-p)]+1,n} - F^{-1}(p))^2]^{1/2} \leq 10^{-d} F^{-1}(p). \qquad (7.9)$$

From Proposition 6.1 we know that we can represent $X_1, \ldots, X_n$ in terms of F and independent random variables $U_1, \ldots, U_n$ uniformly distributed on $(0, 1)$ as $F^{-1}(U_1), \ldots, F^{-1}(U_n)$. In particular, $X_{[n(1-p)]+1,n} = F^{-1}(U_{[n(1-p)]+1,n})$.

Under the assumption that F is differentiable with a density function f we may use a Taylor expansion of F^{-1} around point p to approximate

$$X_{[n(1-p)]+1,n} = F^{-1}(U_{[n(1-p)]+1,n})$$

$$= F^{-1}(p) + \frac{d}{dp} F^{-1}(p)(U_{[n(1-p)]+1,n} - p) + \text{remainder term}$$

$$= F^{-1}(p) + \frac{1}{f(F^{-1}(p))}(U_{[n(1-p)]+1,n} - p) + \text{remainder term}.$$

If we ignore the remainder term and use (7.6), then we may approximate the mean square error of the empirical quantile estimator by

$$E\left[(X_{[n(1-p)]+1,n} - F^{-1}(p))^2\right] \approx \frac{1}{f(F^{-1}(p))^2} E\left[(U_{[n(1-p)]+1,n} - p)^2\right]$$

$$= \frac{1}{f(F^{-1}(p))^2} \frac{p(1-p)}{n}. \qquad (7.10)$$

Can we ignore the error term? By Taylor's formula we find that if f does not vary that much in a neighborhood of $F^{-1}(p)$, then, as $n \to \infty$,

$$E\left[(X_{[n(1-p)]+1,n} - F^{-1}(p))^2\right] \frac{1}{f(F^{-1}(p))^2} \frac{p(1-p)}{n} + O(n^{-2}).$$

The notation $O(n^{-2})$ as $n \to \infty$ means that the remainder term divided by n^{-2} is bounded as $n \to \infty$.

If we want the bounded RMSE in (7.9) and approximate the mean square error by (7.10), then we find that the sample size n should be

$$n \approx \frac{p(1-p)}{f(F^{-1}(p))^2 F^{-1}(p)^2} 10^{2d}.$$

We illustrate the size of n for the exponential and the Pareto distribution. The following example shows that large sample sizes are required to obtain small relative errors for the empirical estimator of high quantiles.

Example 7.6 (Bound on the RMSE). Consider an exponential distribution with $F(x) = 1 - e^{-x}$ for $x > 0$. Then $F^{-1}(p) = -\log(1 - p)$ and $f(F^{-1}(p)) = \exp\{\log(1 - p)\} = 1 - p$. If $p = 0.99$ and $d = 2$ (RMSE of 1% of $F^{-1}(p)$), then we need a sample size n such that

$$n \approx \frac{p}{(1 - p)[\log(1 - p)]^2} 10^4 \approx 4.7 \cdot 10^4.$$

Consider a Pareto distribution with $F(x) = 1 - x^{-3}$ for $x > 1$. Then $F^{-1}(p) = (1 - p)^{-1/3}$ and $f(F^{-1}(p)) = 3(1 - p)^{4/3}$. If $p = 0.99$ and $d = 2$ (a RMSE of 1% of $F^{-1}(p)$), then we need a sample size n such that

$$n \approx \frac{p}{9(1 - p)} 10^4 = 1.1 \cdot 10^5.$$

7.4   Empirical VaR and ES

If X is the value at time 1 of a financial portfolio, then $\mathrm{VaR}_p(X) = F_L^{-1}(1 - p)$, where $L = -X/R_0$, where R_0 is the return on the reference instrument (a risk-free zero-coupon bond, say). Given a sample $L_1, \dots, L_n$ of independent copies of L, the empirical estimate of $\mathrm{VaR}_p(X)$ is therefore given by

$$\widehat{\mathrm{VaR}}_p(X) = L_{[np]+1,n},$$

where $L_{1,n} \geq \cdots \geq L_{n,n}$ is the ordered sample. Note that $\widehat{\mathrm{VaR}}_p(X)$ is simply the empirical $(1 - p)$-quantile of L_k.

   To compute the empirical VaR estimate from a sample of historical prices, we first transform the prices into a sample $\{L_1, \dots, L_n\}$ and then compute the VaR estimate as an empirical quantile. The following example illustrates this procedure.

Example 7.7 (Empirical VaR). Consider a portfolio consisting of long positions in two different assets, one unit of the first asset and two units of the second asset. Historical daily prices per unit of the two assets over the last 20 days are given by S_t^1 and S_t^2 for $t = -20, \dots, 0$. Assume the corresponding pairs of returns

$$\mathbf{R}_t = (R_t^1, R_t^2) = (S_t^1/S_{t-1}^1, S_t^2/S_{t-1}^2), \quad t = -19, \dots, 0,$$

are independent and identically distributed. The value of the portfolio at time 0 is given by $V_0 = S_0^1 + 2S_0^2$, and the value at time 1 is $V_1 = S_1^1 + 2S_1^2$. Suppose we want to compute the empirical $\mathrm{VaR}_{0.05}$ estimate of $X = V_1 - V_0R_0$, where for simplicity we set $R_0 = 1$. We may express X as the value of a function evaluated at the point $\mathbf{R}_1$:

$$X = V_1 - V_0 = (S_1^1 - S_0^1) + 2(S_1^2 - S_0^2)$$
$$= S_0^1(R_1^1 - 1) + 2S_0^2(R_1^2 - 1) = g(\mathbf{R}_1).$$

Under the assumption that the $\mathbf{R}_t$, for $t = -19, \ldots, 0$, are independent copies of $\mathbf{R}_1$, we can easily construct independent copies of X by setting $X_k = g(\mathbf{R}_{-20+k})$ for $k = 1, \ldots, 20$. Setting $L_k = -X_k$ and ordering the sample of L_k as $L_{1,20} \geq \cdots \geq L_{20,20}$, we compute the empirical estimate of $\text{VaR}_{0.05}(X)$ as $\widehat{\text{VaR}}_{0.05}(X) = L_{[20 \cdot 0.05]+1,20} = L_{2,20}$.

Example 7.8 (Thinning versus historical simulation). Suppose today is November 3, 2010, and we have just invested an amount of 100 in the Dow Jones Industrial Average (DJIA) stock market index. We want to analyze the risk we face from holding the position over a period of 20 trading days. The value of the position 20 trading days from today is $V_{20} = 100R_1 \cdots R_{20}$, where $R_1, \ldots, R_{20}$ are the daily returns over the time period under consideration. We want to estimate $\text{VaR}_p(V_{20} - 100)$ (the effect of interest rates are ignored) based on the 801 historical index values of DJIA from August 30, 2007 to November 2, 2010.

If the thinning approach in Example 7.2 is used, then we use every 20th value of the sample of historical DJIA values to obtain the sample $\{R_{-39}^{(20)}, \ldots, R_0^{(20)}\}$ of historical 20-day returns. We set

$$X_k = 100(R_{-40+k}^{(20)} - 1) \quad \text{and} \quad L_k = -X_k \quad \text{for } k = 1, \ldots, 40$$

and estimate $\text{VaR}_p(V_{20} - 100)$ as $L_{[40p]+1,40}$.

If the historical simulation approach in Example 7.3 is used, then we may choose $m = 5,000$ in Example 7.3 and use the sample of historical DJIA values to obtain the sample $\{R_1^{*(20)}, \ldots, R_{5000}^{*(20)}\}$ of fictive 20-day returns. We set

$$X_k = 100(R_k^{*(20)} - 1) \quad \text{and} \quad L_k = -X_k \quad \text{for } k = 1, \ldots, 5000$$

and estimate $\text{VaR}_p(V_{20} - 100)$ as $L_{[5000p]+1,5000}$.

The left plot in Fig. 7.3 shows the left tail of the empirical distribution function of X_k for the two approaches. Because the thinning approach gives a sample of small size 40, the staircase shape is pronounced and the tail estimate is unreliable. The historical simulation approach gives a much smoother, and likely more reliable, estimate of the left tail of the distribution function of $V_{20} - 100$. The right plot in Fig. 7.3 shows the estimates of $\text{VaR}_p(V_{20} - 100)$ as a function of p for the two approaches.

We now present the empirical ES estimator. Recall that the ES at level p of a portfolio with value X at time 1 is given by

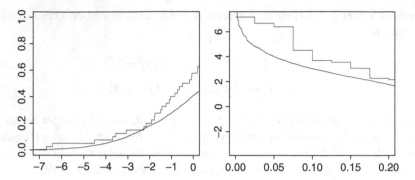

Fig. 7.3 *Above*: observed upper triangle of paid claims; *Below*: unobserved triangle of outstanding claims

$$\mathrm{ES}_p(X) = \frac{1}{p} \int_0^p \mathrm{VaR}_u(X)\,du.$$

The empirical ES estimator is obtained by simply replacing $\mathrm{VaR}_p(X)$ by its empirical estimator $\widehat{\mathrm{VaR}}_p(X) = L_{[np]+1,n}$, where $L_k = -X_k/R_0$ and $L_{1,n} \geq \cdots \geq L_{n,n}$ is the ordered loss sample. This implies

$$\widehat{\mathrm{ES}}_p(X) = \frac{1}{p} \int_0^p L_{[nu]+1,n}\,du$$

$$= \frac{1}{p} \left(\sum_{k=1}^{[np]} \frac{L_{k,n}}{n} + \left(p - \frac{[np]}{n} \right) L_{[np]+1,n} \right). \qquad (7.11)$$

If $[np]$ is an integer, then the expression in the last display reduces to the sample mean of the $[np]$ largest losses. To clarify how one arrives at the expression in (7.11), suppose that $[np] \geq 2$, and notice that in this case

$$\int_0^p L_{[nu]+1,n}\,du$$

$$= \int_0^{1/n} L_{1,n}\,du + \cdots + \int_{([np]-1)/n}^{[np]/n} L_{[np],n}\,du + \int_{[np]/n}^p L_{[np]+1,n}\,du$$

$$= \frac{1}{n} L_{1,n} + \cdots + \frac{1}{n} L_{[np],n} + \left(p - \frac{[np]}{n} \right) L_{[np]+1,n}.$$

Example 7.9 (Empirical ES*).* Consider the historical daily prices of two assets A and B, listed in Table 7.1. You are considering taking a position corresponding to a long position of two units of asset A and three units of asset B.

Table 7.1 Historical daily prices of two assets A and B

Day	−20	−19	−18	−17	−16	−15	−14
Asset A	81.75	81.35	80.4	81.05	83.35	83.00	83.30
Asset B	81.25	81.00	81.5	81.50	81.85	81.25	81.45
Day	−13	−12	−11	−10	−9	−8	−7
Asset A	86.0	85.5	84.50	84.00	84.05	82.35	83.45
Asset B	83.5	83.5	83.75	86.00	85.75	84.60	83.85
Day	−6	−5	−4	−3	−2	−1	0
Asset A	83.50	84.4	86.9	85.90	82.55	83.75	84.75
Asset B	84.55	84.0	84.3	84.75	85.35	87.00	85.75

Table 7.2 Sample of X_k values and corresponding ordered L_k values

Sample of X_k-values: transformation of historical prices

−1.62	−0.39	1.37	5.91	−2.60	1.25	11.97	−0.99	−1.21	5.91
−0.65	−6.88	−0.02	2.25	0.15	5.94		−0.58	−4.79	7.44	−1.67

Ordered sample of corresponding L_k values

6.88	4.79	2.60	1.67	1.62	1.21	0.99	0.65	0.58	0.39
0.02	−0.15	−1.25	−1.37	−2.25	−5.91	−5.91	−5.94	−7.44	−11.97

To evaluate the riskiness of this investment, you want to compute the empirical ES estimate $\mathrm{ES}_p(X)$, where $p = 0.06$ and X is the difference between the value of the position tomorrow and its current value. We may express X as a function of the vector $\mathbf{R}_1 = (R_1^A, R_1^B)$ of returns over the next day as

$$X = V_1 - V_0 = 2S_0^A(R_1^A - 1) + 3S_0^B(R_1^B - 1) = f(\mathbf{R}_1).$$

From Table 7.1 we can compute the corresponding vectors of historical returns. The function f then transforms these vectors into the sample of X_k values shown in Table 7.2 (rounded off to two decimal points). Setting $L_k = -X_k$ and ordering the L_k gives the ordered sample of L_k values in Table 7.2. From (7.11) we find that the ES estimate based on the values $l_{1,n} \geq \cdots \geq l_{n,n}$ is

$$\frac{1}{p}\left(\sum_{k=1}^{[np]} \frac{l_{k,n}}{n} + \left(p - \frac{[np]}{n} \right) l_{[np]+1,n} \right).$$

Here, with $n = 20$ and $p = 0.06$ and the values $l_{k,n}$ in Table 7.2 we get

$$\frac{1}{0.06}\left(\frac{6.88}{20} + (0.06 - 0.05)4.79 \right) \approx 6.53.$$

7.5 Confidence Intervals

Suppose we have observations $x_1, \ldots, x_n$ of independent and identically distributed random variables $X_1, \ldots, X_n$ from an unknown distribution function F and we want to know the value $\theta = \theta(F)$ of some quantity that is determined by the unknown F. Examples include the mean, the variance, some quantile, or some risk measure that depends on F. We may estimate θ by the empirical estimator $\widehat{\theta} = \theta(F_{n,X})$ obtained by computing θ from $F_{n,X}$ instead of F. However, a point estimate is not meaningful unless we have some way of assessing its accuracy. Since we can never know whether the observations $x_1, \ldots, x_n$ are representative outcomes from the unknown distribution F, we can never know whether the empirical estimate $\widehat{\theta}_{\mathrm{obs}} = \theta(F_n)$ based on these observations is close to the true value θ. What we can do is compute a confidence interval for θ.

Let us first recall what a confidence interval is. Given $q \in (0, 1)$, we want to form a stochastic interval (A, B), where $A = f_A(X_1, \ldots, X_n)$ and $B = f_B(X_1, \ldots, X_n)$ for some functions f_A and f_B such that

$$P(A < \theta < B) = q,$$

i.e., the stochastic interval (A, B) covers the value θ with probability q. Clearly, we want q to be close to 1, e.g., $q = 0.95$, and at the same time we want that the length of the interval is likely to be small. The interval (a, b), where $a = f_A(x_1, \ldots, x_n)$ and $b = f_B(x_1, \ldots, x_n)$, is a confidence interval for θ with confidence level q. We may say that we feel confident at level q that the interval (a, b) covers the value θ. Note that q is not the probability that the specific interval (a, b) covers θ (either it does or it does not), but q is the probability that the procedure generating the interval will produce an interval covering θ if fed with a new random sample of the same size from the same probability distribution. Often we want to find a double-sided interval so that

$$P(A < \theta < B) = q, \quad P(A \geq \theta) = P(B \leq \theta) = (1 - q)/2.$$

Since F is unknown, we do not know the functions f_A, f_B, but we can construct approximate confidence intervals. If θ is a quantile of F, i.e., $\theta = F^{-1}(p)$, then we can actually find exact confidence intervals for θ, but not for arbitrary confidence levels q.

7.5.1 Exact Confidence Intervals for Quantiles

Suppose we have observations $x_1, \ldots, x_n$ of outcomes of independent and identically distributed random variables $X_1, \ldots, X_n$ with common unknown continuous distribution function F. Suppose further that we want to construct a confidence interval (a, b) for the quantile $F^{-1}(p)$, where $a = f_A(x_1, \ldots, x_n)$ and $b = f_B(x_1, \ldots, x_n)$ such that

$$P(A < F^{-1}(p) < B) = q, \quad P(A \geq F^{-1}(p)) = P(B \leq F^{-1}(p)) = (1 - q)/2,$$

where q is a confidence level, $A = f_A(X_1, \ldots, X_n)$, and $B = f_B(X_1, \ldots, X_n)$. Since F is unknown, we cannot find a and b. However, we can look for $i > j$ and the smallest $q' \geq q$ such that

$$P(X_{i,n} < F^{-1}(p) < X_{j,n}) = q',$$

$$P(X_{i,n} \geq F^{-1}(p)) \leq (1 - q)/2, \quad P(X_{j,n} \leq F^{-1}(p)) \leq (1 - q)/2. \quad (7.12)$$

It remains to compute the probabilities in (7.12). Let $Y_{F^{-1}(p)}$ be the number of sample points exceeding $F^{-1}(p)$, i.e., the number of indices k for which $X_k > F^{-1}(p)$. It follows immediately that $Y_{F^{-1}(p)}$ is $\text{Bin}(n, r)$-distributed, where $r = P(X_k > F^{-1}(p)) = 1 - F(F^{-1}(p))$. From Proposition 6.1 we know that the continuity of F implies that $F(F^{-1}(u)) = u$ for all $u \in (0, 1)$. In particular, $Y_{F^{-1}(p)}$ is $\text{Bin}(n, 1 - p)$-distributed. The probabilities in (7.12) are easily expressed in terms of the probabilities of $Y_{F^{-1}(p)}$, which are very easily computed (with the assistance of some appropriate software). We have

$$P(X_{1,n} \leq F^{-1}(p)) = P(Y_{F^{-1}(p)} = 0),$$

$$P(X_{2,n} \leq F^{-1}(p)) = P(Y_{F^{-1}(p)} \leq 1),$$

$$\vdots$$

$$P(X_{j,n} \leq F^{-1}(p)) = P(Y_{F^{-1}(p)} \leq j - 1).$$

Similarly, $P(X_{i,n} \geq F^{-1}(p)) = 1 - P(Y_{F^{-1}(p)} \leq i - 1)$. We may now can compute the probabilities $P(X_{j,n} \leq F^{-1}(p))$ and $P(X_{i,n} \geq F^{-1}(p))$ for different i and j until we find indices that satisfy (7.12).

Example 7.10 (Exact intervals for quantiles). Suppose we have a sample $\{X_1, \ldots, X_{200}\}$ of independent and identically distributed random variables with common unknown continuous distribution function F and we want a confidence interval for $F^{-1}(0.95)$ with confidence level $q' \approx q = 0.95$. Since $Y_{F^{-1}(0.95)}$ is $\text{Bin}(200, 0.05)$-distributed, we find that

$$P(X_{5,200} \leq F^{-1}(0.95)) = P(Y_{F^{-1}(0.95)} \leq 4) \approx 0.0264,$$

$$P(X_{17,200} \geq F^{-1}(0.95)) = 1 - P(Y_{F^{-1}(0.95)} \leq 16) \approx 0.0238.$$

Therefore, $P(X_{17,200} < F^{-1}(0.95) < X_{5,200}) \approx 0.95$, so $(x_{17,200}, x_{5,200})$ is a confidence interval for $F^{-1}(0.95)$ with a confidence level of approximately 95%. The length of the confidence interval depends on the sample points, which in turn depends on the unknown distribution function F. Figure 7.4 shows 100 outcomes

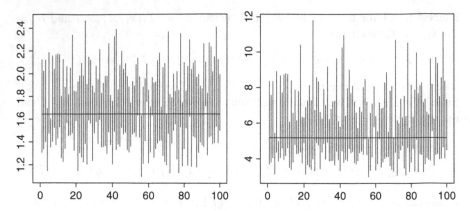

Fig. 7.4 *Each plot* shows empirical confidence intervals $(x_{17,200}, x_{5,200})$ for $F^{-1}(0.95)$ with confidence level 95% for 100 samples of size 200. *Left plot*: empirical confidence intervals for F the standard normal distribution; *right plot*: for F the standard lognormal distribution function

$(x_{17,200}, x_{5,200})$ of the empirical confidence interval for $F^{-1}(0.95)$ for F standard normal (left plot) and for F standard lognormal (right plot). Notice that the 25th confidence interval for the lognormal F says that if we had the 25th lognormal sample, then we could only say that we are rather sure that the 95% quantile of the unknown distribution lies somewhere between 4 and 11.8. This illustrates the difficulty of accurately estimating quantile values.

7.5.2   Confidence Intervals Using the Nonparametric Bootstrap

For quantiles we have seen how to construct exact confidence intervals. However, for risk measures, which unlike VaR are not simply quantile values, and for other quantities such as moments and loss probabilities this approach does not work. We will here investigate a useful method for constructing approximate confidence intervals called the nonparametric bootstrap method.

Suppose we have observations $x_1, \ldots, x_n$ of independent and identically distributed random variables $X_1, \ldots, X_n$ and we want to estimate some quantity $\theta = \theta(F)$ that depends on the unknown distribution F of X_k. For instance, θ could be the p-quantile $\theta = F^{-1}(p)$ giving $\widehat{\theta}_{\text{obs}} = x_{[n(1-p)]+1,n}$ or the mean $\theta = \int x\,dF(x)$ giving $\widehat{\theta}_{\text{obs}} = (x_1 + \cdots + x_n)/n$. We want to construct a confidence interval for θ with confidence level q.

If F were known, we could compute the value θ analytically or approximate it numerically. Alternatively, we could simulate a large sample from F to approximately compute θ as the empirical estimate. The problem here is that we do not know F and we only have one sample $\{x_1, \ldots, x_n\}$ of size n from F.

One way to produce more samples is to randomly draw with replacement n times from the set of observations $x_1, \ldots, x_n$ to produce a sample $\{X_1^*, \ldots, X_n^*\}$. The sample points X_k^* are independent and F_n-distributed (uniformly distributed on the set of the original observations $x_1, \ldots, x_n$). Some of the X_k^* are likely to be equal, even if the x_k are all different. The probability that $X_j^* \neq X_k^*$ for all $j \neq k$ is very small; the probability that none of the x_ks is drawn twice among the n tries is $n!/n^n$. Write F_n^* for the empirical distribution of $X_1^*, \ldots, X_n^*$ and $\widehat{\theta}^* = \theta(F_n^*)$ for the estimate of θ based on the sample $\{X_1^*, \ldots, X_n^*\}$. Even though $\{X_1^*, \ldots, X_n^*\}$ is not a sample from F, it has most of the characteristics of a sample from F as long as n is sufficiently large. In particular, the probability distribution of $\widehat{\theta}^*$ is likely to be close to the probability distribution of $\widehat{\theta}$. Whereas the probability distribution of $\widehat{\theta}$ is unknown (since F is not known), the probability distribution of $\widehat{\theta}^*$ can be approximated arbitrarily well by repeated resampling N times for N large enough.

An approximative confidence interval $I_{\theta,q}$ for θ of confidence level q using the nonparametric bootstrap method is constructed as follows.

- For each j in the set $\{1, \ldots, N\}$ draw with replacement n times from the sample $\{x_1, \ldots, x_n\}$ to obtain the sample $\{X_1^{*(j)}, \ldots, X_n^{*(j)}\}$ and the corresponding empirical distribution function $F_n^{*(j)}$.
- Compute the estimates $\widehat{\theta}_j^* = \theta(F_n^{*(j)})$ of θ and the residuals $R_j^* = \widehat{\theta}_{\text{obs}} - \widehat{\theta}_j^*$ for $j = 1, \ldots, N$.
- Form the interval

$$I_{\theta,q} = \left(\widehat{\theta}_{\text{obs}} + R^*_{[N(1+q)/2]+1,N}, \widehat{\theta}_{\text{obs}} + R^*_{[N(1-q)/2]+1,N}\right),$$

where $R^*_{1,N} \geq \cdots \geq R^*_{N,N}$ is the ordering of the sample $\{R_1^*, \ldots, R_N^*\}$.

Why is the interval $I_{\theta,q}$ a reasonable approximative confidence interval for θ? Here is one way of motivating the procedure.

Let G denote the distribution function of $\theta - \widehat{\theta}$. Then

$$q = P(G^{-1}((1-q)/2) < \theta - \widehat{\theta} < G^{-1}((1+q)/2))$$

$$= P(\widehat{\theta} + G^{-1}((1-q)/2) < \theta < \widehat{\theta} + G^{-1}((1+q)/2)).$$

Therefore, $(\widehat{\theta}_{\text{obs}} + G^{-1}((1-q)/2), \widehat{\theta}_{\text{obs}} + G^{-1}((1+q)/2))$ is a confidence interval for θ of level q. The problem is that we do not know the distribution function G.

The success of the bootstrap relies on the validity of the bootstrap principle, which says that G can be well approximated by the empirical distribution G_N^* of $R_1^*, \ldots, R_N^*$. Then the quantiles $G^{-1}((1-q)/2)$ and $G^{-1}((1+q)/2)$ can be well approximated by the empirical counterparts $R^*_{[N((1+q)/2)]+1,N}$ and $R^*_{[N(1-q)/2]+1,N}$, which leads to the interval $I_{\theta,q}$. We need n to be sufficiently large to make it plausible that the bootstrap principle holds so $\theta - \widehat{\theta}$ and $\widehat{\theta}_{\text{obs}} - \widehat{\theta}^*$ are approximately equally distributed. This requirement is investigated in the following example.

Example 7.11 (Bootstrap intervals for quantiles). Suppose we want to construct confidence intervals for $\theta = \text{VaR}_{0.05}(V_1 - V_0)$, where V_0 and V_1 are respectively the current value and the value tomorrow of a long position in some stock index. Since the time period here is only 1 day, we ignore the impact of interest rates. We may express V_1 in terms of the return R_1 of the stock index as $V_1 = V_0 R_1$, and we assume that $\log R_1$ is normally distributed with zero mean and standard deviation 0.01. For simplicity we also assume that $V_0 = 1$. This implies that

$$\text{VaR}_{0.05}(V_1 - V_0) = F_{V_0 - V_1}^{-1}(0.95) = V_0\big(1 - F_{R_1}^{-1}(0.05)\big) \approx 0.016314.$$

In reality we would not know with certainty the return distribution or, therefore, the value of $\text{VaR}_{0.05}(V_1 - V_0)$. However, we may—under the right circumstances—believe that the past n index returns can be seen as sample points from the distribution of the future return R_1 and in that case transform the historical returns into outcomes $l_1, \ldots, l_n$ of $L_1, \ldots, L_n$ that are independent copies of $L = V_0 - V_1$. The problem we investigate here is how to construct and evaluate confidence intervals for $F_L^{-1}(0.95)$ given the sample $\{l_1, \ldots, l_n\}$. We have already seen how we can construct confidence intervals for quantiles, and this approach is applicable here since $\text{VaR}_{0.05}(V_1 - V_0) = F_L^{-1}(0.95)$. However, the aim here is to investigate the nonparametric bootstrap approach to construct approximative confidence intervals and evaluate it by comparing it to the approach for quantiles.

Recall that the accuracy of the nonparametric bootstrap approach for constructing confidence intervals is likely to be good if $\theta - \widehat{\theta}$, where $\widehat{\theta} = L_{[0.05n]+1,n}$, and $\widehat{\theta}_{\text{obs}} - \widehat{\theta}^*$ have approximately the same probability distribution. The upper left plot in Fig. 7.5 shows a histogram from 2,000 simulations of $\theta - \widehat{\theta}$. The upper right and middle plots in Fig. 7.5 show histograms of 2,000 bootstrap simulations of $\widehat{\theta}_{\text{obs}} - \widehat{\theta}^*$ based on resampling from three different outcomes of $L_1, \ldots, L_{500}$. Based on these plots, we definitely see a resemblance between the distribution of $\theta - \widehat{\theta}$ and $\widehat{\theta}_{\text{obs}} - \widehat{\theta}^*$, but it is clear that much information has been lost in the bootstrap world. One might suspect that increasing the number N of resampling runs could improve things. The middle left and lower left plots show bootstrap simulations of $\widehat{\theta}_{\text{obs}} - \widehat{\theta}^*$ based on the same sample $\{l_1, \ldots, l_{500}\}$, where the number N is 2,000 for the middle left plot and 10,000 for the lower left plot. The same comparison is shown in the middle right and the lower right plots but based on another sample $\{l_1, \ldots, l_{500}\}$. We observe that increasing the number N of bootstrap simulations to a very large number does not improve things much.

Finally, we compute 50 confidence intervals for θ of confidence level 0.95 with the exact method (left plot in Fig. 7.6) and with the nonparametric bootstrap method (right plot in Fig. 7.6). We observe that the results are very similar. The differences among the confidence intervals are to a much greater extent due to the differences among the 50 outcomes of the random sample $\{L_1, \ldots, L_{500}\}$ than to the particular method used. We conclude that here the bootstrap method performs rather well.

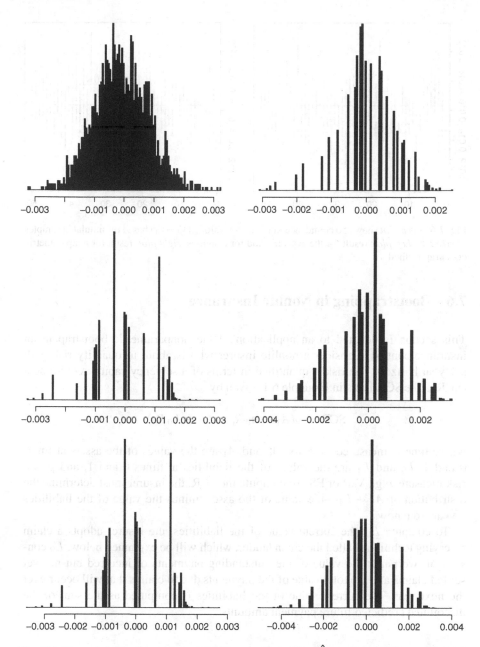

Fig. 7.5 *Upper left plot*: histogram of 2,000 outcomes of $\theta - \hat{\theta}$ based on 2,000 outcomes of $\{L_1, \ldots, L_{500}\}$. Each of the remaining plots shows centered bootstrap estimates $\hat{\theta}_{\text{obs}} - \hat{\theta}^*$. *Upper right* and *middle plots*: histograms based on $N = 2,000$ resampling runs for three different outcomes of the sample $\{L_1, \ldots, L_{500}\}$. *Middle* and *lower left plots*: based on the same original sample, the number of resampling runs is 10,000 for the *lower left plot* instead of 2,000. Similarly for *middle* and *lower right plots*

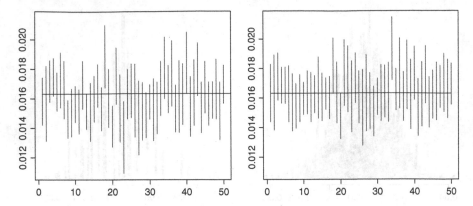

Fig. 7.6 *Each plot* shows 50 confidence intervals for $\mathrm{VaR}_{0.05}(V_1 - V_0)$ based on simulated samples of size 500. *Left plot*: result for the exact method for quantiles; *right plot*: result for nonparametric bootstrap method

7.6 Bootstrapping in Nonlife Insurance

This section is devoted to an application of the nonparametric bootstrap in an insurance context. Consider a nonlife insurer who is about to quantify risk with a 1-year horizon. The risk is quantified in terms of a solvency capital requirement (SCR). The SCR is, as in Example 6.1, given by

$$\mathrm{SCR} = \rho(A_1 - A_0 R_0 - L_1 + L_0 R_0),$$

where time is measured in years, A_0 and A_1 are the values of the assets at times 0 and 1, L_0 and L_1 are the values of the liabilities at times 0 and 1, and ρ is a risk measure, e.g., VaR or ES. To compute the SCR, the insurer must determine the distribution of $A_1 - L_1$—the value of the assets minus the value of the liabilities 1 year from now.

To compute L_0, the current value of the liabilities, the insurer adopts a claim reserving technique called the chain ladder, which will be explained below. L_0 consists of two parts: the value of the outstanding payments of incurred but not yet settled claims and the total value of the payments due to claims that will occur over the next year. The current value of the liabilities is computed as the sum of the discounted predicted future payment amounts.

7.6.1 Claims Reserve Prediction Via the Chain Ladder

The prediction of the future payment amounts is based on a historical record of paid claims. It is assumed that all claims that occurred at least $n + 1$ years ago

Table 7.3 *Above*: observed upper triangle of paid claims; *below*: unobserved triangle of outstanding claims

Origin	Development year					
	0	1	2	$\cdots$	$n-1$	n
$-n-1$	$C_{-n-1,0}$	$C_{-n-1,1}$	$C_{-n-1,2}$	$\cdots$	$C_{-n-1,n-1}$	$C_{-n-1,n}$
$-n$	$C_{-n,0}$	$C_{-n,1}$	$C_{-n,2}$	$\cdots$	$C_{-n,n-1}$	
$\vdots$	$\vdots$	$\vdots$				
-2	$C_{-2,0}$	$C_{-2,1}$				
-1	$C_{-1,0}$					
0						

Origin	Development year					
	0	1	2	$\cdots$	$n-1$	n
$-n-1$						
$-n$						$C_{-n,n}$
$-n+1$					$C_{-n+1,n-1}$	$C_{-n+1,n}$
$\vdots$					$\vdots$	$\vdots$
-1		$C_{-1,1}$	$C_{-1,2}$	$\cdots$	$C_{-1,n-1}$	$C_{-1,n}$
0	$C_{0,0}$	$C_{0,1}$	$C_{0,2}$	$\cdots$	$C_{0,n-1}$	$C_{0,n}$

are completely settled, but all claims that occurred at most n years ago are not completely settled. The historical record of paid claims is displayed in the form of a claims triangle. The triangle of paid claims, called the upper triangle, displays on each row the amounts of paid claims for claims incurred with the same origin year. The columns represent the development years—the difference between the year a claim was settled and the year it was incurred. The entry $C_{-k,l}$ represents the amount paid for claims that were incurred in year $-k$ and paid l year later, in year $-k+l$, for $l = 0, \ldots, k-1$. The upper triangle is illustrated in the top half of Table 7.3.

The insurer relies on the assumption that the payment pattern over the development years is repetitive. Even if the number of accidents and claim amounts may differ significantly from year to year, the payment patterns over the development years look similar. Based on this assumption the upper triangle of paid claims can be used to predict unobserved future payments. The unknown future payments are represented in the lower triangle of outstanding claims, with entries $C_{-k,l}$ for $l = k, \ldots, n$. The lower triangle is illustrated in the bottom half of Table 7.3.

To formulate the idea that the payment patterns are repetitive, one possibility, which is the one we follow here, is to assume a multiplicative structure for the cumulative claims. Consider, for $k = 1, \ldots, n+1$ and $l = 0, \ldots, k-1$, the cumulative amounts paid for claims that occurred in year $-k$,

$$D_{-k,l} = \sum_{j=0}^{l} C_{-k,j}.$$

Suppose the expected cumulative payments can be written as

$$E[D_{-k,l+1}] = E[D_{-k,l}]f_l, \quad l = 0, \ldots, n-1,$$

where $f_0, \ldots, f_{n-1}$ are called development factors. Then the expected amounts paid are given by

$$E[C_{-k,0}] = E[D_{-k,0}],$$

$$E[C_{-k,l}] = E[D_{-k,l} - D_{-k,l-1}] = E[C_{-k,0}]f_0 \cdots f_{l-2}(f_{l-1} - 1), \quad l = 1, \ldots, n.$$

A simple model for $C_{-k,l}$ is obtained by assuming that the observed payments have the representation $C_{-k,l} = E[C_{-k,l}]R_{-k,l}$, where $\{R_{-k,l}\}_{k,l=0}^{n}$ are independent and identically distributed with $E[R_{-k,l}] = 1$.

A standard method for predicting the lower triangle (outstanding claims) is called the chain ladder method. In the chain ladder method, the development factors are estimated by

$$\widehat{f}_l = \frac{\sum_{k=l+2}^{n+1} D_{-k,l+1}}{\sum_{k=l+2}^{n+1} D_{-k,l}}, \quad l = 0, \ldots, n-1. \tag{7.13}$$

The expected amounts of paid claims, $E[C_{-k,l}]$, in the upper triangle can be estimated by

$$\widehat{C}_{-k,0} = \widehat{D}_{-k,0} = \frac{D_{-k,k-1}}{\widehat{f}_0 \ldots \widehat{f}_{k-2}}, \tag{7.14}$$

$$\widehat{C}_{-k,l} = \widehat{D}_{-k,l} - \widehat{D}_{-k,l-1} = \frac{D_{-k,k-1}}{\widehat{f}_l \ldots \widehat{f}_{k-2}} \left(1 - \frac{1}{\widehat{f}_{l-1}} \right) \tag{7.15}$$

for $k = 1, \ldots, n+1$, $l = 1, \ldots, k-1$, and the residuals are computed as

$$R_{-k,l} = \frac{C_{-k,l}}{\widehat{C}_{-k,l}}, \quad k = 1, \ldots, n+1, \ l = 0, \ldots, k-1.$$

The predictions for the unobserved cumulative claim amounts in the lower triangle are given by

$$\widehat{D}_{-k,l} = D_{-k,k-1}\widehat{f}_{k-1} \cdots \widehat{f}_{l-1} \tag{7.16}$$

for $k = 1, \ldots, n$ and $l = k, \ldots, n$. The corresponding predictions for the unobserved future payments in the lower triangle are

$$\widehat{C}_{-k,l} = \widehat{D}_{-k,l+1} - \widehat{D}_{-k,l}. \tag{7.17}$$

The last row in the lower triangle of outstanding payments, corresponding to $k = 0$, represents amounts for claims that will occur during the next year. Therefore, it does not contain any observations in the upper triangle and cannot be predicted with the standard chain ladder. We will predict this row by predicting the initial payment

$D_{0,0}$ by the mean of the predictions for the previous years and then apply the chain ladder method for the predictions of $D_{0,l}$ for $l \geq 1$. More precisely, the predictions for the last row may be constructed as follows:

$$\widehat{D}_{0,0} = \frac{1}{n+1} \sum_{k=1}^{n+1} \frac{D_{-k,k-1}}{\widehat{f}_0 \cdots \widehat{f}_{k-2}}, \qquad \widehat{D}_{0,l} = \widehat{D}_{0,0}\widehat{f}_0 \cdots \widehat{f}_{l-1},$$

$$\widehat{C}_{0,0} = \widehat{D}_{0,0}, \qquad \widehat{C}_{0,l} = \widehat{D}_{0,l+1} - \widehat{D}_{0,l}.$$

When the prediction of all future payments $\widehat{C}_{-k,l}$, $k = 0, \ldots, n$, $l = k, \ldots, n$, is completed, the present value L_0 of the outstanding claims is computed as

$$L_0 = \sum_{k=0}^{n}\sum_{l=k}^{n} \widehat{C}_{-k,l} e^{-r_{l-k+1}(l-k+1)},$$

where $\mathbf{r}^{\mathsf{T}} = (r_1, \ldots, r_n)$ is the vector of current zero rates.

At time 1, new information is available as the values of the diagonal entries $C_{-k,k}$, for $k = 0, \ldots, n$, are observed. The value L_1 of the liabilities at time 1 will be computed similarly to L_0. First, the new observations $C_{-k,k}$, for $k = 0, \ldots, n$, are entered into the upper triangle of observed payments. Then, the development factors are updated by

$$\widehat{f}_l^{(1)} = \frac{\sum_{k=l+1}^{n+1} D_{-k,l+1}}{\sum_{k=l+1}^{n+1} D_{-k,l}} \qquad \text{for } l = 0, \ldots, n-1,$$

and the predictions, denoted by $\widehat{C}_{-k,l}^{(1)}$, are updated accordingly by entering the updated development factors in (7.16) and (7.17). Assuming the zero rates at time 1 are $\mathbf{r} + \Delta\mathbf{r}$, the value of the liabilities at time 1 can be expressed as

$$L_1 = \sum_{k=0}^{n} C_{-k,k} + \sum_{k=0}^{n-1}\sum_{l=k+1}^{n} \widehat{C}_{-k,l}^{(1)} e^{-(r_{l-k} + \Delta r_{l-k})(l-k)}.$$

Note that L_1 is completely determined by the random variables $C_{-k,k}$, for $k = 0, \ldots, n$, and $\Delta\mathbf{r}$, all observed at time 1.

To protect the value of the liabilities against changes in the zero rates, it is assumed that the insurer has purchased a bond portfolio. Rewrite the current value of the liabilities by summing along the diagonals as

$$L_0 = \sum_{m=1}^{n+1} \left(\sum_{j=m-1}^{n-m+1} \widehat{C}_{-(m-j-1),j} \right) e^{-r_m m}.$$

Then we see that a good choice of the bond portfolio is obtained by buying $\sum_{j=m-1}^{n-m+1} \widehat{C}_{-(m-j-1),j}$ zero-coupon bonds with maturity m years from now. If the zero rate changes are independent of the claim amounts $C_{-k,k}$, for $k = 0, \ldots, n$, then this bond portfolio is the quadratic hedge of the value of the liabilities at time 1.

To compute the SCR, we need to apply the risk measure ρ to the quantity $A_1 - A_0 R_0 - L_1 + L_0 R_0$. Under the assumption that $A_0 = L_0$, it is sufficient to consider the distribution of $A_1 - L_1$. If the true development factors were known, the distribution of L_1 could be sampled by sampling the diagonal elements $C_{-k,k}$, for $k = 0, \ldots, n$, updating the prediction of the lower triangle, sampling the zero rate changes $\Delta \mathbf{r}$, and computing the outcome of L_1 by discounting the future payments in each simulated scenario. When adopting an empirical approach, the diagonal elements are sampled by sampling the residuals $R_{-k,k}$, for $k = 0, \ldots, m$, from the empirical distribution of the residuals and putting $C_{-k,k} = \widehat{C}_{-k,k} R_{-k,k}$. Similarly, the zero rate changes may be sampled from the empirical distribution of historical zero rate changes.

A problem with the empirical approach is that it does not account for parameter uncertainty in the development factors. The development factors, used for predicting the diagonal means $\widehat{C}_{-k,k}$, are not known but merely estimated from the upper triangle. Since the amount of data is rather limited, the parameter uncertainty may be substantial. To account for the parameter uncertainty, a bootstrap procedure can be implemented as outlined below. The algorithm below generates a sample from the so-called predictive distribution of $A_1 - L_1$, in which the parameter uncertainty is taken into account. The input to the algorithm is an upper triangle of amounts paid, as in the left table in Table 7.3. The algorithm proceeds as follows.

1. Compute the estimates $\widehat{f}_0, \ldots, \widehat{f}_{n-1}$ of the development factors by (7.13).
2. Compute the estimates $\widehat{C}_{-k,l}$ of $\mathrm{E}[C_{-k,l}]$ for $k = 1, \ldots, n+1, l = 0, \ldots, k-1$, by (7.14) and (7.15).
3. Compute the residuals $R_{-k,l} = C_{-k,l}/\widehat{C}_{-k,l}$, for $k = 1, \ldots, n+1, l = 0, \ldots, k-1$.
4. For each bootstrap iteration, $j = 1, \ldots, N$, repeat the following:

    (a) Draw with replacement bootstrapped residuals $R^*_{-k,l}$, for $k = 1, \ldots, n+1$, $l = 0, \ldots, k-1$, from the set $\{R_{-k,l}, k = 1, \ldots, n+1, l = 0, \ldots, k-1\}$.
    (b) Compute a bootstrapped upper triangle with entries $C^*_{-k,l} = \widehat{C}_{-k,l} R^*_{-k,l}$ for $k = 1, \ldots, n+1, l = 0, \ldots, k-1$.
    (c) Compute the development factors $f^*_0, \ldots, f^*_{n-1}$ of the bootstrapped upper triangle as in (7.13).
    (d) Compute one-step predictions $\widehat{C}^*_{-k,k}$, for $k = 0, \ldots, n$, using the bootstrapped upper triangle.
    (e) Draw with replacement the outcomes of diagonal residuals $R^{**}_{-k,k}$, for $k = 0, \ldots, n$, from the set of residuals $\{R_{-k,l}, k = 1, \ldots, n+1, l = 0, \ldots, k-1\}$.
    (f) Add the diagonal $C^{**}_{-k,k} = \widehat{C}^*_{-k,k} R^{**}_{-k,k}$, for $k = 0, \ldots, n$, to the bootstrapped upper triangle to form a sample of the upper triangle at time 1.

Table 7.4 Current zero rates used in Example 7.12

Maturity (years)	1	2	3	4	5	6	7	8	9	10
Zero rate (%)	0.82	1.57	2.16	2.54	2.82	3.04	3.23	3.37	3.49	3.58

(g) Compute the development factors $\widehat{f}_0^{**}, \ldots, \widehat{f}_{n-1}^{**}$ of the upper triangle at time 1.

(h) Compute the predictions $\widehat{C}_{-k,l}^{**}$, for $k = 1, \ldots, n, l = k, \ldots, n$, for the lower triangle at time 1.

(i) Draw one outcome $\Delta \mathbf{r}$ of zero rate changes from the set of historical zero rate changes.

(j) Compute the value of the liabilities at time 1 as

$$L_1 = \sum_{k=0}^{n} C_{-k,k}^{**} + \sum_{k=0}^{n-1} \sum_{l=k+1}^{n} \widehat{C}_{-k,l}^{**} e^{-(r_{l-k} + \Delta r_{l-k})(l-k+1)}$$

and the value of the bond portfolio as

$$A_1 = \sum_{k=0}^{n} \widehat{C}_{-k,k} + \sum_{k=0}^{n-1} \sum_{l=k+1}^{n} \widehat{C}_{-k,l} e^{-(r_{l-k} + \Delta r_{l-k})(l-k+1)}$$

and store the difference $A_1 - L_1$.

Example 7.12 (Sampling from the predictive distribution). Consider a nonlife insurer with upper triangle of paid claim amounts as in Table 7.5. The claims are assumed to be completely settled 9 years after the incident year. The objective is to determine the predictive distribution of the value of the assets minus the value of the liabilities, $A_1 - L_1$, 1 year from now. The bootstrapping algorithm outlined above is run. A historical sample of quarterly zero rate changes serves as the basis for generating annual zero rate changes $\Delta \mathbf{r}$. Each annual zero rate scenario is constructed by sampling four quarterly scenarios, with replacement, and adding them up. The current zero rates are given in Table 7.4.

A histogram of $N = 10,000$ samples from the predictive distribution of $A_1 - L_1$ is given in Fig. 7.7.

7.7 Notes and Comments

An introduction to the bootstrap and related resampling procedures, including statistical applications, is given in the classic book [11] by Bradley Efron and Robert Tibshirani.

Stochastic claims reserving techniques, extending the chain ladder, have been developed in the actuarial literature in recent decades by Thomas Mack [28] and

Fig. 7.7 Histogram in
Example 7.12 for predictive
distribution of the value of
assets minus the value of the
liabilities in 1 year from now

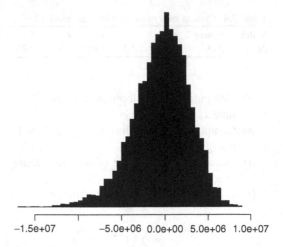

$$-1.5e+07 \qquad\qquad -5.0e+06 \;\; 0.0e+00 \;\; 5.0e+06 \;\; 1.0e+07$$

many others. A comprehensive treatment of such techniques is the book [46] by
Mario Wütrich and Michael Merz. Our approach to bootstrapping the chain ladder
method is a slight variation of the method presented by Peter England and Richard
Verrall [14]. The upper claims triangle in Table 7.5 used in Example 7.12 originates
from a paper by G.C. Taylor and F.R. Ashe [45].

7.8 Exercises

Exercise 7.1 (Empirical VaR). A unit within a bank is required to report an
empirical estimate of $\mathrm{VaR}_{0.01}(X)$, where X is the portfolio value the next day from
its trading activities. The empirical estimate $\widehat{\mathrm{VaR}}_{0.01}(X)$ is based on market prices
from the previous $n + 1$ days that are transformed into a sample of size n from
the distribution of X and the sample points are assumed to be independent and
identically distributed. Compute the probability

$$\mathrm{P}\left(\widehat{\mathrm{VaR}}_{0.01}(X) > \mathrm{VaR}_{0.01}(X)\right)$$

as a function of n and determine its minimum and maximum for n in $\{100, \dots, 300\}$.

Exercise 7.2 (Empirical tail conditional median). The tail conditional median
$\mathrm{TCM}_p(X) = \mathrm{median}[L \mid L \geq \mathrm{VaR}_p(X)]$, where $L = -X/R_0$, has been proposed
as a more robust alternative to $\mathrm{ES}_p(X)$ since $\mathrm{TCM}_p(X)$ is not as sensitive as
$\mathrm{ES}_p(X)$ to the behavior of the left tail of the distribution of X.

   Let Y have a standard Student's t distribution with ν degrees of freedom, and set
$X = e^{0.01Y} - 1$. Consider the empirical estimators $\widehat{\mathrm{TCM}}_{0.05}(X)$ and $\widehat{\mathrm{ES}}_{0.05}(X)$ based
on a sample of size 200 from the distribution of $L = -X$. Generate histograms
based on samples of size 10^5 from the distributions of $\widehat{\mathrm{TCM}}_{0.05}(X)$ and $\widehat{\mathrm{ES}}_{0.05}(X)$
for $\nu = 2$ and $\nu = 10$.

Table 7.5 Upper triangle of paid claims and development factors in Example 7.12

Origin year	Development year									
	0	1	2	3	4	5	6	7	8	9
-10	357,848	766,940	610,542	482,940	527,326	574,398	146,342	139,950	227,229	67,948
-9	352,118	884,021	933,894	1,183,289	445,745	320,996	527,804	266,172	425,046	
-8	290,507	1,001,799	926,219	1,016,654	750,816	146,923	495,992	280,405		
-7	310,608	1,108,250	776,189	1,562,400	272,482	352,053	206,286			
-6	443,160	693,190	991,983	769,488	504,851	470,639				
-5	396,132	937,085	847,498	805,037	705,960					
-4	440,832	847,631	1,131,398	1,063,269						
-3	359,480	1,061,648	1,443,370							
-2	376,686	986,608								
-1	344,014									
-0										
Development factor	3.491	1.747	1.457	1.174	1.104	1.086	1.054	1.077	1.018	1.000

Exercise 7.3 (Empirical expected shortfall). Let $\{Z_1, \ldots, Z_n\}$ be a sample of independent and identically distributed historical log returns that are distributed as the log return $\log(S_T/S_0)$ of an asset from today until time $T > 0$. Show that if the risk-free return over the investment period is 1, then the empirical estimator of $\mathrm{ES}_p(S_T - S_0)$ is given by

$$\min_c -c + \frac{1}{np} \sum_{k=1}^{n} (c + S_0 - S_0 e^{Z_k}) I\{Z_k \leq \log(1 + c/S_0)\}.$$

Exercise 7.4 (Empirical spectral risk measure). Let $\{Z_1, \ldots, Z_n\}$ be a sample of independent and identically distributed historical log returns that are distributed as the log return $\log(S_T/S_0)$ of an asset from today until time $T > 0$. Show that if the risk-free return over the investment period is 1 and if ρ_ϕ is a spectral risk measure with risk aversion function ϕ, then the empirical estimator of $\rho_\phi(S_T - S_0)$ is given by

$$S_0\left(1 - \sum_{k=1}^{n} \phi_k e^{Z_{k,n}}\right), \quad \text{where } \phi_k = \int_{(n-k)/n}^{(n-k+1)/n} \phi(u) du.$$

Project 7 (Total returns). Consider a 5-year investment in a portfolio of dividend-paying stocks. The yearly portfolio returns S_{t+1}/S_t and dividends D_{t+1} paid at time $t + 1$ are modeled as

$$\frac{S_{t+1}}{S_t} = e^{\mu + 0.2 X_{t+1}} \quad \text{and} \quad \frac{D_{t+1}}{S_t} = 0.05 e^{-0.05^2/2 + 0.05 Y_{t+1}},$$

where $X_1, \ldots, X_5, Y_1, \ldots, Y_5$ are independent and standard normally distributed.

(a) Consider the value in 5 years of investing \$1 million in a portfolio of stocks and reinvesting the dividends in the portfolio of stocks. Determine the function f such that the value V_5 in 5 years of the investment strategy can be expressed as $V_5 = f(\mu, X_1, \ldots, X_5, Y_1, \ldots, Y_5)$.

(b) Simulate a sample of suitable size from the distribution of $(X_1, \ldots, X_5, Y_1, \ldots, Y_5)$ and use this sample to determine the empirical distribution of V_5 for a range of values of the parameter μ. Estimate the smallest value of μ for which the probability that V_5 exceeds the value in 5 years of an investment of \$1 million in a 5-year zero-coupon bond with zero rate 5% per year is 0.75.

Project 8 (Pension savings). Consider a yearly investment of \$1,000 in long positions in a portfolio of stocks and a risk-free, 1-year, zero-coupon bond over a 30-year period. The yearly returns on the portfolio of stocks in year k is modeled as $R_k = e^{\mu + \sigma Z_k}$, where Z_k is standard normally distributed. The yearly returns are assumed to be independent. The yearly return on the risk-free bond is assumed to be $e^{0.01}$. The fraction of the yearly amount invested in the portfolio of stocks at the beginning of year k is $p(1 - c(k - 1)/30)$, where $p, c \in [0, 1]$.

(a) Determine a function f such that the value of the pension savings in 30 years can be expressed as $V_{30} = f(\mu, \sigma, p, c, Z_1, \ldots, Z_{30})$.

Simulate a sample of suitable size n from the distribution of $(Z_1, \ldots, Z_{30})$ and use this sample to determine the empirical distribution F_n of V_{30} for a range of values of the parameters μ, σ, p, c.

(b) Set $\mu = 0.06$ and $\sigma = 0.2$ and investigate the effects on the empirical distribution $F_n(p, c)$ of V_{30} of varying p and c. Suggest a suitable criterion for selecting the optimal empirical distribution $F_n(p, c)$ and determine the optimizer (p, c).

Chapter 8
Parametric Models and Their Tails

In this chapter we consider approaches to selecting a parametric family of
distributions for a random variable and approaches to estimating the parameters. We
also present techniques for analyzing the tails of the chosen probability distribution
and the effect of the tails on the estimation of risk measures. Finally, we consider
a semiparametric approach to the estimation of tail probabilities. It provides an
alternative to relying on a full parametric model in order to produce estimates of
tail probabilities beyond the range of the sample data.

A common situation is that we want to model the future value V of a portfolio
of financial instruments and that we may express this random value as $V = g(\mathbf{Z})$,
where $\mathbf{Z}$ is, for instance, a vector of log returns of financial assets or changes in
zero rates over the next time period and g is a known function that depends only on
the current positions and the prices of the different instruments. Even when $Z = \mathbf{Z}$
is univariate, the problem of assigning a good parametric model to Z is far from
straightforward. Therefore, we only consider the univariate problem here and return
to the multivariate case in the following chapter.

A parametric family of distribution functions is a set $\{F_\theta : \theta \in \mathbf{\Theta}\}$ of distribution
functions, where θ is the parameter and $\mathbf{\Theta} \subset \mathbb{R}^k$ the parameter space. One example
is the family of normal distribution functions with parameter (μ, σ^2), parameter
space $\mathbb{R} \times (0, \infty) \subset \mathbb{R}^2$, and distribution function $\Phi((x - \mu)/\sigma)$. Another example
is Student's t location-scale family with parameter (μ, σ, ν), parameter space $\mathbb{R} \times
(0, \infty) \times (0, \infty) \subset \mathbb{R}^3$, and distribution function $t_\nu((x - \mu)/\sigma)$, where $t_\nu(x)$ is
the distribution function for a standard Student's t distribution with ν degrees of
freedom.

In most cases, the parameter θ is at least partly estimated from past observations
$z_{-n+1}, \ldots, z_0$ that are obtained by transforming historical prices of the relevant
asset. If we feel comfortable assuming that the z_{-k} are observations of random
variables $Z_{-n+1}, \ldots, Z_0$ with the same distribution as the random variable Z_1
that we want to model, then we may assign the distribution function $F_{\widehat{\theta}}$ to Z_1,
where the parameter $\widehat{\theta}$ is estimated from the observations z_{-k}. Maximum-likelihood
estimation or least-squares estimation are natural choices of estimation techniques

H. Hult et al., *Risk and Portfolio Analysis: Principles and Methods*, Springer Series
in Operations Research and Financial Engineering, DOI 10.1007/978-1-4614-4103-8_8,
© Springer Science+Business Media New York 2012

that often perform well. The quality of the parameter estimation is best if we know the correct time series model for $Z_{-n+1}, \ldots, Z_0$ and use this knowledge in the estimation procedure. Here we will not go deeper into time series models. It will be assumed that $Z_{-n+1}, \ldots, Z_0$ are roughly independent and identically distributed. Of course, there may be situations where we choose the parameter partly based on the historical observations and partly based on other relevant information that is not present in the historical data.

In an attempt to be objective, it is common to take the parametric family as given and to estimate the parameters of the model based on historical data. Although this may be both sensible and useful, it is important to note that this is nevertheless not an objective approach to choosing a model. The choice of the parametric family, the belief in the explanatory power of the historical data, and the choice of sample size of the historical data sample used to fit the parameters are some of the factors that influence the results, and they are always subjective choices by the modeler. Parametric modeling usually consists of three steps:

1. Select a parametric family of distributions.
2. Estimate the parameters.
3. Validate the resulting distribution.

If in the third step the validation fails (for instance, we assumed a normal family but the estimated normal distribution gave a poor fit to the historical data), then we start over from the first step with a different parametric family. If the validation is considered successful, then the fitted parametric distribution may be used as a subjective model for future values.

Model selection, parameter estimation, and model validation are presented and analyzed in Sect. 8.1.

The choice of parametric model will be crucial for quantifying the riskiness of a portfolio by a risk measure such as value-at-risk (VaR) or expected shortfall (ES). Of particular importance to a risk manager is the probability of extreme outcomes, as these outcomes may lead to large portfolio losses. Therefore, model selection cannot only be based on the data in the center of the distribution, but one must take into account the effect of the modeling of the tails. Throughout the chapter we will emphasize the tail behavior of the parametric models. That is, the shape of the distribution far from the center, where few observations are available. Section 8.2 presents different kinds of tail behavior and analyzes their effects on the potential benefits from diversification for a financial or insurance portfolio. A key message is that if the probability mass of the distribution decays slowly, then the idea of diversification, that works so well in the presence of, for example, normal distributions, may work poorly in the sense that the risk is not reduced by much, if at all.

8.1 Model Selection and Parameter Estimation

In this section, we review some useful techniques for selecting a parametric model to a data sample. It will be assumed that the historical data have been transformed into n observations $z_{-n+1}, \ldots, z_0$ that are the outcomes of the random variables $Z_{-n+1}, \ldots, Z_0$, which we believe are close to independent and identically distributed. A standard approach to selecting an appropriate model is the three-step procedure of first selecting a parametric family, then estimating the parameters, and finally validating the chosen parametric model.

The starting point for model selection is to plot the raw data, make a histogram, and inspect the plots. Based on this graphical inspection, the next step is to compare the empirical quantiles of the data with the quantiles of a few reference distributions that may be suitable by making so-called quantile–quantile (q–q) plots. This approach is presented below. Once a good candidate for a parametric family has been chosen, the next step is to estimate the parameters of the model. The aim here is to find the distribution function F_θ (or density f_θ) in the family $\{F_\theta : \theta \in \Theta\}$ that best represents the distribution of the historical data.

The methods presented below differ in the notion of "best." There is not a single criterion that is always best, but one must understand the properties of the different model selection and estimation techniques and select one that is appropriate for the problem at hand.

8.1.1 Examples of Parametric Distributions

In this section we list some commonly encountered one-dimensional parametric probability distributions used to model financial and insurance data. There are definitely many more distributions that may be considered, but here we restrict ourselves to a short list. In Sect. 8.1.6, a method for constructing new parametric models that provide a good fit to the data is described.

A standard model for financial data is the normal distribution $N(\mu, \sigma^2)$ with density

$$\frac{1}{\sigma}\phi\left(\frac{x-\mu}{\sigma}\right) = \frac{1}{\sqrt{2\pi}\sigma}\exp\left\{-\frac{(x-\mu)^2}{2\sigma^2}\right\} \quad \text{for all } x. \tag{8.1}$$

The normal distribution typically gives a poor fit to financial log-return data regardless of how one sets the parameters. The reason for the poor fit is that its probability mass decays faster as one moves away from its mean than what is indicated by data of daily log returns. An approach to overcoming this problem is to replace the standard deviation parameter σ in the representation $X \stackrel{d}{=} \mu + \sigma Z$ by the random variable $\sigma(\nu/S_\nu)^{1/2}$, where S_ν has a χ^2-distribution with ν degrees of freedom. Recall that if ν is an integer, then S_ν is distributed as the sum of ν squared independent standard normals. The density function of $Y = \mu + \sigma(\nu/S_\nu)^{1/2}Z$ is given by

$$\frac{\Gamma((\nu+1)/2)}{\sqrt{\nu\pi}\,\Gamma(\nu/2)}\left(1 + \frac{(x-\mu)^2}{\nu\sigma^2}\right)^{-(\nu+1)/2} \quad \text{for all } x, \tag{8.2}$$

where Γ denotes the gamma function. This can be recognized as the density of a Student's t distribution with ν degrees of freedom, and location and scale parameters

μ and σ, written $t_\nu(\mu, \sigma^2)$. The standard t_ν distribution corresponds to $(\mu, \sigma^2) = (0, 1)$. For a standard t-distributed random variable, the expected value is 0 (if $\nu > 1$) and the variance is $\nu/(\nu - 2)$ (if $\nu > 2$).

We now take a look at some models for random variables taking only positive values. A random variable X with a lognormal distribution is simply the exponential function of a normally distributed random variable, X is $\mathrm{LN}(\mu, \sigma^2)$-distributed if $\log X$ is $\mathrm{N}(\mu, \sigma^2)$-distributed and the density function of X is given by

$$\frac{1}{\sqrt{2\pi}\sigma x} \exp\left\{-\frac{(\log x - \mu)^2}{2\sigma^2}\right\} \quad \text{for } x > 0.$$

The random variable Y has an exponential distribution with parameter $\lambda > 0$, written $\mathrm{Exp}(\lambda)$, if it has distribution function $\mathrm{P}(Y \leq x) = 1 - e^{-\lambda x}$ for $x \geq 0$. The random variable X has a Pareto distribution with parameters $\alpha > 0$, written $\mathrm{Pa}(\alpha)$, if it has distribution function $\mathrm{P}(X \leq x) = 1 - x^{-\alpha}$ for $x \geq 1$. Since

$$\mathrm{P}(\log X > y) = \mathrm{P}(X > e^y) = e^{-\alpha y},$$

we find that the logarithm of a $\mathrm{Pa}(\alpha)$-distributed X is $\mathrm{Exp}(\alpha)$-distributed. There are two two-parameter versions of the Pareto distribution corresponding to the distribution functions $1 - (c/x)^\alpha$ for $c > 0$ and $x \geq c$ and the distribution function $1 - k^\alpha/(k + x)^\alpha$ for $k > 0$ and $x \geq 0$. We now turn to the Gamma distribution $\Gamma(\alpha, \beta)$ whose density function is given by

$$\frac{\beta^\alpha}{\Gamma(\alpha)} x^{\alpha-1} \exp\{-\beta x\} \quad \text{for } x, \alpha, \beta > 0.$$

If α is an integer, then $\Gamma(\alpha, \beta)$ is the distribution of the sum of α independent $\mathrm{Exp}(1/\beta)$-distributed random variables. Finally, the Weibull distribution $\mathrm{WBL}(\alpha, \beta)$ allows for a wide range of tail probabilities by varying the parameter α. That a random variable X is $\mathrm{WBL}(\alpha, \beta)$-distributed is equivalent to βX^α having a standard exponential distribution $\mathrm{Exp}(1)$. Note that all the probability distributions presented here are related in some rather simple way to the standard normal or the exponential distribution.

Example 8.1 (Normal tails). Let Y be $\mathrm{N}(\mu, \sigma^2)$-distributed. We want to investigate the behavior of its left tail $F_{\mu,\sigma}(x) = \Phi((x - \mu)/\sigma)$ as $x \to -\infty$. In other words, we want to understand how the normal model for a log return assigns probability mass to events corresponding to negative log returns with very large absolute values. We claim that, as $x \to -\infty$,

$$\Phi\left(\frac{x-\mu}{\sigma}\right) \sim \frac{\sigma}{(-x)} \phi\left(\frac{x-\mu}{\sigma}\right) = \frac{1}{\sqrt{2\pi}(-x)} \exp\left\{-\frac{(x-\mu)^2}{2\sigma^2}\right\},$$

where ϕ is the standard normal density and $\sim$ means that the quotient of the left- and right-hand sides tends to one. An application of l'Hôpital's rule gives

$$\lim_{x \to -\infty} \frac{\Phi((x-\mu)/\sigma)}{\phi((x-\mu)/\sigma)\sigma/(-x)} = \lim_{x \to -\infty} \frac{\frac{d}{dx}\Phi((x-\mu)/\sigma)}{\frac{d}{dx}\phi((x-\mu)/\sigma)\sigma/(-x)},$$

provided that the right-hand side converges to some positive limit. We need to show that the limit is one. Computing the derivatives and using the relation $\phi'(y) = -y\,\phi(y)$ give

$$\frac{d}{dx}\Phi\left(\frac{x-\mu}{\sigma}\right) = \frac{1}{\sigma}\phi\left(\frac{x-\mu}{\sigma}\right),$$

$$\frac{d}{dx}\frac{\sigma}{(-x)}\phi\left(\frac{x-\mu}{\sigma}\right) = \frac{\sigma}{x^2}\phi\left(\frac{x-\mu}{\sigma}\right) + \frac{1}{(-x)}\phi'\left(\frac{x-\mu}{\sigma}\right)$$

$$= \frac{1}{\sigma}\phi\left(\frac{x-\mu}{\sigma}\right)\left(1 + \frac{\mu}{(-x)} + \frac{\sigma^2}{x^2}\right),$$

from which the claim follows.

The tail $F_{\mu,\sigma}(x) = \Phi((x-\mu)/\sigma)$ decays faster than an exponential rate in the sense that

$$\lim_{x \to -\infty} \frac{F_{\mu,\sigma}(x)}{e^{-\lambda(-x)}} = 0 \quad \text{for every } \lambda > 0.$$

In addition, the location parameter μ matters for the asymptotic behavior of the tail in the sense that $F_{\mu,\sigma}(x)/F_{\tilde{\mu},\sigma}(x)$ does not converge as $x \to -\infty$ if $\tilde{\mu} \neq \mu$.

Example 8.2 (Student's t tails). Consider a random variable Y with a location-scale Student's t distribution with ν degrees of freedom. The distribution function of Y is given by $t_\nu((x-\mu)/\sigma)$, and the density function is given by

$$g_\nu\left(\frac{x-\mu}{\sigma}\right) = C\left(1 + \frac{(x-\mu)^2}{\nu\sigma^2}\right)^{-(\nu+1)/2} \quad \text{with} \quad C = \frac{\Gamma((\nu+1)/2)}{\sqrt{\nu\pi}\,\Gamma(\nu/2)}.$$

Applying l'Hôpital's rule gives

$$\lim_{x \to -\infty} \frac{t_\nu((x-\mu)/\sigma)}{(-x)^{-\nu}} = \lim_{x \to -\infty} \frac{g_\nu((x-\mu)/\sigma)/\sigma}{\nu(-x)^{-(\nu+1)}}$$

$$= \frac{C}{\nu}\lim_{x \to -\infty}\left(\frac{1}{x^2}\left(1 + \frac{x^2 - 2\mu x + \mu^2}{\nu\sigma^2}\right)\right)^{-(\nu+1)/2}$$

$$= C\nu^{(\nu-1)/2}\sigma^{\nu+1}.$$

We conclude that

$$t_\nu\left(\frac{x-\mu}{\sigma}\right) \sim \frac{\Gamma((\nu+1)/2)}{\sqrt{\nu\pi}\,\Gamma(\nu/2)}\nu^{(\nu-1)/2}\sigma^{\nu+1}(-x)^{-\nu} \quad \text{as } x \to -\infty.$$

The polynomial rate of decay $(-x)^{-\nu}$ is slow in the sense that it is slower than any exponential:

$$\lim_{x \to -\infty} \frac{t_\nu((x-\mu)/\sigma)}{e^{-\lambda(-x)^\tau}} = \infty \quad \text{for every } \lambda, \tau > 0.$$

8.1.2 Quantile–Quantile Plots

In this section, we will consider some useful practical methods to study the distributional properties of a data set. In particular, we will emphasize the extreme values of the data.

Suppose that we have observations $z_1, \dots, z_n$ of independent and identically distributed random variables $Z_1, \dots, Z_n$ with an unknown common distribution function that we would like to determine. A common approach is to suggest a candidate reference distribution F and to test whether it is reasonable to assume that the observations form a sample from F. The q–q plot provides a useful graphical test. First recall that we write $z_{1,n} \geq \cdots \geq z_{n,n}$ for the ordered sample. A q–q plot is a plot of the points

$$\left\{ \left(F^{-1}\left(\frac{n-k+1}{n+1} \right), z_{k,n} \right) : k = 1, \dots, n \right\}. \tag{8.3}$$

This is a plot of the empirical quantiles against the quantiles of the reference distribution. At first sight this is not quite clear. However, notice that with $p = (n-k+1)/(n+1)$ we have $F_n^{-1}(p) = z_{[n(1-p)]+1,n}$, where

$$[n(1-p)] = \left[n\left(1 - \frac{n-k+1}{n+1} \right) \right] = \left[\frac{nk}{n+1} \right] < k,$$

$$n(1-p) = \frac{nk}{n+1} > \frac{nk}{n+1} + \frac{k-1}{n+1} - \frac{n}{n+1} = k - 1.$$

Therefore, (8.3) may equivalently be written as

$$\left\{ \left(F^{-1}\left(\frac{n-k+1}{n+1} \right), F_n^{-1}\left(\frac{n-k+1}{n+1} \right) \right) : k = 1, \dots, n \right\}.$$

If the data are generated by a probability distribution similar to the reference distribution, then the q–q plot is approximately linear. An important property is that the q–q plot remains approximately linear if the data are transformed by an affine transformation, i.e., if the distribution of the data is approximately in the same location-scale family as the reference distribution. If the data form a sample from the reference distribution F, then the q–q plot should be approximately linear with intercept 0 and slope 1. If the data form a sample from $F_{\mu,\sigma}(x) = F((x-\mu)/\sigma)$, then the q–q plot is still approximately linear since $F_{\mu,\sigma}^{-1}(p) = \mu + \sigma F^{-1}(p)$. Moreover, the parameters μ and σ can be estimated from the intercept and slope of the q–q plot.

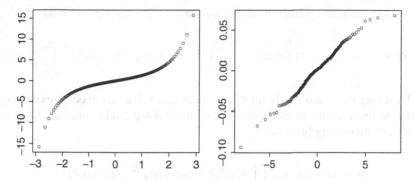

Fig. 8.1 *Left plot*: q–q plot of standard Student's t distribution with two degrees of freedom (y-axis) against standard normal distribution (x-axis). *Right plot*: q–q plot of Nasdaq Composite log-return data (y-axis) against standard Student's t distribution with three degrees of freedom (x-axis)

The q–q plot is particularly useful for studying the tails of a distribution. Suppose that a reference distribution F is given. If the empirical distribution of the data has a heavier right tail than the reference distribution, meaning that $(1 - F_n(x))/(1 - F(x))$ is increasing for x sufficiently large, then the q–q plot will curve up (convex shape) for large values on the x-axis. If the empirical distribution of the data has a heavier left tail than F, meaning that $F_n(x)/F(x)$ is increasing for x sufficiently small, then the q–q plot will curve down (concave shape) for small values on the x-axis. This behavior is illustrated in the left plot in Fig. 8.1. If the empirical tails are lighter than those of the reference distribution, then the q–q plot shows the opposite behavior (an S-shaped q–q plot).

Quantile–quantile plots are often used to analyze the goodness of fit of a reference distribution. If a parametric family has been selected and parameters estimated with some statistical technique, then a q–q plot can be used to check that the suggested distribution actually gives a good fit; the q–q plot should look linear. If it does not look linear, then one should reconsider the choice of parametric family or the estimation technique.

8.1.3   Maximum-Likelihood Estimation

Consider observations $z_1, \ldots, z_n$ of independent and identically distributed random variables $Z_1, \ldots, Z_n$ with the density function f_{θ_0}, where the parameter θ_0 is unknown. In maximum-likelihood estimation (MLE), the unknown parameter θ_0 is estimated as the parameter value θ that maximizes the probability of the observed data. More precisely, the maximum-likelihood estimator $\widehat{\theta}$ is given by

$$\widehat{\theta} = \operatorname{argmax}_\theta \prod_{k=1}^n f_\theta(Z_k).$$

Strictly speaking, the probability of any outcome $z_1, \ldots, z_n$ is zero. However,

$$P(Z_k \in (z_k - \varepsilon, z_k + \varepsilon) \text{ for all } k) = \prod_{k=1}^{n} \int_{z_k - \varepsilon}^{z_k + \varepsilon} f_\theta(x_k) dx_k \approx (2\varepsilon)^n \prod_{k=1}^{n} f_\theta(z_k),$$

and by taking $\varepsilon > 0$ arbitrarily small, the parameter value that maximizes the right-hand side also maximizes the probability on the left-hand side. Since the logarithm is a strictly increasing function

$$\widehat{\theta} = \operatorname{argmax}_\theta \log \prod_{k=1}^{n} f_\theta(Z_k) = \operatorname{argmax}_\theta \sum_{k=1}^{n} \log f_\theta(Z_k).$$

The sum on the right-hand side is the log-likelihood function $\log L(\theta; Z_1, \ldots, Z_n)$.

To see that the maximum-likelihood estimator is a good estimator, we want $\widehat{\theta} = \operatorname{argmax}_\theta L(\theta; Z_1, \ldots, Z_n)$ to converge to θ_0 with probability one as $n \to \infty$. We do not give a full proof here but an argument that can be refined into a rigorous proof (under additional rather mild assumptions). It follows from the law of large numbers that, with probability one,

$$\frac{1}{n} \log L(\theta; Z_1, \ldots, Z_n) - \frac{1}{n} \log L(\theta_0; Z_1, \ldots, Z_n) = \frac{1}{n} \sum_{k=1}^{n} \log \frac{f_\theta(Z_k)}{f_{\theta_0}(Z_k)}$$

$$\to \operatorname{E}\left[\log \frac{f_\theta(Z_1)}{f_{\theta_0}(Z_1)} \right]$$

as $n \to \infty$. Notice that $\widehat{\theta}$ is the maximizer of the left-hand side. We claim that θ_0 is the maximizer of the right-hand side. Indeed, from Jensen's inequality,

$$\operatorname{E}\left[\log \frac{f_\theta(Z_1)}{f_{\theta_0}(Z_1)} \right] \le \log \operatorname{E}\left[\frac{f_\theta(Z_1)}{f_{\theta_0}(Z_1)} \right] = \log \int f_\theta(x) dx = 0,$$

with equality if and only if $f_\theta = f_{\theta_0}$.

Example 8.3 (MLE for the normal distribution). For the normal distribution $N(\mu, \sigma^2)$ the log-likelihood function is given by

$$\log L(\mu, \sigma^2; z_1, \ldots, z_n) = \sum_{k=1}^{n} \left(-\log(2\pi\sigma^2)^{1/2} - \frac{1}{2\sigma^2}(z_k - \mu)^2 \right)$$

$$= -\frac{n}{2} \log(2\pi) - \frac{n}{2} \log(\sigma^2) - \frac{1}{2\sigma^2} \sum_{k=1}^{n} (z_k - \mu)^2.$$

For notational convenience write $g(\mu, \sigma^2)$ for the log-likelihood function. It can be checked that $g(\mu, \sigma^2)$ is concave in (μ, σ^2). Therefore, the maximum-likelihood

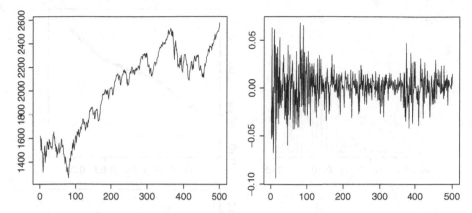

Fig. 8.2 *Left plot*: values of Nasdaq Composite index from November 11, 2008 until November 4, 2010. *Right plot*: corresponding log returns

estimator is obtained by computing the partial derivatives of g with respect to μ and with respect to σ^2, setting these two expressions to zero, and solving the equation system for (μ, σ^2). We obtain

$$\frac{\partial g}{\partial \mu}(\mu, \sigma^2) = \frac{1}{\sigma^2} \sum_{k=1}^{n}(z_k - \mu) = \frac{n}{\sigma^2}(\bar{z} - \mu)$$

$$\frac{\partial g}{\partial \sigma^2}(\mu, \sigma^2) = -\frac{n}{2\sigma^2} + \frac{1}{2(\sigma^2)^2} \sum_{k=1}^{n}(z_k - \mu)^2$$

$$= \frac{1}{2(\sigma^2)^2}\left(\sum_{k=1}^{n}(z_k - \mu)^2 - n\sigma^2 \right),$$

from which it follows that the maximum-likelihood estimator for (μ, σ^2) is

$$(\widehat{\mu}, \widehat{\sigma}^2) = \left(\bar{z}, \frac{1}{n}\sum_{k=1}^{n}(z_k - \bar{z})^2 \right). \tag{8.4}$$

[It would be more appropriate to write $\widehat{(\mu, \sigma^2)}$ instead of $(\widehat{\mu}, \widehat{\sigma}^2)$ but that hurts the eye.] The optimization procedure ignored the constraint $\sigma^2 \geq 0$, but this turned out not to be a problem.

Example 8.4 (Nasdaq data I). Consider a sample consisting of $n = 500$ log returns $z_1, \dots, z_n$ from the Nasdaq Composite index (index values from November 11, 2008 through November 4, 2010). Here we want to fit a parametric probability distribution to the log-return data and use this distribution as a model for future log returns. The historical index values and the log returns are shown as time series in Fig. 8.2. An

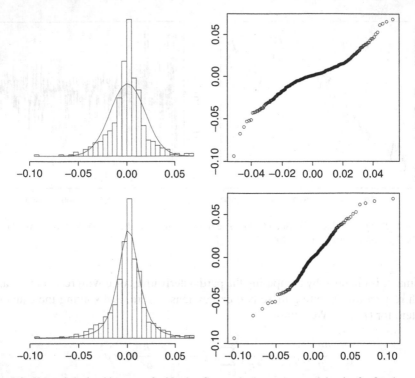

Fig. 8.3 *Upper left plot*: histogram for Nasdaq Composite log returns and density for fitted normal distribution. *Upper right plot*: q–q plot of Nasdaq Composite log returns (*y*-axis) against fitted normal quantiles (*x*-axis). *Lower left plot*: histogram for Nasdaq Composite log returns and density for fitted Student's *t* distribution. *Lower right plot*: q–q plot of Nasdaq Composite log returns (*y*-axis) against fitted Student's *t* quantiles (*x*-axis)

initial eyeball inspection leaves it unclear as to whether the observed log returns can be seen as outcomes of independent and identically distributed random variables $Z_1, \ldots, Z_n$. For instance, we observe a somewhat higher volatility in the beginning of the sample representing a time period of financial turmoil. Nevertheless, we make the assumption that the observed log returns are outcomes of independent and identically distributed random variables. First we propose a normal distribution $N(\mu, \sigma^2)$ as a model for the log returns. The density is given by (8.1), and the maximum-likelihood estimator of (μ, σ^2) is given in (8.4). Inserting the numerical values gives the estimate

$$(\widehat{\mu}, \widehat{\sigma}) \approx (9.33 \cdot 10^{-4}, 1.82 \cdot 10^{-2}).$$

The upper left plot in Fig. 8.3 of the density with estimated parameters shows that the fit to the log-return data is not particularly good. There is too little mass in the center and the tails. The q–q plot in the upper right of Fig. 8.3 also confirms that the center and the tails are not well modeled by the normal distribution. Due to the poor fit of the normal distribution, we consider fitting a location-scale Student's t distribution, $t_\nu(\mu, \sigma^2)$, to the log-return data. The density function is given by (8.2), and the parameter vector to be estimated is $\theta = (\mu, \sigma, \nu)$. Numerical maximization of the likelihood function gives

$$(\widehat{\mu}, \widehat{\sigma}, \widehat{\nu}) \approx (1.57 \cdot 10^{-3}, 1.12 \cdot 10^{-2}, 2.65).$$

The lower left plot in Fig. 8.3 shows the Student's t density with estimated parameters. The plot shows a better, but not perfect, fit than for the normal distribution, and the q–q plot in the lower right shows that the center and the tails are better modeled with the location-scale Student's t than with the normal distribution. One should not take the numerical estimates too seriously. The log-likelihood function is typically rather flat, and small changes in the data can result in big changes in the parameter estimates. Moreover, for typical financial log-return data two different numerical optimizers may produce rather different parameter estimates. For Student's t model increasing both the scale parameter σ and the degrees of freedom parameter ν (the latter corresponds to making the tails lighter) may give the same value for the log-likelihood function as the obtained optimal parameters.

There is another problem with MLE in this context: maximizing the log-likelihood function is not the same as choosing the parameters so that the q–q plot is as linear as possible.

It is important to realize that the model selection can have a dramatic impact on the estimate of a quantity such as a risk measure that depends on the shape of the tail of the distribution. This is a problem because usually we have very little information about the shape of the tail. By definition, there are few observations there. In the following example, we illustrate the sensitivity of ES to model selection. We fit three models to log-return data: an empirical distribution, a normal distribution, and a Student's t distribution. Although all three give a reasonable fit and very similar values of the VaR at level 0.05, the estimates for ES are quite different. This is because ES depends heavily on the shape of the tail in regions far out where we have no data. Consequently, methods of portfolio optimization, where the optimal portfolio allocations are largely determined by the tail of the distribution, should be regarded with some skepticism.

Example 8.5 (Nasdaq data II, continutation of Example 8.4). Consider a long position of value 100 in the Nasdaq Composite index. We want to analyze the risk we face by holding this position until tomorrow by comparing the estimates of VaR and ES at the 5% level for the position held over 1 day (due to a change in the closing price from today until tomorrow).

We begin with VaR and consider the change X in the value of the position over the next day (the effect of interest rates is ignored). Recall that $X = V_1 - V_0 R_0$ is seen as the net value tomorrow of a position obtained by borrowing V_0 today, taking the index position with value V_1 tomorrow. The value of the loan does not change much over 1 day, so we approximate $V_0 R_0 \approx V_0$. If Z_1 is the log return on the index from today until tomorrow, then we may write

$$X = V_1 - V_0 = 100(\exp\{Z_1\} - 1).$$

Proposition 6.3, with $g(z) = 100(\exp\{z\} - 1)$, and Proposition 6.4 imply that

$$\mathrm{VaR}_{0.05}(X) = F^{-1}_{-g(Z_1)}(1 - 0.05)$$

$$= -g(F^{-1}_{Z_1}(0.05))$$

$$= -100(\exp\{F^{-1}_{Z_1}(0.05)\} - 1),$$

where Z_1 is either normally or Student's t distributed. If Z_1 is normally distributed, then $F^{-1}_{Z_1}(0.05) = \mu + \sigma\Phi^{-1}(0.05)$ and if Z_1 is t distributed, then $F^{-1}_{Z_1}(0.05) = \mu + \sigma t^{-1}_\nu(0.05)$. The estimates of $\mathrm{VaR}_{0.05}(X)$ for the two models are obtained by plugging in the estimated parameters from Example 8.4. The empirical estimate of $\mathrm{VaR}_{0.05}(X)$ is obtained by setting $l_k = -100(\exp\{z_k\} - 1)$, ordering the l_k, and finally taking $l_{[500 \cdot 0.05]+1,500} = l_{26,500}$ as the empirical estimate. The estimates are

$$\widehat{\mathrm{VaR}}_{0.05}(X) \approx \begin{cases} 3.01, \text{ empirical,} \\ 2.85, \text{ normal,} \\ 2.59, \text{ Student's } t. \end{cases}$$

We now turn to estimation of ES. The empirical estimate of $\mathrm{ES}_{0.05}(X)$ is obtained by setting $l_k = -100(\exp\{z_k\} - 1)$, ordering the l_k, and finally taking $(l_{1,500} + \cdots + l_{25,500})/25$ as the empirical estimate. If Z_1 is normally distributed, then Example 6.15 provides the formula

$$\mathrm{ES}_p(X) = V_0 \left(1 - \frac{\Phi(\Phi^{-1}(p) - \sigma)e^{\mu+\sigma^2/2}}{p} \right).$$

For the location-scale Student's t model Example 6.15 provides the formula

$$\mathrm{ES}_p(X) = V_0 \left(1 - \frac{1}{p} \int_0^p \exp\{\mu + \sigma t^{-1}_\nu(u)\})du \right).$$

With $p = 0.05$ and μ, σ, ν replaced by the estimates in Example 8.4, numerical integration gives the result. The ES estimates are

$$\widehat{\mathrm{ES}}_{0.05}(X) \approx \begin{cases} 4.20, \text{ empirical,} \\ 3.59, \text{ normal,} \\ 4.52, \text{ Student's } t. \end{cases}$$

Note that the estimate for ES for the Student's t model is approximately 25% larger than that for the normal model, even though the VaR estimate for the Student's t model is approximately 10% smaller than that for the normal model. An explanation for the sizes of the estimates follows from the plot in Fig. 8.4. The left plot shows

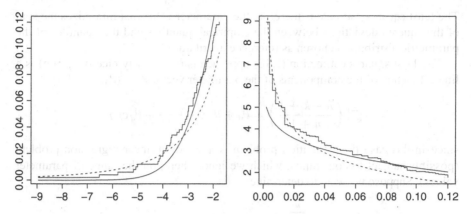

Fig. 8.4 *Left plot*: tail of net worth X in Example 8.4. The empirical tail is the *staircase curve*. The fitted normal tail is the *solid curve*, and the fitted Student's t tail is the *dashed curve*. *Right plot*: $\text{VaR}_p(X)$ as a function of p. The estimate of $\text{ES}_{0.05}(X)$ is the "area under the graph" from 0 to 0.05 divided by 0.05

the left tail of the distribution of X for the three models (empirical, normal, t). We see that for $x_0 \approx -3$

$$\Phi((x - \widehat{\mu}_n)/\widehat{\sigma}_n) > t_{\widehat{\nu}}((x - \widehat{\mu}_t)/\widehat{\sigma}_t) \text{ for } x > x_0,$$
$$\Phi((x - \widehat{\mu}_n)/\widehat{\sigma}_n) < t_{\widehat{\nu}}((x - \widehat{\mu}_t)/\widehat{\sigma}_t) \text{ for } x < x_0,$$

where the subscripts indicate the model (normal or t). The fitted normal tail dominates the fitted Student's t tail for moderate x-values, whereas the opposite holds in the far-out tail. The right plot in Fig. 8.4 shows $\text{VaR}_p(X)$ as a function of p for the three models. Since the Student's t $\text{VaR}_p(X)$ estimates are much larger than those for the normal model for small values of p, the estimates of $\text{ES}_p(X)$ for the Student's t model dominate those for the normal model for rather moderate values of p. It looks like the Student's t distribution gives a much better fit to the tail and the quantile function than the normal distribution. But one should be aware that the limited amount of data available in the tail makes it is impossible to say how the true distribution looks beyond the range of the data.

8.1.4 Least-Squares Estimation

Consider observations $z_1, \ldots, z_n$ of independent and identically distributed random variables $Z_1, \ldots, Z_n$ with the distribution function F_{θ_0}, where the parameter θ_0 is unknown. The least-squares estimator $\widehat{\theta}$ of the unknown parameter θ_0 is given by

$$\widehat{\theta} = \operatorname{argmin}_\theta \sum_{k=1}^n \left(z_{k,n} - F_\theta^{-1}\left(\frac{n-k+1}{n+1}\right) \right)^2.$$

The least-squares estimate is therefore the parameter value that minimizes the sum of the squared deviations between the empirical quantiles and the quantiles of the parametric distribution chosen as reference distribution.

The least-squares estimation (LSE) approach is particularly nice if $F_\theta^{-1}(p)$ is a linear function of the components of the parameter vector $\theta = (\theta_0, \ldots, \theta_d)^T$,

$$F_\theta^{-1}\left(\frac{n-k+1}{n+1}\right) = \theta_0 + \theta_1 c_{k,1} + \cdots + \theta_d c_{k,d},$$

since in this case the estimation problem is a standard linear regression problem (possibly subject to constraints, which we ignore here, on the range of parameter values). Suppose we want to determine

$$\widehat{\theta} = \mathrm{argmin}_\theta \sum_{k=1}^n (z_{k,n} - \theta_0 - \theta_1 c_{k,1} - \cdots - \theta_d c_{k,d})^2,$$

where $c_{k,l}$ are some given numbers. Write

$$\mathbf{z} = \begin{pmatrix} z_{1,n} \\ \vdots \\ z_{n,n} \end{pmatrix} \quad \text{and} \quad \mathbf{C} = \begin{pmatrix} 1 & c_{1,1} & \cdots & c_{1,d} \\ \vdots & & & \vdots \\ 1 & c_{n,1} & \cdots & c_{n,d} \end{pmatrix}.$$

We look for the minimizer $\widehat{\theta}$ of $(\mathbf{z} - \mathbf{C}\theta)^T(\mathbf{z} - \mathbf{C}\theta)$, which is a convex function of θ. We know from Chap. 2 that a solution to the linear equation system $\nabla(\mathbf{z} - \mathbf{C}\theta)^T(\mathbf{z} - \mathbf{C}\theta) = \mathbf{0}$ is the least-squares estimator $\widehat{\theta}$. Computing the partial derivatives of $(\mathbf{z} - \mathbf{C}\theta)^T(\mathbf{z} - \mathbf{C}\theta)$ with respect to $\theta_0, \ldots, \theta_d$ gives $\nabla(\mathbf{z} - \mathbf{C}\theta)^T(\mathbf{z} - \mathbf{C}\theta) = \mathbf{C}^T(\mathbf{z} - \mathbf{C}\theta)$, which in turn gives the solution

$$\widehat{\theta} = (\mathbf{C}^T\mathbf{C})^{-1}\mathbf{C}^T\mathbf{z} \tag{8.5}$$

if $\mathbf{C}^T\mathbf{C}$ is invertible.

Example 8.6 (LSE for a normal distribution). Here we consider the normal distribution $N(\mu, \sigma^2)$ and take $\theta = (\mu, \sigma)$ and

$$F_\theta^{-1}\left(\frac{n-k+1}{n+1}\right) = \mu + \sigma\Phi^{-1}\left(\frac{n-k+1}{n+1}\right).$$

With the notation above we have $d = 1$, $\theta_0 = \mu$, $\theta_1 = \sigma$, and $c_{k,1} = \Phi^{-1}((n-k+1)/(n+1))$. Moreover,

$$(\mathbf{C}^T\mathbf{C})^{-1} = \begin{pmatrix} n & \sum_{k=1}^n \Phi^{-1}\left(\frac{n-k+1}{n+1}\right) \\ \sum_{k=1}^n \Phi^{-1}\left(\frac{n-k+1}{n+1}\right) & \sum_{k=1}^n \Phi^{-1}\left(\frac{n-k+1}{n+1}\right)^2 \end{pmatrix}^{-1}$$

$$= \begin{pmatrix} n & 0 \\ 0 & \sum_{k=1}^n \Phi^{-1}\left(\frac{n-k+1}{n+1}\right)^2 \end{pmatrix}^{-1}$$

and

$$\mathbf{C}^{\mathrm{T}}\mathbf{z} = \begin{pmatrix} \sum_{k=1}^{n} z_k \\ \sum_{k=1}^{n} z_{k,n} \, \Phi^{-1}\left(\frac{n-k+1}{n+1}\right) \end{pmatrix}.$$

Therefore, the least-squares estimator of (μ, σ) is given by

$$(\widehat{\mu}, \widehat{\sigma}) = \left(\bar{z}, \frac{\sum_{k=1}^{n} z_{k,n} \, \Phi^{-1}\left(\frac{n-k+1}{n+1}\right)}{\sum_{k=1}^{n} \Phi^{-1}\left(\frac{n-k+1}{n+1}\right)^2} \right).$$

Example 8.7 (LSE for Student's t distribution). Here we consider Student's t distribution $t_\nu(\mu, \sigma^2)$ and take $\theta = (\mu, \sigma, \nu)$ and

$$F_\theta^{-1}\left(\frac{n-k+1}{n+1}\right) = \mu + \sigma t_\nu^{-1}\left(\frac{n-k+1}{n+1}\right).$$

The least-squares estimate $(\widehat{\mu}, \widehat{\sigma}, \widehat{\nu})$ is the parameter triple that minimizes the sum of the squared deviations between the empirical and Student's t quantiles:

$$(\widehat{\mu}, \widehat{\sigma}, \widehat{\nu}) = \operatorname{argmin}_{(\mu, \sigma \nu)} \sum_{k=1}^{n} \left(z_{k,n} - \mu - \sigma t_\nu^{-1}\left(\frac{n-k+1}{n+1}\right) \right)^2$$

subject to the constraint $\sigma, \nu > 0$. Since the model quantiles are nonlinear functions of the parameter ν, the estimation problem cannot be reduced to linear regression and must be solved numerically.

Example 8.8 (Nasdaq data III, continuation of Example 8.5). In Example 8.4 we fitted a normal distribution and a location-scale Student's t distribution to the log returns of the Nasdaq Composite index using maximum likelihood. The corresponding least-squares estimate is

$$(\widehat{\mu}, \widehat{\sigma}) \approx (9.33 \cdot 10^{-4}, 1.79 \cdot 10^{-2})$$

for the normal distribution and

$$(\widehat{\mu}, \widehat{\sigma}, \widehat{\nu}) \approx (9.33 \cdot 10^{-4}, 1.28 \cdot 10^{-2}, 3.51)$$

for Student's t distribution. The least-squares parameter estimates lead to different estimates of $\mathrm{VaR}_{0.05}(X)$ and $\mathrm{ES}_{0.05}(X)$ compared to the case in Example 8.5 with maximum-likelihood parameter estimates. Here the $\mathrm{VaR}_{0.05}(X)$ estimates are

$$\widehat{\mathrm{VaR}}_{0.05}(X) \approx \begin{cases} 3.01, & \text{empirical,} \\ 2.80, & \text{normal,} \\ 2.71, & \text{Student's } t, \end{cases}$$

and the $\text{ES}_{0.05}(X)$ estimates are

$$\widehat{\text{ES}}_{0.05}(X) \approx \begin{cases} 4.20, \text{ empirical,} \\ 3.52, \text{ normal,} \\ 4.22, \text{ Student's } t. \end{cases}$$

We can compare the estimates of VaR and ES when the parametric model is estimated using least squares, as in the last example, to those in Example 8.5, where maximum likelihood was used. We observe that the parameter estimation method has an impact on the estimates of $\text{VaR}_{0.05}(X)$ and $\text{ES}_{0.05}(X)$. To evaluate the accuracy of the estimation method, we may apply the parametric bootstrap to obtain approximate confidence intervals for VaR and ES. The procedure is illustrated in the following section.

8.1.5 Parametric Bootstrap

The parametric bootstrap works similarly to the nonparametric bootstrap presented in Sect. 7.5.2, with the difference that the parametric form of the distribution is taken into account. Let us illustrate how the method works to construct an approximate confidence interval for a quantile.

Suppose we have a sample $\{x_1, \dots, x_n\}$ of size n from a distribution F_θ, which depends on an unknown parameter θ. We have an estimator $\widehat{\theta} = \widehat{\theta}(X_1, \dots, X_n)$, which is used to estimate θ, and $\widehat{\theta}_{\text{obs}} = \widehat{\theta}(x_1, \dots, x_n)$ is the point estimate of θ. An approximative confidence interval $I_{\theta,q}$ for $F_\theta^{-1}(p)$ of confidence level q using the parametric bootstrap method is constructed as follows:

- For each j in the set $\{1, \dots, N\}$ draw with replacement n times from the sample $\{x_1, \dots, x_n\}$ to obtain the sample $\{X_1^{*(j)}, \dots, X_n^{*(j)}\}$.
- Compute the estimates $\widehat{\theta}_j^* = \widehat{\theta}(X_1^{*(j)}, \dots, X_n^{*(j)})$ of θ and the corresponding quantiles $F_{\widehat{\theta}_j^*}^{-1}(p)$ for $j = 1, \dots, N$.
- Form the residuals $R_j^* = F_{\widehat{\theta}_{\text{obs}}}^{-1}(p) - F_{\widehat{\theta}_j^*}^{-1}(p)$.
- Compute the interval

$$I_{\theta,q} = (F_{\widehat{\theta}_{\text{obs}}}^{-1}(p) + R_{[N(1+q)/2]+1,N}^*, F_{\widehat{\theta}_{\text{obs}}}^{-1}(p) + R_{[N(1-q)/2]+1,N}^*),$$

where $R_{1,N}^* \geq \cdots \geq R_{N,N}^*$ is the ordering of the sample $\{R_1^*, \dots, R_N^*\}$.

Example 8.9 (Nasdaq data IV, cont. of Example 8.8). In Examples 8.4, 8.5, and 8.8 we estimated VaR and ES at level 0.05 for an investment in the Nasdaq Composite stock index using two different estimation methods. To evaluate the accuracy of the estimates, the parametric bootstrap is applied, where in the procedure outlined

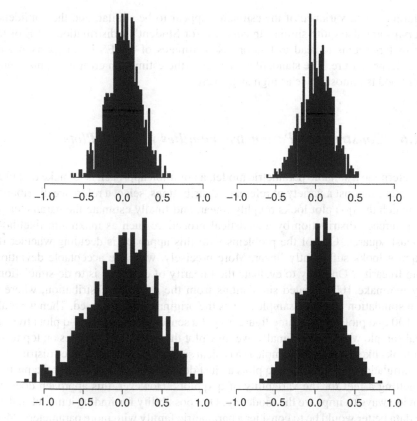

Fig. 8.5 *Upper plots*: residuals R_j^*, for $j = 1, \ldots, 1,000$, from parametric bootstrap for computing ES at level 0.05 based on a normal distribution (*upper left*: residuals when maximum likelihood is used; *upper right*: residuals when LSE is used). *Lower plots*: residuals when ES is computed using Student's t distribution (*lower left*: maximum likelihood; *lower right*: LSE)

previously the quantile $F_\theta^{-1}(p)$ is replaced by $ES_{0.05}$. The parametric bootstrap procedure is performed with 1,000 resamples to compute approximate confidence intervals for the ES. The residuals R_j^*, for $j = 1, \ldots, 1000$, from the parametric bootstrap procedure are illustrated in Fig. 8.5. The resulting confidence intervals at level $q = 0.95$ are

$$I_{N,MLE} \approx (3.22, 3.99) \quad \text{and} \quad I_{N,LSE} \approx (3.17, 3.90)$$

for the normal distribution and

$$I_{t,MLE} \approx (3.70, 5.29) \quad \text{and} \quad I_{t,LSE} \approx (3.80, 4.79)$$

for Student's t distribution. We observe that for the normal model estimating the parameters with maximum likelihood or least squares does not make much of a difference. The variance of the estimates appear to be similar, and the confidence intervals are also quite similar. In contrast, for Student's t distribution MLE of the parameters seems to lead to less precise estimates of the ES in comparison with the LSE procedure. The standard deviation of the estimation error using maximum likelihood is almost twice as high as for LSE.

8.1.6   Constructing Parametric Families with q–q Plots

To determine a suitable parametric model, a common approach is to make q–q plots of the data against a variety of reference distributions, select a reference distribution for which the q–q plot looks roughly linear, and finally estimate the parameters of the reference distribution by a statistical procedure such as maximum likelihood or least squares. One of the problems with this approach is deciding whether the q–q plot looks sufficiently linear. More precisely, what are acceptable deviations from linearity? One way to evaluate the linearity of q–q plots is to do simulations. Say we make 100 repeated simulations from the reference distribution, where in each simulation the same sample size as the original sample is used. Then we make the 100 q–q plots, in the same figure, to get a sense of how linear q–q plots from an ideal sample would look. Finally, we may plot the empirical quantiles on top to see if it looks like the original sample is a typical sample from the reference distribution.

Simulations of repeated q–q plots as just described may be good for the purpose of getting a feel for the variability of q–q plots. However, this approach does not provide ways to improve the modeling. One possibility for finding a model that fits the data better would be to consider a parametric family with more parameters. More parameters implies more flexibility and typically gives a better fit. We prefer to keep models as simple as possible and keep the number of parameters small.

An alternative approach starts with the observation that if $Z = g(Y)$ for a nondecreasing left-continuous function g, then $F_Z^{-1}(p) = g(F_Y^{-1}(p))$ (this is the statement of Proposition 6.3). Assign some simple standard distribution to Y; for instance, a standard normal if the data take both positive and negative values or a standard exponential if the data take only positive values. Make a q–q plot of the empirical quantiles against the quantiles for Y. Then find a suitable function g such that $F_n^{-1}(p) \approx g(F_Y^{-1}(p))$. The quality of the fit certainly depends on the choice of g. If $g(x) = F_n^{-1}(F_Y(x))$, then the probability distribution of $g(Y)$ is the empirical distribution of the z_k. In particular, the q–q plot is linear and a perfect fit is obtained. However, we have obtained a model with too many parameters, $\theta = (z_1, \ldots, z_n)$. Loosely speaking, we want the simplest model that gives a sufficiently good fit to the data.

Example 8.10 (Polynomial normal model). Consider the 500 observed log returns from the Nasdaq Composite index. The lower left plot in Fig. 8.7 shows a q–q plot of the empirical quantiles against the quantiles of the standard normal distribution. The q–q plot looks like the graph of a third-degree polynomial. Therefore, it seems plausible that the log-return sample can be seen as outcomes of the random variable $g(Y; \boldsymbol{\theta})$, where Y is standard normally distributed, $\boldsymbol{\theta} = (\theta_0, \theta_1, \theta_2, \theta_3)$, and

$$g(y; \boldsymbol{\theta}) = \theta_0 + \theta_1 y + \theta_2 y^2 + \theta_3 y^3.$$

We call this model the polynomial normal model. Since a third-degree polynomial $g(y; \boldsymbol{\theta})$ that fits a q–q plot of empirical quantiles against the standard normal quantiles is an increasing function, Proposition 6.3 implies that the quantile function of $g(Y; \boldsymbol{\theta})$ is given by

$$F_{g(Y;\boldsymbol{\theta})}^{-1}(p) = \theta_0 + \theta_1 \Phi^{-1}(p) + \theta_2 \Phi^{-1}(p)^2 + \theta_3 \Phi^{-1}(p)^3.$$

In particular, LSE of $\boldsymbol{\theta}$ is given by

$$(\widehat{\theta}_0, \dots, \widehat{\theta}_3) = \operatorname{argmin}_{(\theta_1, \dots, \theta_4)} \sum_{k=1}^{n} \left(z_{k,n} - \sum_{l=0}^{3} \theta_l \Phi^{-1} \left(\frac{n-k+1}{n+1} \right)^l \right)^2,$$

subject to the constraint $g'(y; \boldsymbol{\theta}) = \theta_1 + 2\theta_2 y + 3\theta_3 y^2 \geq 0$. Except for the constraint, this is just ordinary linear regression. We may ignore the constraint $g'(y; \boldsymbol{\theta}) \geq 0$, solve the linear regression problem, and verify that $g'(y; \widehat{\boldsymbol{\theta}}) \geq 0$. The solution $\widehat{\boldsymbol{\theta}}$ to the linear regression problem is given by (8.5), which here yields $\widehat{\boldsymbol{\theta}} = (\mathbf{C}^{\mathsf{T}} \mathbf{C})^{-1} \mathbf{C}^{\mathsf{T}} \mathbf{z}$, where

$$\mathbf{C} = \begin{pmatrix} 1 & c_{1,1} & c_{1,2} & c_{1,3} \\ \vdots & & & \vdots \\ 1 & c_{n,1} & c_{n,2} & c_{n,3} \end{pmatrix} \quad \text{with} \quad c_{k,l} = \Phi^{-1} \left(\frac{n-k+1}{n+1} \right)^l.$$

Notice that $g'(y; \boldsymbol{\theta}) = \theta_1 + 2\theta_2 y + 3\theta_3 y^2 \geq 0$ for all y if $\theta_3 \geq 0$ and if $g'(y^*; \boldsymbol{\theta}) \geq 0$ for the minimizer y^* of $g'(y; \boldsymbol{\theta})$. The minimizer y^* solves $g''(y; \boldsymbol{\theta}) = 0$ and is given by $y^* = -\theta_2/(3\theta_3)$. Therefore, $g(y; \boldsymbol{\theta})$ is an increasing function if $\theta_3 \geq 0$ and $3\theta_1 \theta_3 - \theta_2^2 \geq 0$.

If $3\theta_1 \theta_3 - \theta_2^2 > 0$ so that $g(y; \boldsymbol{\theta})$ is a strictly increasing function, then

$$P(g(Y; \boldsymbol{\theta}) \leq x) = P(Y \leq g^{-1}(x; \boldsymbol{\theta})) = \Phi(g^{-1}(x; \boldsymbol{\theta})),$$

from which the chain rule gives the density function

$$f_{g(Y;\boldsymbol{\theta})}(x) = \frac{d}{dx} \Phi(g^{-1}(x; \boldsymbol{\theta})) = \phi(g^{-1}(x; \boldsymbol{\theta})) \frac{d}{dx} g^{-1}(x; \boldsymbol{\theta})$$

$$= \frac{\phi(g^{-1}(x; \boldsymbol{\theta}))}{g'(g^{-1}(x; \boldsymbol{\theta}); \boldsymbol{\theta})}.$$

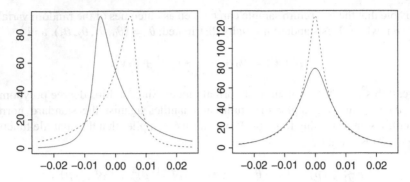

Fig. 8.6 Density functions for polynomial normal model. Parameter vectors: $(\theta_0, \theta_1, \theta_2, \theta_3) = (-3, 5, 3, 1.19089) \cdot 10^{-3}$ *(left plot, solid curve)*, $(\theta_0, \theta_1, \theta_2, \theta_3) = (3, 5, -3, 1.19089) \cdot 10^{-3}$ *(left plot, dashed curve)*, $(\theta_0, \theta_1, \theta_2, \theta_3) = (0, 5, 0, 1.44949) \cdot 10^{-3}$ *(right plot, solid curve)*, $(\theta_0, \theta_1, \theta_2, \theta_3) = (0, 3, 0, 2.038181) \cdot 10^{-3}$ *(right plot, dashed curve)*

The number $g^{-1}(x; \boldsymbol{\theta})$ is the unique real root of the (strictly increasing) third-order polynomial $\theta_3 y^3 + \theta_2 y^2 + \theta_1 y + \theta_0 - x$ for which an explicit formula exists. In particular, we may study the effect on the density $f_{g(Y;\theta)}$ of varying the parameter vector $\boldsymbol{\theta}$. If $\theta_2 = 0$, then the density is symmetric around θ_0. If θ_2 is negative, then the density is left-skewed, whereas it is right-skewed if θ_2 is positive. To study the effect of the parameters on the density, we fix the mean and standard deviation and vary the parameter values. Notice that

$$E[g(Y; \boldsymbol{\theta})] = \theta_0 + \theta_2,$$
$$E[g(Y; \boldsymbol{\theta})^2] = \theta_0^2 + (2\theta_0\theta_2 + \theta_1^2) E[Y^2] + (2\theta_1\theta_3 + \theta_2^2) E[Y^4] + \theta_3^2 E[Y^6].$$

and therefore, since $E[Y^2] = 1$, $E[Y^4] = 3$, and $E[Y^6] = 15$,

$$\mathrm{Var}(g(Y; \boldsymbol{\theta})) = \theta_1^2 + 2\theta_2^2 + 6\theta_1\theta_3 + 15\theta_3^2.$$

Figure 8.6 shows densities for the polynomial normal model with zero mean and standard deviation 0.01. The left plot illustrates how the value of θ_2 affects the skewness. The right plot illustrates the effect of simultaneously increasing θ_3 and decreasing θ_1.

Example 8.11 (Nasdaq data V, continuation of Example 8.5). Consider the polynomial normal model in Example 8.10. For the Nasdaq log returns we get the least-squares parameter estimate

$$(\widehat{\theta_0}, \widehat{\theta_1}, \widehat{\theta_2}, \widehat{\theta_3}) \approx (1.46, 12.3, -0.54, 1.98) \cdot 10^{-3}.$$

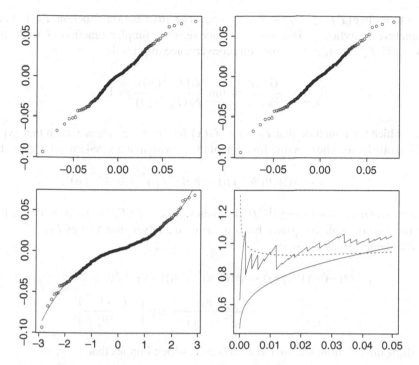

Fig. 8.7 *Upper plots, lower left plot:* q–q plots for Nasdaq log returns: empirical quantiles against fitted Student's t quantiles (*upper left*), empirical quantiles against a fitted third-degree polynomial of standard normal quantiles (*upper right*), and empirical quantiles against standard normal quantiles and graph of fitted third-degree polynomial (*lower left*). *Lower right plot:* empirical/normal/Student's t $\mathrm{VaR}_p(X)$ estimates divided by those of the polynomial normal model (jagged/solid/dotted)

The q–q plots of the empirical quantiles against those of the least-squares-fitted Student's t distribution and against the polynomial normal quantiles are shown in Fig. 8.7. Both models give a good fit, and it is difficult to see any difference between the two. The estimates of $\mathrm{VaR}_{0.05}(X)$ and $\mathrm{ES}_{0.05}(X)$ for the polynomial normal model are

$$\widehat{\mathrm{VaR}}_{0.05}(X) \approx 2.87 \quad \text{and} \quad \widehat{\mathrm{ES}}_{0.05}(X) \approx 4.44.$$

The ratios of the $\mathrm{VaR}_p(X)$ estimates for the normal, Student's t, and empirical method against those of the polynomial normal model as functions of p in $(0, 0.05)$ are shown in Fig. 8.7. Only for very small values of p is there a significant difference between the $\mathrm{VaR}_p(X)$ estimates of the Student's t model and the polynomial normal model.

Example 8.12 (Polynomial normal tails). Let Y be standard normally distributed and consider the polynomial normal random variable $g(Y) = \theta_0 + \theta_1 Y + \theta_2 Y^2 + \theta_3 Y^3$ with $\theta_3 > 0$. We want to investigate the behavior of its left tail

$F_\theta(x) = P(g(Y) \leq x)$ as $x \to -\infty$. The distribution function F_θ is hard to analyze analytically. However, we may select a simpler function G such that $\lim_{p \to 0} G(F_\theta^{-1}(p))/p = 1$ since this convergence implies that

$$\lim_{x \to -\infty} \frac{G(x)}{F_\theta(x)} = \lim_{p \to 0} \frac{G(F_\theta^{-1}(p))}{F_\theta(F_\theta^{-1}(p))} = 1,$$

from which we conclude that $F_\theta(x) \approx G(x)$ for negative values x such that $|x|$ is sufficiently large. The quantile function of the polynomial normal model is given by

$$F_\theta^{-1}(p) = \theta_0 + \theta_1 \Phi^{-1}(p) + \theta_2 \Phi^{-1}(p)^2 + \theta_3 \Phi^{-1}(p)^3.$$

Therefore, $G(x) = \Phi(-(-x)^{1/3}/\theta_3)$ satisfies $\lim_{p \to 0} G(F_\theta^{-1}(p))/p = 1$. The left tail of G is not explicitly given, but it is easier to analyze than that of F_θ.

We claim that, as $x \to -\infty$,

$$\Phi(-(-x)^{1/3}/\theta_3) \sim \phi((-x)^{1/3}/\theta_3)(-x)^{-1/3}\theta_3$$

$$= \frac{\theta_3}{\sqrt{2\pi}(-x)^{1/3}} \exp\left\{-\frac{(-x)^{2/3}}{2\theta_3^2}\right\}.$$

To show this relation, we use l'Hôpital's rule, which implies that

$$\lim_{x \to -\infty} \frac{\Phi(-(-x)^{1/3}/\theta_3)}{\phi((-x)^{1/3}/\theta_3)(-x)^{-1/3}\theta_3} = \lim_{x \to -\infty} \frac{\frac{d}{dx}\Phi(-(-x)^{1/3}/\theta_3)}{\frac{d}{dx}\phi((-x)^{1/3}/\theta_3)(-x)^{-1/3}\theta_3}.$$

If we compute the derivatives and use that $\phi'(y) = -y\phi(y)$ and $\phi(-y) = \phi(y)$, then we have

$$\frac{d}{dx}\Phi(-(-x)^{1/3}/\theta_3) = \phi((-x)^{1/3}/\theta_3)(-x)^{-2/3}/(3\theta_3),$$

$$\frac{d}{dx}\phi((-x)^{1/3}/\theta_3)(-x)^{-1/3}\theta_3 = -\phi'((-x)^{1/3}/\theta_3)(-x)^{-1}/3$$

$$+ \phi((-x)^{1/3}/\theta_3)(-x)^{-4/3}\theta_3/3$$

$$= \phi((-x)^{1/3}/\theta_3)(-x)^{-2/3}/(3\theta_3)$$

$$+ \phi((-x)^{1/3}/\theta_3)(-x)^{-4/3}\theta_3/3.$$

Since

$$\phi((-x)^{1/3}/\theta_3)(-x)^{-2/3}\left(\frac{1}{3\theta_3} + \frac{\theta_3}{3}(-x)^{-2/3}\right) \sim \phi((-x)^{1/3}/\theta_3)(-x)^{-2/3}/(3\theta_3)$$

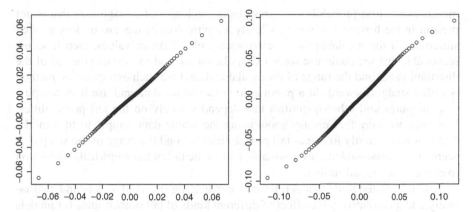

Fig. 8.8 Quantiles $0.012\Phi^{-1}(p) + 0.002\Phi^{-1}(p)^3$ (*y*-axes) of polynomial normal model against Student's *t* quantiles $0.012t^{-1}_{3.15}(p)$ (*x*-axes) for $p = 1/201,\ldots,200/201$ (*left plot*) and $p = 1/1{,}001,\ldots,1{,}000/1{,}001$ (*right plot*)

as $x \to -\infty$, we have verified the claim. We conclude that the distribution function F of the polynomial normal random variable $\theta_0 + \theta_1 Y + \theta_2 Y^2 + \theta_3 Y^3$ satisfies

$$F_\theta(x) \sim \frac{\theta_3}{\sqrt{2\pi}(-x)^{1/3}} \exp\left\{ -\frac{(-x)^{2/3}}{2\theta_3^2} \right\} \quad \text{as } x \to -\infty.$$

Notice that the asymptotic behavior of the left tail of F_θ depends on θ only via the coefficient $\theta_3 > 0$ of the third-degree term. Notice also that F_θ decays slower than an exponential rate in the sense that

$$\lim_{x \to -\infty} \frac{F_\theta(x)}{e^{-\lambda(-x)}} = \infty \quad \text{for every } \lambda > 0.$$

From Example 8.2 we know that the tail probabilities of Student's *t* distribution has a slower rate of decay compared to the polynomial normal model. However, Fig. 8.8 shows that this difference in tail behavior cannot be observed by comparing a few hundred sample points from the two distributions. Given a sample of daily log returns that shows no signs of asymmetry, the preference of one of the two models in favor of the other is ultimately a subjective choice of the modeler that cannot be justified by only the log-return data.

8.2 Extreme Values and Tail Probabilities

Given historical loss data, a risk manager typically wants to estimate the probability of future large losses and extreme events in order to assess the risk of holding a certain portfolio. It is rather common that a risk manager is asked to compute risk measures and loss probabilities corresponding to losses of a magnitude that is not present in the historical samples. Clearly, empirical estimates are useless in such situations. If the available data are representative of future values, then it seems reasonable that we could use some extrapolation method to estimate the tail of the distribution beyond the range of the available data. One such extrapolation method is rather straightforward: fit a parametric model to the data and use it to compute risk measures and other quantities that depend strongly on the tail probabilities. However, we may feel uneasy about using the whole data sample to fit a model that is used primarily to assess tail probabilities beyond the range of the sample. It seems more reasonable to use a suitable part of the tail of the empirical distribution to extrapolate the tail further.

Another related topic that is relevant for managing risk is to understand qualitatively and quantitatively the effect of different kinds of tail probabilities for models of returns or log returns on the tail of a distribution for the future value of a portfolio. For example, we should understand the interplay between the characteristics of the tails of claim size distributions and the potential benefits from diversification by pooling independent risks.

8.2.1   Heavy Tails and Diversification

Empirical investigations such as that for the Nasdaq log returns have shown that daily log returns of financial assets typically have distributions with heavy left tails. A log-return distribution with a distribution function F has a heavy left tail if the tail probability $F(x)$ decays slowly as x decreases ($x \to -\infty$). Although there is no definition of "heavy tail," it is common to consider the left tail $F(x)$, for $-x$ large, heavy if

$$\lim_{x \to -\infty} \frac{F(x)}{e^{-\lambda(-x)}} = \infty \quad \text{for every } \lambda > 0.$$

From Examples 8.1, 8.2, and 8.12 we know that the polynomial normal and Student's t distributions have heavy tails, whereas the normal distribution has light tails.

Sometimes we consider random variables representing losses or claim sizes that take only positive values. For such random variables we want to understand the behavior of the right tail $\overline{F}(x) = 1 - F(x)$, where x is large and F denotes the distribution function. The distribution function has a heavy right tail if

$$\lim_{x \to \infty} \frac{\overline{F}(x)}{e^{-\lambda x}} = \infty \quad \text{for every } \lambda > 0, \tag{8.6}$$

i.e., if it is heavier than the right tail of every exponential distribution.

Example 8.13 (Indications of heavy tails). Suppose we have observations of independent and identically distributed random variables with an unknown distribution.

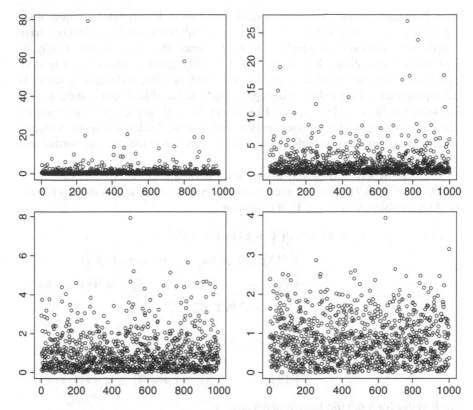

Fig. 8.9 Simulated samples of size 1,000 from Pa(2)-distribution (*upper left*), lognormal distribution (*upper right*), exponential distribution (*lower left*), and distribution of absolute value of standard normally distribution random variable (*lower right*)

The first step in trying to make sense of the tail behavior of the unknown distribution is to look at the plot of the sample points. Figure 8.9 shows plots of simulated samples of size 1,000 from the Pa(2)-distribution (upper left plot), the standard lognormal distribution (upper right plot), the standard exponential distribution (lower left plot), and the distribution of the absolute value of a standard normally distributed random variable (lower right plot). If the largest handful of sample points is substantially larger than the remaining sample points (upper two plots in Fig. 8.9), then we have reasons to believe that the unknown distribution has heavy tails.

We will now study the family of heavy-tailed distributions called subexponential distributions. Consider a nonnegative random variable X, $n \geq 2$, and independent copies $X_1, \ldots, X_n$ of X. If

$$\lim_{x \to \infty} \frac{P(X_1 + \cdots + X_n > x)}{P(X > x)} = n, \qquad (8.7)$$

then X is said to have a subexponential distribution. If this limit relation holds for some $n \geq 2$, then it holds for all $n \geq 2$. Subexponential distributions have distribution functions for which (8.6) holds, hence the name of this family of probability distributions. It is not necessary to only consider nonnegative random variables. A random variable Y that can take both positive and negative values is subexponential if there exists a nonnegative random variable X that is subexponential such that $\lim_{x \to \infty} P(Y > x)/P(X > x) = 1$. Examples of subexponential distributions are the (third-degree) polynomial normal, Student's t, lognormal, and Pareto. Examples of distributions that are not subexponential are the normal and exponential distributions.

What can we say qualitatively about the approximation $P(X_1 + \cdots + X_n > x) \approx n\, P(X > x)$ in (8.7)? For any nonnegative independent and identically distributed random variables $X, X_1, \ldots, X_n$ it holds that

$$P(X_1 + \cdots + X_n > x) = n\, P(X > x)\, P(X \leq x)^{n-1}$$

$$+ P(X_k > x \text{ and } X_l > x \text{ for some } k \neq l)$$

$$+ P(X_1 + \cdots + X_n > x \text{ and } X_k \leq x \text{ for every } k)$$

$$= n\, P(X > x)\, P(X \leq x)^{n-1}$$

$$+ \binom{n}{2} P(X > x)^2$$

$$+ P(X_1 + \cdots + X_n > x \text{ and } X_k \leq x \text{ for every } k).$$

We find that for a not too large n and a large x

$$\frac{P(X_1 + \cdots + X_n > x)}{P(X > x)} \approx n + \frac{P(X_1 + \cdots + X_n > x \text{ and } X_k \leq x \text{ for every } k)}{P(X > x)}.$$

Moreover, we find that typically $P(X_1 + \cdots + X_n > x) > n\, P(X > x)$. Notice that subexponentiality implies that

$$\lim_{x \to \infty} \frac{P(X_1 + \cdots + X_n > x \text{ and } X_k \leq x \text{ for every } k)}{P(X > x)} = 0.$$

However, subexponentiality does not say anything about the speed of convergence, which may be rather slow.

Notice also that $\max(X_1, \ldots, X_n) > x$ implies that $X_1 + \cdots + X_n > x$ and that

$$P(\max(X_1, \ldots, X_n) > x) = P(X_1 > x) + P(X_1 \leq x, X_2 > x)$$

$$+ \cdots + P(X_1 \leq x, \ldots, X_{n-1} \leq x, X_n > x)$$

$$= P(X > x) \sum_{m=0}^{n-1} P(X \leq x)^m.$$

Therefore,

$$\sum_{m=0}^{n-1} P(X \le x)^m \le \frac{P(X_1 + \cdots + X_n > x)}{P(X > x)}$$

$$< n + \frac{P(X_1 + \cdots + X_n > x \text{ and } X_k \le x \text{ for every } k)}{P(X > x)},$$

where the last inequality follows from the inequality $P(X_k > x$ for some $k) \le n\,P(X > x)$.

Finally, if the random variables $X_1, \ldots, X_n$ are subexponential, then $\lim_{x \to \infty} P(X_k > x \mid X_1 + \cdots + X_n > x) = 1/n$ (which follows immediately from the definition) and

$$\lim_{x \to \infty} \frac{P(\max(X_1, \ldots, X_n) > x)}{P(X_1 + \cdots + X_n > x)} = 1.$$

The interpretation is that the sum takes a very large value due to precisely one of the terms taking a very large value and the sum of the remaining terms being small.

The following proposition states that we may add a constant to a random variable with a subexponential distribution without affecting the tail probabilities asymptotically. The situation here is quite different from a light-tailed distribution like the normal distribution (Example 8.1).

Proposition 8.1. *If X is a nonnegative random variable with a subexponential distribution, then* $\lim_{x \to \infty} P(X > x - y)/P(X > x) = 1$ *for every* y.

Proof. Denote by F the distribution function of X and let X_1 and X_2 be two independent copies of X. For $x \ge y > 0$,

$$\frac{P(X_1 + X_2 > x)}{P(X > x)} = \frac{1}{\overline{F}(x)} \int_0^\infty P(X_1 + t > x) dF(t)$$

$$= \int_0^y \frac{\overline{F}(x-t)}{\overline{F}(x)} dF(t) + \int_y^x \frac{\overline{F}(x-t)}{\overline{F}(x)} dF(t)$$

$$+ \int_x^\infty \frac{\overline{F}(x-t)}{\overline{F}(x)} dF(t)$$

$$\ge F(y) + \frac{\overline{F}(x-y)}{\overline{F}(x)} (F(x) - F(y)) + 1.$$

Therefore,

$$1 \le \frac{\overline{F}(x-y)}{\overline{F}(x)} \le \left(\frac{P(X_1 + X_2 > x)}{P(X > x)} - 1 - F(y) \right) \frac{1}{F(x) - F(y)}.$$

Since the right-hand side converges to 1 as $x \to \infty$, we conclude that $\lim_{x\to\infty} P(X > x - y)/P(X > x) = 1$ for $y > 0$. For $y < 0$, $z = -y$, and $v = x + z$,

$$1 = \left(\lim_{v\to\infty} \frac{\overline{F}(v-z)}{\overline{F}(v)}\right)^{-1} = \left(\lim_{x\to\infty} \frac{\overline{F}(x)}{\overline{F}(x+z)}\right)^{-1} = \lim_{x\to\infty} \left(\frac{\overline{F}(x)}{\overline{F}(x+z)}\right)^{-1}$$

$$= \lim_{x\to\infty} \frac{\overline{F}(x-y)}{\overline{F}(x)},$$

which shows that $\lim_{x\to\infty} P(X > x - y)/P(X > x) = 1$ for $y < 0$. $\square$

We will now introduce the family of heavy-tailed distributions with regularly varying tails. A distribution function F has a regularly varying right tail $\overline{F} = 1 - F$ if there exists a number ρ such that

$$\lim_{t\to\infty} \frac{\overline{F}(tx)}{\overline{F}(t)} = x^\rho \quad \text{for every } x > 0. \tag{8.8}$$

Since $\overline{F}(x)$ is decreasing in x, it necessarily holds that $\rho \leq 0$, and we may set $\rho = -\alpha$ for $\alpha \geq 0$. We may formulate the regular variation property (8.8) as

$$\lim_{t\to\infty} P(X > tx \mid X > t) = x^{-\alpha} \quad \text{for every } x > 1.$$

The regular variation property for the left tail of F is defined similarly to (8.8).

Example 8.14 (Pareto and Student's t tails). The canonical example of a distribution with a regularly varying right tail is the Pareto distribution with distribution function $F(x) = 1 - (c/x)^\alpha$ for $c > 0$ and $x \geq c$. If $tx, t > c$, then

$$\frac{\overline{F}(tx)}{\overline{F}(t)} = x^{-\alpha}.$$

Example 8.2 shows that the Student's t distribution function has a regularly varying left tail with index $-\nu$:

$$\lim_{t\to-\infty} \frac{F(tx)}{F(t)} = x^{-\nu}.$$

Since Student's t distribution is symmetric, the same holds for the right tail.

Proposition 8.2. *Consider a nonnegative random variable X with distribution function F. If F has a regularly varying right tail, then X has a subexponential distribution.*

Proof. Consider two independent copies X_1 and X_2 of X. If $\varepsilon \in (0, 1/2)$, then

$$P(X_1 + X_2 > x) = 2\,P(X_1 + X_2 > x, X_1 \leq \varepsilon x)$$
$$+ P(X_1 + X_2 > x, X_1 > \varepsilon x, X_2 > \varepsilon x)$$
$$\leq 2\,P(X_2 > (1 - \varepsilon)x) + P(X_1 > \varepsilon x)^2.$$

Moreover,

$$P(X_1 + X_2 > x) \geq P(X_1 > x \text{ or } X_2 > x) = 2\,P(X_1 > x) - P(X_1 > x)^2.$$

Therefore,

$$\underbrace{\frac{2\,P(X_1 > x) - P(X_1 > x)^2}{P(X_1 > x)}}_{g(\alpha,\epsilon,x)} \leq \frac{P(X_1 + X_2 > x)}{P(X_1 > x)}$$

$$\leq \underbrace{\frac{2\,P(X_2 > (1 - \varepsilon)x) + P(X_1 > \varepsilon x)^2}{P(X_1 > x)}}_{h(\alpha,\epsilon,x)}.$$

We have $\lim_{x \to \infty} g(\alpha, \epsilon, x) = 2$ and

$$\lim_{x \to \infty} h(\alpha, \epsilon, x) = 2 \lim_{x \to \infty} \frac{P(X_1 > (1 - \varepsilon)x)}{P(X_1 > x)} = 2(1 - \varepsilon)^{-\alpha}.$$

Since $\varepsilon > 0$ can be chosen arbitrary small, we conclude that

$$\lim_{x \to \infty} \frac{P(X_1 + X_2 > x)}{P(X_1 > x)} = 2. \qquad \square$$

Example 8.15 (Diversification and heavy tails). Consider two independent nonnegative random variables X_1 and X_2 with common distribution function F with a regularly varying right tail $\overline{F} = 1 - F$. The random variables represent aggregate claim amounts during a 1-year period for an insurance product sold to two groups of customers in different geographical areas. The insurance company wants to compare the risk of shutting down its business in one geographical area and doubling its business in the other area versus the status quo. The right tail of the aggregate claim amount distribution in the former case is $P(2X_1 > x)$ and $P(X_1 + X_2 > x)$ in the latter case.

From the subexponential property (8.7) and the regular variation property (8.8) we find that

$$\lim_{x \to \infty} \frac{P(X_1 + X_2 > x)}{P(2X_1 > x)} = \lim_{x \to \infty} \frac{P(X_1 + X_2 > x)}{P(X_1 > x)} \frac{P(X_1 > x)}{P(2X_1 > x)} = 2^{1-\alpha}.$$

The interpretation is that for $\alpha < 1$ (very heavy tails) diversification does not give a portfolio with smaller probability of large losses. However, for $\alpha > 1$ (the aggregate claim amount distribution has finite mean) diversification reduces the probability of large losses and the diversification effect increases with α. How can we interpret the findings for $\alpha < 1$? As an example we may consider proportional reinsurance of nuclear power plants. The potential loss from a loss-generating event may be enormous, and insuring twice as much of the potential losses from one nuclear power plant may be less risky than insuring two different power plants.

Example 8.16 (Nonsubadditivity of the quantile function). Take $\alpha \in (0, 1)$ and let X_1 and X_2 be as in the previous example. We saw that for x sufficiently large $P(X_1 + X_2 > x) > P(2X_1 > x)$. Therefore, for $p \in (0, 1)$ sufficiently large

$$F_{X_1}^{-1}(p) + F_{X_2}^{-1}(p) = 2F_{X_1}^{-1}(p) = F_{2X_1}^{-1}(p)$$
$$= \min\{x : P(2X_1 > x) \le 1 - p\}$$
$$< \min\{x : P(X_1 + X_2 > x) \le 1 - p\}$$
$$= F_{X_1+X_2}^{-1}(p).$$

We conclude that the sum of the quantiles for two independent and identically distributed random variables is not necessarily greater than the quantile of the sum.

The last example can be modified to show that VaR is not subadditive, which we already know from Example 6.10.

Example 8.17 (The heavier tail wins). Let X and Y be random variables representing, for example, losses in two lines of business (losses due to fire and car accidents, say) of an insurance company. Suppose that X has a distribution function with a regularly varying right tail with index $\alpha > 0$ and that $|Y|$ has a finite moment of order $\alpha + \delta$ for some $\delta > 0$, i.e., $E[|Y|^{\alpha+\delta}] < \infty$. The insurance company wants to investigate the probability $P(X + Y > x)$ of very large aggregate losses.

For every $\varepsilon \in (0, 1)$ and $x > 0$,

$$P(X + Y > x) = P(X + Y > x, X > (1 - \varepsilon)x)$$
$$+ P(X + Y > x, X < -(1 - \varepsilon)x)$$
$$+ P(X + Y > x, |X| \le (1 - \varepsilon)x)$$
$$\le P(X > (1 - \varepsilon)x)$$
$$+ P(Y > (2 - \varepsilon)x)$$
$$+ P(|Y| > \varepsilon x)$$
$$\le P(X > (1 - \varepsilon)x) + 2P(|Y| > \varepsilon x).$$

Therefore,

$$1 \le \frac{P(X+Y > x)}{P(X > x)}$$

$$\le \frac{P(X > (1-\varepsilon)x)}{P(X > x)} + \frac{2\,P(|Y| > \varepsilon x)}{P(X > x)}$$

$$\le \frac{P(X > (1-\varepsilon)x)}{P(X > x)} + \frac{2\,E[|Y|^{\alpha+\delta}]}{(\varepsilon x)^{\alpha+\delta}\,P(X > x)}$$

$$\to (1-\varepsilon)^{-\alpha} + 0$$

as $x \to \infty$, where Markov's inequality was used in the second-to-last step above. Choosing ε arbitrarily small gives

$$\lim_{x\to\infty} \frac{P(X+Y > x)}{P(X > x)} = 1,$$

which shows that only the loss variable with the heaviest right tail matters for probabilities of very large losses.

Example 8.18 (Diversification in proportional reinsurance). Diversification is the key principle for insurers to deal with risks in large portfolios. Here we investigate potential diversification benefits for a reinsurance company selling proportional excess loss reinsurance. Let $Y_1, \ldots, Y_n$ be independent and identically distributed random variables representing aggregate losses of some kind during a 1-year period for some insurers. The reinsurer offers to pay $\lambda_k(Y_k - y_k)_+$, the fraction $\lambda_k \in [0, 1]$ of the aggregate claim amount $(Y_k - y_k)_+$ exceeding the so-called retention level y_k, and in return demands a certain premium from the insurer buying the protection. For simplicity we set $y_k = y$ for all k and compare the aggregate claim amount

$$X_D = \frac{1}{n}\sum_{k=1}^{n}(Y_k - y)_+$$

for the reinsurer, corresponding to a presumed optimally diversified reinsurance portfolio, to the aggregate claim amount $X_1 = (Y_1 - y)_+$, corresponding to a concentrated reinsurance portfolio. We want to compare the quantiles $F_{X_D}^{-1}(p)$ and $F_{X_1}^{-1}(p)$ under the assumption that the common distribution of the Y_k is subexponential. Proposition 8.1 implies that

$$\lim_{x\to\infty} \frac{P((Y_k - y)_+ > x)}{P(Y_k > x)} = \lim_{x\to\infty} \frac{P(Y_k > x + y)}{P(Y_k > x)} = 1,$$

which in turn implies that $X_k = (Y_k - y)_+$ has a subexponential distribution (see Sect. 8.3 for details). From the subexponentiality of the X_k we find that, for p large,

$$
\begin{aligned}
F_{X_D}^{-1}(p) &= \min\{x : F_{X_D}(x) \geq p\} \\
&= \min\{x : \overline{F}_{X_D}(x) \leq 1 - p\} \\
&\approx \min\{x : n\overline{F}_{X_1}(nx) \leq 1 - p\} \\
&= \min\{x : n\overline{F}_{Y_1}(nx + y) \leq 1 - p\} \\
&\approx \min\{x : n\overline{F}_{Y_1}(nx) \leq 1 - p\} \\
&= \frac{1}{n} F_{Y_1}^{-1}\left(1 - \frac{1 - p}{n}\right),
\end{aligned}
\tag{8.9}
$$

where the last equality assumes that there exists an x such that $n\overline{F}_{Y_1}(nx) = 1 - p$ (this is not a major problem since we could always increase p until we found an x solving the equation). The approximation of $F_{X_D}^{-1}(p)$ by (8.9) for p large can be made precise in the sense that $F_{X_D}^{-1}(p)$ divided by (8.9) converges to one as $p \to 1$.

If $\overline{F}_{Y_1}$ is regularly varying with index $-\alpha$, then $\lim_{x\to\infty} \overline{F}_{Y_1}(nx)/\overline{F}_{Y_1}(x) = n^{-\alpha}$, which yields the approximation

$$
F_{X_D}^{-1}(p) \approx F_{Y_1}^{-1}\left(1 - \frac{1 - p}{n^{1-\alpha}}\right).
\tag{8.10}
$$

From (8.10) we observe that for p large and α close to 1 there is no diversification benefit since $F_{X_D}^{-1}(p) \approx F_{Y_1}^{-1}(p)$.

What can we say qualitatively about the approximation of $F_{X_D}^{-1}(p)$ by (8.9)? On the one hand, $\overline{F}_{X_D}(x) \geq n\overline{F}_{X_1}(nx)$, which contributes to underestimating $F_{X_D}^{-1}(p)$ by (8.9). On the other hand $\overline{F}_{Y_1}(nx + y) \leq \overline{F}_{Y_1}(nx)$, which contributes to overestimating $F_{X_D}^{-1}(p)$ by (8.9) if y is large.

To test the accuracy of the approximation, we let $n = 10$ and consider Y_k that are Pa(2)-distributed, $\overline{F}_{Y_1}(x) = x^{-2}$ for $x > 1$, simulate samples of size 10^6 from the distribution of X_D for different values of y, and approximate $F_{X_D}^{-1}(p)$ by the empirical quantile. The quotients of the empirical quantiles estimates and the estimates from (8.10) are shown in Fig. 8.10.

Example 8.19 (The central limit theorem and heavy tails). The central limit theorem says that the sum of n independent and identically distributed random variables $S_n = X_1 + \cdots + X_n$ is approximately normally distributed in the sense that

$$
\lim_{n\to\infty} \mathrm{P}\left(\frac{S_n - n\mu}{\sqrt{n}\sigma} \leq x\right) = \Phi(x),
$$

Fig. 8.10 *Upper curve:* empirical estimates of p-quantiles of X_D for Pa(2)-distributed Y_k, $n = 10$, and $y = 0$, based on sample of size 10^6, divided by corresponding values of expression in (8.10). The remaining curves (*upper to lower*) correspond to $y = 1$, 2, and 5. All curves are based on the same samples from the Pa(2) distribution

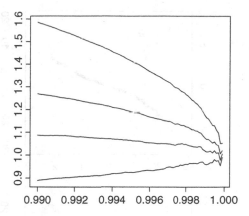

where μ and σ denote the mean and standard deviation of the X_k. On the other hand, if the distribution function of the X_k has a regularly varying left tail with index $-\alpha$, then, as $x \to -\infty$,

$$P\left(\frac{S_n - n\mu}{\sqrt{n}\sigma} \le x\right) \sim n\,P(X_1 \le \sqrt{n}\sigma x + n\mu) \sim n^{1-\alpha/2}\sigma^{-\alpha}\,P(X_1 \le x),$$

so the far-out left tail is unaffected by the summation and stays regularly varying with index $-\alpha$.

For n and $-x$ large and $-\alpha$ small but greater than 2, it is not obvious a priori which one of the two approximations yields the best estimate for the tail probabilities $P(X_1 + \cdots + X_n \le x)$ of the sum. One possible interpretation of the sum $X_1 + \cdots + X_n$ would be an n-day log return, which is the sum of n 1-day log returns. Empirical investigations show that the empirical distribution of log returns of stock prices and index values over longer time periods tends to be better approximated by the normal distribution, whereas this is not true for log returns over shorter time periods. However, empirical investigations also show that there is nonnegligible dependence between the absolute values of log returns for consecutive 1-day periods, so this interpretation is not necessarily consistent with data.

The convergence toward the normal distribution as the number of terms n increases for Student's t-distributed X_k with three degrees of freedom is illustrated in Fig. 8.11. The upper left plot shows the empirical quantiles of $S_1 = X_1$ based on a sample of size 500 (y-axis) against the standard normal quantiles (x-axis). The upper right plot shows the empirical quantiles of $S_{10} = X_1 + \cdots + X_{10}$ based on a sample of size 500 (y-axis) against the standard normal quantiles (x-axis). The lower left plot shows the empirical quantiles of $S_{20} = X_1 + \cdots + X_{20}$ based on a sample of size 500 (y-axis) against the standard normal quantiles (x-axis). The lower right plot shows the empirical quantiles of $S_{40} = X_1 + \cdots + X_{40}$ based on a sample of size 500 (y-axis) against the standard normal quantiles (x-axis).

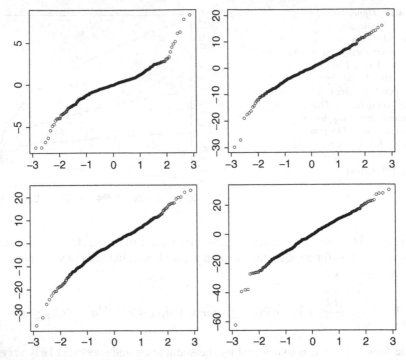

Fig. 8.11 Simulations of 40 samples of size 500 from Student's t distribution with three degrees of freedom are illustrated in four q–q plots. *Upper left*: empirical quantiles based on first sample against standard normal quantiles. *Upper right/lower left/lower right*: empirical quantiles based on cumulative sum of first 10/20/40 samples against standard normal quantiles

Consider the lower right plot in Fig. 8.11. We see that the smallest value is approximately -60. Denote the sample points by $Y_1, \ldots, Y_{500}$ and note that Y_k and $X_1 + \cdots + X_{40}$ are equally distributed. The subexponentiality of Student's t distribution yields the approximation

$$P(\min(Y_1, \ldots, Y_{500}) < -60) \approx 500\, P(Y_1 < -60)$$

$$= 500\, P(X_1 + \cdots + X_{40} < -60)$$

$$\approx 500 \cdot 40\, P(X_1 < -60)$$

$$\approx 0.102.$$

This estimate is an overestimate of the true probability but is reasonably close. The central limit theorem yields the approximation

$$P(\min(Y_1, \ldots, Y_{500}) < -60) = 1 - P(Y_1 \geq -60)^{500}$$

$$= 1 - P(X_1 + \cdots + X_{40} \geq -60)^{500}$$

$$= 1 - \mathrm{P}(X_1 + \cdots + X_{40} \leq 60)^{500}$$

$$\approx 1 - \Phi(60/\sqrt{120})^{500}$$

$$\approx 1.08 \cdot 10^{-5}.$$

This estimate is an underestimate of the true probability and is far from the true value.

8.2.2   Peaks Over Threshold Method

The asymptotic properties of the tail of a distribution function can never be determined from a sample of the distribution. Every sample has a finite sample size, and the tail behavior outside the range of the sample is a subjective assessment of the modeler. However, if the sample is made up of outcomes of independent and identically distributed random variables with an unknown distribution, then the large values contain relevant information that may be used to extrapolate the empirical tail outside the range of the sample. Here we present one such extrapolation approach. It is important to bear in mind that if there is significant dependence between the random variables whose outcomes form the sample, then it may be impossible to use this approach.

Suppose we have observations of independent and identically distributed random variables $X_1, \ldots, X_n$ with common unknown distribution function F with a regularly varying right tail $\overline{F}(x) = \mathrm{P}(X_k > x)$. It turns out that the distribution of appropriately scaled excesses $X_k - u$ over a high threshold u is typically well approximated by a distribution called the generalized Pareto distribution. This fact can be used to construct estimates of tail probabilities and quantiles.

For $\gamma > 0$ and $\beta > 0$ the generalized Pareto distribution function $G_{\gamma,\beta}$ is given by

$$G_{\gamma,\beta}(x) = 1 - (1 + \gamma x/\beta)^{-1/\gamma} \quad \text{for } x \geq 0.$$

Suppose that X is a random variable with distribution function F that has a regularly varying right tail so that $\lim_{u \to \infty} \overline{F}(\lambda u)/\overline{F}(u) = \lambda^{-\alpha}$ for all $\lambda > 0$ and some $\alpha > 0$. Then

$$\lim_{u \to \infty} \mathrm{P}\left(\frac{X - u}{u/\alpha} > x \mid X > u\right) = \lim_{u \to \infty} \frac{\mathrm{P}(X > u(1 + x/\alpha))}{\mathrm{P}(X > u)}$$

$$= (1 + x/\alpha)^{-\alpha}$$

$$= \overline{G}_{1/\alpha,1}(x).$$

The excess distribution function of X over the threshold u is given by

$$F_u(x) = \mathrm{P}(X - u \leq x \mid X > u) \quad \text{for } x \geq 0.$$

Notice that

$$\overline{F}_u(x) = \frac{\overline{F}(u+x)}{\overline{F}(u)} = \frac{\overline{F}(u(1+x/u))}{\overline{F}(u)}. \tag{8.11}$$

Since $\overline{F}$ is regularly varying with index $-\alpha < 0$, it holds that $\overline{F}(\lambda u)/\overline{F}(u) \to \lambda^{-\alpha}$ uniformly in $\lambda \geq 1$ as $u \to \infty$, i.e.,

$$\lim_{u\to\infty} \sup_{\lambda\geq 1} |\overline{F}(\lambda u)/\overline{F}(u) - \lambda^{-\alpha}| = 0.$$

Hence, from expression (8.11) we see that

$$\lim_{u\to\infty} \sup_{x>0} |\overline{F}_u(x) - \overline{G}_{\gamma,\beta(u)}(x)| = 0, \tag{8.12}$$

where $\gamma = 1/\alpha$ and $\beta(u) \sim u/\alpha$ as $u \to \infty$. We now demonstrate how these findings lead to natural tail and quantile estimators based on the sample points $X_1, \ldots, X_n$. Choose a high threshold u and let

$$N_u = \#\{i \in \{1, \ldots, n\} : X_i > u\}$$

be the number of exceedances of u by $X_1, \ldots, X_n$. Recall from (8.11) that

$$\overline{F}(u+x) = \overline{F}(u)\overline{F}_u(x). \tag{8.13}$$

If u is not too far out into the tail, then the empirical approximation $\overline{F}(u) \approx \overline{F}_n(u) = N_u/n$ is accurate. Moreover, (8.12) shows that the approximation

$$\overline{F}_u(x) \approx \overline{G}_{\gamma,\beta(u)}(x) \approx \overline{G}_{\widehat{\gamma},\widehat{\beta}}(x) = \left(1 + \widehat{\gamma}\frac{x}{\widehat{\beta}}\right)^{-1/\widehat{\gamma}},$$

where $\widehat{\gamma}$ and $\widehat{\beta}$ are the estimated parameters, makes sense. Relation (8.13) then suggests estimating the tail of F by estimating $\overline{F}_u(x)$ and $\overline{F}(u)$ separately. We consider the estimator for $\overline{F}(u+x)$ given by

$$\widehat{\overline{F}(u+x)} = \frac{N_u}{n}\left(1 + \widehat{\gamma}\frac{x}{\widehat{\beta}}\right)^{-1/\widehat{\gamma}}. \tag{8.14}$$

Expression (8.14) immediately leads to the following estimator of the quantile $F^{-1}(p)$:

$$\widehat{F^{-1}}(p) = \min\{x : \widehat{\overline{F}}(x) \leq 1-p\}$$

$$= \min\{u + x : \widehat{\overline{F}(u+x)} \leq 1-p\}$$

$$= u + \min \left\{ x : \frac{N_u}{n} \left(1 + \widehat{\gamma} \frac{x}{\widehat{\beta}} \right)^{-1/\widehat{\gamma}} \le 1 - p \right\}$$

$$= u + \frac{\widehat{\beta}}{\widehat{\gamma}} \left(\left(\frac{n}{N_u} (1 - p) \right)^{-\widehat{\gamma}} - 1 \right). \tag{8.15}$$

The peaks over threshold (POT) method for estimating tail probabilities and quantiles can be summarized by the following procedure. Each step will be discussed further below.

(i) Choose a high threshold u using some statistical method and count the number N_u of exceedances $X_k > u$.
(ii) Given the sample $Y_1, \ldots, Y_{N_u}$ of excesses $X_k - u$ if $X_k > u$, estimate the parameters γ and β.
(iii) Combine steps (i) and (ii) to get estimates of the form (8.14) and (8.15).

The rest of this section will be devoted to steps (i) and (ii). How do we choose a high threshold u in a suitable way? How should we estimate the parameters γ and β?

The choice of a suitable high threshold u is important but difficult. If we choose u too large, then we will have few observations to use for parameter estimation, resulting in poor estimates with large variance. If the threshold is too low, then we have more data, but on the other hand, the approximation $\overline{F}_u(x) \approx \overline{G}_{\gamma, \beta(u)}(x)$ will be questionable. The main idea when choosing the threshold u is to look at the tail of the empirical distribution tail and choose u such that the tail above this level looks somewhat like the tail of a Pareto distribution. Many different algorithmic approaches to the problem of choosing the suitable threshold value have been suggested. Here we take a less formal approach to this problem:

- Inspect q–q plots for the empirical quantiles against the quantiles of suitable reference distributions (different Pareto distributions, say).
- Select a not-too-high threshold value u and make a q–q plot of the empirical quantiles of the N_u excesses against the quantiles of a generalized Pareto distribution whose parameters are estimated by maximum likelihood. From the expression for the distribution function of the generalized Pareto distribution we get the quantile function

$$G_{\gamma, \beta}^{-1}(p) = \frac{\beta}{\gamma} \left((1 - p)^{-\gamma} - 1 \right).$$

- Try a slightly larger threshold value u, repeat the procedure above, and observe whether the smaller sample of excesses gives a better fit to a generalized Pareto distribution.

Given the threshold u, we may estimate the parameters γ and β based on the observations of excesses $Y_1, \ldots, Y_{N_u}$ over u. Denote the observations by $y_1, \ldots, y_{n_u}$. The parameters γ and β can be estimated by least squares by minimizing

$$
\sum_{k=1}^{n_u} \left(y_{k,n_u} - G_{\gamma,\beta}^{-1}\left(\frac{n_u - k + 1}{n_u + 1} \right) \right)^2
$$

$$
= \sum_{k=1}^{n_u} \left(y_{k,n_u} - \frac{\beta}{\gamma}\left(\left(1 - \frac{n_u - k + 1}{n_u + 1} \right)^{-\gamma} - 1 \right) \right)^2,
$$

where $y_{1,n_u} \geq \cdots \geq y_{n_u,n_u}$ are the ordered excesses.

Alternatively, the parameters of the generalized Pareto distribution can be estimated by maximum likelihood. The likelihood function is given by

$$
L(\gamma, \beta; y_1, \ldots, y_{n_u}) = \prod_{k=1}^{n_u} g_{\gamma,\beta}(y_k), \quad g_{\gamma,\beta}(y) = \frac{1}{\beta}\left(1 + \gamma\frac{y}{\beta} \right)^{-1/\gamma - 1},
$$

which gives the log-likelihood function

$$
\log L(\gamma, \beta; y_1, \ldots, y_{n_u}) = -n_u \ln\beta - \left(\frac{1}{\gamma} + 1 \right)\sum_{k=1}^{n_u} \log\left(1 + \frac{\gamma}{\beta}y_k \right).
$$

To understand the difference between the maximum-likelihood and least-squares approaches to fitting the generalized Pareto distribution to the excesses over the threshold u, we simulate 1,000 samples of size 1,000 from the Pa(3) distribution, and for each sample we estimate the parameters using both least squares and maximum likelihood with the threshold u chosen as the 101st largest outcome $X_{101,1000}$. The theoretical values are $(\gamma, \beta) \approx (0.33, 0.72)$ in the sense that $\gamma = 1/\alpha = 1/3$ and $\beta = \gamma X_{101,1000} = \gamma F_{n,X}^{-1}(0.9)$, where $\gamma F^{-1}(0.9) = (1 - 0.9)^{-1/3}/3 \approx 0.72$. For the least-squares estimator the sample mean and covariance matrix of

$$
\begin{pmatrix} \widehat{\gamma} \\ \widehat{\beta} \end{pmatrix} \quad \text{are} \quad \begin{pmatrix} 0.44 \\ 0.66 \end{pmatrix} \quad \text{and} \quad \begin{pmatrix} 8.60 & -5.35 \\ -5.35 & 4.71 \end{pmatrix} \cdot 10^{-2}.
$$

The corresponding correlation coefficient is -0.841. The estimator appears to be biased. The estimated median values for (γ, β) are $(0.41, 0.68)$. For the maximum-likelihood estimator the sample mean and covariance matrix of

$$
\begin{pmatrix} \widehat{\gamma} \\ \widehat{\beta} \end{pmatrix} \quad \text{are} \quad \begin{pmatrix} 0.31 \\ 0.74 \end{pmatrix} \quad \text{and} \quad \begin{pmatrix} 2.21 & -1.17 \\ -1.17 & 1.59 \end{pmatrix} \cdot 10^{-2}.
$$

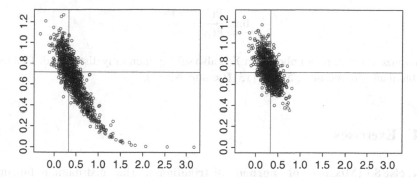

Fig. 8.12 *Each scatter plot* shows 1,000 pairs of parameter estimates. *Left plot*: least-squares estimates; *right plot*: maximum-likelihood estimates. The 101st largest outcome of each sample was used as the threshold value

The corresponding correlation coefficient is -0.622. The estimated median values for (γ, β) are $(0.31, 0.73)$. The maximum-likelihood estimator performs much better here than the least-squares estimator. This is also clear from Fig. 8.12, which shows scatter plots of the 1,000 outcomes of $(\widehat{\gamma}, \widehat{\beta})$ for the two estimation approaches.

There is another reason for preferring MLE in this context. LSE is a natural choice for parameter estimation if the tails of the model have significant influence on the estimates. Here we want all of the samples of excesses to fit nicely to the generalized Pareto distribution, not primarily the tails.

8.3 Notes and Comments

Parameter estimation and principles of statistical inference are treated in many books; see, for instance, the book [9] by George Casella and Roger Berger. A more comprehensive treatment of parameter estimation, including robust and Bayesian techniques, with an emphasis on applications to financial asset allocation, is found in Attilio Meucci's book [33].

There is an extensive literature on statistical inference for and modeling of extreme events and heavy-tailed phenomena. Three comprehensive yet accessible accounts are the books [13] by Paul Embrechts, Claudia Klüppelberg, and Thomas Mikosch, [20] by Laurens de Haan and Ana Ferreira, and [37] by Sidney Resnick. Here we only touched upon a few select topics from the wide variety of topics and applications considered in these books. Our presentation in Sect. 8.2 is to a large extent based on [13]. The POT method is much more general than the version presented here. See [13, 20], or [37] for a more general version of the method with a wider range of possible applications.

In Example 8.18 we used the fact that if random variable X is subexponentially distributed and if Y is a random variable such that

$$\lim_{x\to\infty} \frac{P(Y > x)}{P(X > x)}$$

is a nonzero and finite number, then Y is also subexponentially distributed. This fact
is stated and proved on p. 572 in [13] (Lemma A3.15).

8.4   Exercises

Exercise 8.1 (Mixture of normal distributions). The distribution function
$F(x) = p\Phi(x/\sigma_1) + (1 - p)\Phi(x/\sigma_2)$ of a mixture of the two normal distributions
$N(0, \sigma_1^2)$ and $N(0, \sigma_2^2)$ corresponds to drawing a value with probability p from the
$N(0, \sigma_1^2)$ distribution and with probability $1 - p$ from the $N(0, \sigma_2^2)$ distribution.

(a) Use maximum likelihood to estimate the parameters p, σ_1, σ_2 based on the
    sample $\{t_4^{-1}(k/201) : k = 1, \ldots, 200\}$.
(b) Plot the density function of the mixture distribution with the parameters
    estimated in (a) and compare it to the density function of the standard Student's
    t distribution with four degrees of freedom.
(c) Plot the quantiles of the Student's t distribution with four degrees of freedom
    against the quantiles of the mixture distribution with the parameters estimated
    in (a).
(d) Determine the asymptotic behavior of $F(x)$ as $x \to -\infty$ in terms of an
    explicitly given function G such that $\lim_{x\to-\infty} F(x)/G(x) = 1$.

Exercise 8.2 (Parameter estimation). Consider Student's t location-scale family
with parameter vector (μ, σ, ν).

(a) Determine the log-likelihood function and estimate the parameters based on the
    sample $\{t_4^{-1}(k/201) : k = 1, \ldots, 200\}$.

Simulate 3,000 samples of size 200 from the standard Student's t distribution
with four degrees of freedom.

(b) For each sample compute the maximum-likelihood estimate of the parameter
    vector (μ, σ, ν). Make a scatter plot of the 3,000 parameter estimates $(\hat{\sigma}, \hat{\nu})$ and
    interpret the plot.
(c) For each sample compute the least-squares estimate of the parameter vector
    (μ, σ, ν). Make a scatter plot of the 3,000 parameter estimates $(\hat{\sigma}, \hat{\nu})$, interpret
    the plot, and compare the plot to that in (b).
(d) For each sample compute the sample standard deviation and divide the sample
    by the sample standard deviation. Consider each rescaled sample to be a sample
    from a Student's t distribution with unit variance and estimate the degrees-
    of-freedom parameter by maximum likelihood. Transform the estimates into
    estimates of the parameter pair (σ, ν) for a centered Student's t distribution
    with scale parameter σ. Make a scatter plot of the 3,000 parameter estimates
    $(\hat{\sigma}, \hat{\nu})$, interpret the plot, and compare the plot to that in (b).

Exercise 8.3 (Lognormal tail). Let X be $LN(\mu, \sigma^2)$-distributed.

(a) Show that, as $x \to \infty$,

$$P(X > x) \sim \frac{\sigma}{\sqrt{2\pi}(\log x - \mu)} \exp\left\{-\frac{(\log x - \mu)^2}{2\sigma^2}\right\}.$$

(b) Use the result in (a) to show that, for any $\lambda, \alpha > 0$,

$$\lim_{x \to \infty} \frac{P(X > x)}{e^{-\lambda x}} = \infty \quad \text{and} \quad \lim_{x \to \infty} \frac{P(X > x)}{x^{-\alpha}} = 0.$$

Project 9 (Estimation of high quantiles). Consider the following four random variables:

$$X_1 = e^{\Phi^{-1}(t_5(Y))}, \quad X_2 = a(4 + Y)^2, \quad X_3 = e^{bY}, \quad X_4 = c(1 - t_5(Y))^{-2/5},$$

where Y has a standard Student's t distribution with five degrees of freedom. All three have distributions with heavy right tails.

(a) Determine a, b, and c such that $F_{X_k}^{-1}(0.995) = F_{X_1}^{-1}(0.995)$ for $k = 2, 3, 4$.
(b) Determine which of the random variables $X_1, \ldots, X_4$ have a regularly varying right tail.
(c) Simulate a sample of size 500 from the distribution of Y and transform this sample into samples from the distributions of $X_1, \ldots, X_4$. For each k estimate the quantile $F_{X_k}^{-1}(0.995)$ based on the sample from the distribution of X_k. For each k use the POT method to estimate the quantile $F_{X_k}^{-1}(0.995)$ based on the sample from the distribution of X_k using the empirical estimate of $F_{X_k}^{-1}(0.9)$ as the threshold value.

Repeat the procedure until the samples of quantile estimators are sufficiently representative of the unknown distributions of the quantile estimators. Illustrate the results in histograms.

Chapter 9
Multivariate Models

In this chapter, we consider multivariate models for the joint distribution of several risk factors such as returns or log returns for different assets, zero rate changes for different maturity times, changes in implied volatility, and losses due to defaults on risky loans. Our aim is to specify a good model for the future value $g(\mathbf{X})$ of a portfolio, where the function g is known and its argument $\mathbf{X}$ is a random vector of, for instance, log returns and zero rate changes over a given future time period. Since the function g is known, what remains is to make a good choice of probability distribution for random vector $\mathbf{X}$.

The first sections, Sects. 9.1–9.3, present spherical and elliptical distributions and their applicability in a wide range of problems in risk management. Elliptical distributions provide convenient and flexible multivariate models. This set of models includes the multivariate normal model but allows for a much wider range of tail behavior and dependence properties.

Elliptical distributions have the following important property: if $\mathbf{X}$ has an elliptical distribution, then the distribution of any linear combination $\mathbf{w}^{\mathrm{T}}\mathbf{X}$ of its components is known. This property is useful because if $\mathbf{X}$ represents the returns of the financial assets in a portfolio, then we know the distribution of every linear portfolio. The property is useful even if we do not model the returns directly with an elliptical distribution. Suppose that $\mathbf{X}$ represents a vector of log returns, zero rate changes, etc. and is modeled by an elliptical distribution. If the portfolio value at some future time is given by $g(\mathbf{X})$, then a first order Taylor approximation of g around the mean vector $\boldsymbol{\mu} = \mathrm{E}[\mathbf{X}]$ leads to the first-order approximation

$$g(\mathbf{X}) \approx g(\boldsymbol{\mu}) + \sum_{k=1}^{d} \frac{\partial g}{\partial x_k}(\boldsymbol{\mu})(X_k - \mu_k).$$

The right-hand side is a linear combination of the components of $\mathbf{X}$ whose distribution therefore is known. Thus, whenever linearization of the nonlinear function g is justified, we can approximate the probability distribution of $g(\mathbf{X})$ analytically.

H. Hult et al., *Risk and Portfolio Analysis: Principles and Methods*, Springer Series in Operations Research and Financial Engineering, DOI 10.1007/978-1-4614-4103-8_9, © Springer Science+Business Media New York 2012

An important property of spherically distributed random vectors is that they can be decomposed into a product of a radial part and an independent angular part that is uniformly distributed on a sphere. This property makes it easy to simulate from a spherical (elliptical) distribution in any dimension. In particular, we can approximate the probability distribution of $g(\mathbf{X})$ arbitrarily well by simulating a large enough sample from $\mathbf{X}$ and consider the resulting empirical distribution of the simulated outcomes of $g(\mathbf{X})$.

A series of applications of elliptical distribution in risk management, including risk aggregation, solvency computations for an insurance company, and hedging of options, is presented in Sect. 9.3.

Then we turn our attention to multivariate models for random vectors that do not show signs of elliptical symmetry, and the notion of copula is introduced in Sect. 9.4. On the one hand, the copula is just a multivariate distribution function appearing in the representation of a multivariate distribution function in terms of its (continuous) marginal distribution functions. On the other hand, the copula may be identified as the dependence structure of a multivariate distribution, and by varying the copula for a random vector $\mathbf{X}$ for which the distributions of the components X_k are held fixed, we may understand better the effect of the dependence between the X_k on the distribution for the future portfolio value $g(\mathbf{X})$. We rarely have sufficient information to accurately specify the copula of a random vector $\mathbf{X}$, and by varying the copula within a set of copula functions, we may study the robustness of the distribution of the portfolio value $g(\mathbf{X})$ to misspecifications of the dependence between the components of $\mathbf{X}$. Moreover, the representation of a multivariate model for $\mathbf{X}$ in terms of a copula and distribution functions for the X_k is useful for simulation from the distribution of $\mathbf{X}$: an outcome from $\mathbf{X}$ is constructed as an outcome from the copula together with an application of the quantile transform.

Finally, in Sect. 9.5, we consider the effect of dependence modeling for large homogeneous portfolios. We consider a high-dimensional random vector $\mathbf{X}$ with equally distributed components and study the effect of the dependence between the components on the distribution of the sum of the components of $\mathbf{X}$.

9.1 Spherical Distributions

A random vector $\mathbf{Y}$ has a spherical distribution in $\mathbb{R}^d$ if its distribution is spherically symmetric. In other words, its distribution is invariant under rotations and reflections. Linear transformations that represent rotations and reflections correspond to multiplication by orthogonal matrices. Recall that a matrix $\mathbf{O}$ is orthogonal if it has real entries and $\mathbf{O}\mathbf{O}^{\mathrm{T}} = \mathbf{I}$, where $\mathbf{I}$ is the identity matrix. Formally, $\mathbf{Y}$ has a spherical distribution if

$$\mathbf{O}\mathbf{Y} \stackrel{\mathrm{d}}{=} \mathbf{Y} \quad \text{for every orthogonal matrix } \mathbf{O}. \tag{9.1}$$

Figure 9.1 shows scatter plots of samples from two spherical distributions.

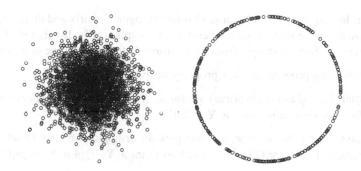

Fig. 9.1 *Left plot*: sample of size 3,000 from bivariate standard normal distribution. *Right plot*: sample of size 300 from uniform distribution on the unit circle

Three examples of spherical distributions are presented below. Before presenting the examples, let us recall the definition of the multivariate normal distribution.

(1) A random vector $\mathbf{Z}$ has standard normal distribution $N_d(\mathbf{0}, \mathbf{I})$ if $\mathbf{Z} = (Z_1, \ldots, Z_d)^T$, where $Z_1, \ldots, Z_d$ are independent and $N(0, 1)$-distributed.
(2) A random vector $\mathbf{X}$ is $N_d(\boldsymbol{\mu}, \boldsymbol{\Sigma})$-distributed if $\mathbf{X} \stackrel{d}{=} \boldsymbol{\mu} + \mathbf{AZ}$, where $\mathbf{AA}^T = \boldsymbol{\Sigma}$ and $\mathbf{Z}$ is $N_d(\mathbf{0}, \mathbf{I})$-distributed.

Example 9.1 (Standard normal distribution). The first example of a spherical distribution is the standard normal distribution $N_d(\mathbf{0}, \mathbf{I})$. Let $\mathbf{Z}$ have a $N_d(\mathbf{0}, \mathbf{I})$ distribution, and let $\mathbf{O}$ be an arbitrary orthogonal matrix. By property (2) above, $\mathbf{OZ}$ has the distribution $N_d(\mathbf{0}, \mathbf{OO}^T)$. Since $\mathbf{OO}^T = \mathbf{I}$, we conclude that $\mathbf{Z}$ satisfies (9.1). The left plot in Fig. 9.1 shows a scatter plot of a sample from $N_d(\mathbf{0}, \mathbf{I})$.

Example 9.2 (Standard normal variance mixture). Another example of a spherical distribution is obtained by multiplying a $N_d(\mathbf{0}, \mathbf{I})$-distributed random vector $\mathbf{Z}$ by an independent nonnegative random variable W. Notice that, for any orthogonal matrix $\mathbf{O}$,

$$\mathbf{OWZ} = W\mathbf{OZ} \stackrel{d}{=} W\mathbf{Z},$$

where the last equality follows since $\mathbf{Z}$ is spherically distributed.

The uniform distribution on the unit sphere $\mathbb{S}^{d-1} = \{\mathbf{x} \in \mathbb{R}^d : |\mathbf{x}| = 1\}$, where $|\mathbf{x}|^2 = \mathbf{x}^T\mathbf{x}$, assigns equal probability to any two subsets of $\mathbb{S}^{d-1}$ with the same surface area.

Example 9.3 (Uniform distribution on the unit sphere). A third example of a spherical distribution is the uniform distribution on the unit sphere, i.e., the probability mass is distributed uniformly on the unit sphere $\mathbb{S}^{d-1}$. Let $\mathbf{U}$ be uniformly distributed on the unit sphere and consider a subset A of the unit sphere. For any orthogonal matrix $\mathbf{O}$ it holds that

$$P(\mathbf{OU} \in A) = P(\mathbf{U} \in \mathbf{O}^{-1}A) = P(\mathbf{U} \in \mathbf{O}^T A) = P(\mathbf{U} \in A),$$

where the last equality holds because $\mathbf{O}$ is an orthogonal matrix and therefore A and $\mathbf{O}^{\mathrm{T}} A$ have the same surface area. Therefore, $\mathbf{U}$ is spherically distributed. The right plot in Fig. 9.1 shows a sample from the uniform distribution on the unit circle.

The following property is a key property of spherical distributions.

Proposition 9.1. *If $\mathbf{a}$ is an arbitrary vector in $\mathbb{R}^d$ and $\mathbf{Y}$ is spherically distributed and of the same dimension, then $\mathbf{a}^{\mathrm{T}}\mathbf{Y} \overset{\mathrm{d}}{=} |\mathbf{a}|Y_1$.*

Proof. Take $\mathbf{a} \neq \mathbf{0}$, let $\mathbf{u} = \mathbf{a}/|\mathbf{a}|$, and pick an orthogonal matrix $\mathbf{O}$ whose first row is equal to $\mathbf{u}^{\mathrm{T}}$. Since $\mathbf{OY} \overset{\mathrm{d}}{=} \mathbf{Y}$, it follows that $\mathbf{a}^{\mathrm{T}}\mathbf{Y} = |\mathbf{a}|\mathbf{u}^{\mathrm{T}}\mathbf{Y} = |\mathbf{a}|(\mathbf{OY})_1 \overset{\mathrm{d}}{=} |\mathbf{a}|Y_1$. $\qquad\square$

The following property is another key property of spherical distributions.

Proposition 9.2. *If $\mathbf{Y}$ is spherically distributed, then $\mathbf{Y} \overset{\mathrm{d}}{=} R\mathbf{U}$, where $R \overset{\mathrm{d}}{=} |\mathbf{Y}|$, $\mathbf{U}$ is uniformly distributed on the unit sphere and R and $\mathbf{U}$ are independent. Moreover, $\mathrm{P}(\mathbf{Y}/|\mathbf{Y}| \in \cdot \mid |\mathbf{Y}| > 0) = \mathrm{P}(\mathbf{U} \in \cdot)$.*

The proposition provides a way to simulate from a spherical distribution. First draw a vector from the uniform distribution on the unit sphere by sampling from a standard normal distribution and dividing by its norm. Then draw the radial part by sampling from the distribution of $|\mathbf{Y}|$.

To prove Proposition 9.2, we first state and prove the following lemma.

Lemma 9.1. *The uniform distribution on the unit sphere is the unique spherical distribution on the unit sphere.*

Proof. Let $\mathbf{Z}$ have a spherical distribution on the unit sphere. For any orthogonal matrix $\mathbf{O}$ and subset A of $\mathbb{S}^{d-1}$ it holds that $\mathrm{P}(\mathbf{Z} \in \mathbf{O}A) = \mathrm{P}(\mathbf{O}^{\mathrm{T}}\mathbf{Z} \in \mathbf{O}^{\mathrm{T}}\mathbf{O}A) = \mathrm{P}(\mathbf{Z} \in A)$ since $\mathbf{Z}$ is spherically distributed and $\mathbf{O}^{\mathrm{T}}$ is an orthogonal matrix. If $\mathbf{Z}$ were not uniformly distributed on $\mathbb{S}^{d-1}$, then there would exist a subset A_0 of $\mathbb{S}^{d-1}$ and an orthogonal matrix $\mathbf{O}_0$ such that $\mathrm{P}(\mathbf{Z} \in A) \neq \mathrm{P}(\mathbf{Z} \in \mathbf{O}_0 A_0)$, which contradicts that $\mathbf{Z}$ is spherically distributed. $\qquad\square$

Proof of Proposition 9.2. It is sufficient to show that $\mathrm{P}(|\mathbf{Y}| > r, \mathbf{Y}/|\mathbf{Y}| \in A) = \mathrm{P}(|\mathbf{Y}| > r)\,\mathrm{P}(\mathbf{U} \in A)$ for any $r \geq 0$ and any subset A of $\mathbb{S}^{d-1}$, where $\mathbf{U}$ is uniformly distributed on the unit sphere.

We claim that, for any $r \geq 0$, $I\{|\mathbf{Y}| > r\}\mathbf{Y}/|\mathbf{Y}|$ is spherically distributed. To prove the claim, note that for any orthogonal matrix $\mathbf{O}$ it holds that $|\mathbf{OY}| = |\mathbf{Y}|$ and $\mathbf{OY} \overset{\mathrm{d}}{=} \mathbf{Y}$ and therefore

$$\mathbf{O}I\{|\mathbf{Y}| > r\}\mathbf{Y}/|\mathbf{Y}| = I\{|\mathbf{OY}| > r\}\mathbf{OY}/|\mathbf{OY}| \overset{\mathrm{d}}{=} I\{|\mathbf{Y}| > r\}\mathbf{Y}/|\mathbf{Y}|.$$

To complete the proof of Proposition 9.2, we may without loss of generality take r such that $\mathrm{P}(|\mathbf{Y}| > r) > 0$ and note that

$$P(\mathbf{Y}/|\mathbf{Y}| \in A \mid |\mathbf{Y}| > r) = P(I\{|\mathbf{Y}| > r\}\mathbf{Y}/|\mathbf{Y}| \in A)/\,P(|\mathbf{Y}| > r)$$
$$= P(I\{|\mathbf{Y}| > r\}\mathbf{Y}/|\mathbf{Y}| \in \mathbf{O}A)/\,P(|\mathbf{Y}| > r)$$
$$= P(\mathbf{Y}/|\mathbf{Y}| \in \mathbf{O}A \mid |\mathbf{Y}| > r).$$

It now follows from Lemma 9.1 that $P(\mathbf{Y}/|\mathbf{Y}| \in A \mid |\mathbf{Y}| > r) = P(\mathbf{U} \in A)$, and therefore $P(|\mathbf{Y}| > r, \mathbf{Y}/|\mathbf{Y}| \in A) - P(|\mathbf{Y}| > r)\,P(\mathbf{U} \in A)$. $\qquad\square$

9.2 Elliptical Distributions

The multivariate normal distribution is very useful in the construction of multivariate models. Its popularity derives primarily from the fact that it is tractable, allowing for explicit calculations, and it that can be motivated asymptotically by the central limit theorem. For univariate data that show clear signs of symmetry the univariate normal distribution does not necessarily give a good fit to the data. Typically normal tails do not match empirical tails particularly well. Similarly, the multivariate normal distribution is often at best a reasonable first approximation for samples of multivariate observations with clear signs of elliptical symmetry.

A random vector $\mathbf{X}$ has a $N_d(\mu, \Sigma)$ distribution if

$$\mathbf{X} \stackrel{d}{=} \mu + \mathbf{AZ}, \tag{9.2}$$

where $\mathbf{AA}^\mathsf{T} = \Sigma$ and $\mathbf{Z}$ has a $N_d(\mathbf{0}, \mathbf{I})$ distribution. An easy way to obtain a richer class of multivariate distributions, which share many of the tractable properties of the multivariate normal distribution, is to replace the standard normal vector $\mathbf{Z}$ in (9.2) by an arbitrary spherically distributed random vector $\mathbf{Y}$. Formally, a random vector $\mathbf{X}$ has an elliptical distribution if there exist a vector μ, a matrix $\mathbf{A}$, and a spherically distributed vector $\mathbf{Y}$ such that

$$\mathbf{X} \stackrel{d}{=} \mu + \mathbf{AY}. \tag{9.3}$$

The matrix $\mathbf{A}$ and the spherical distribution of $\mathbf{Y}$ in (9.3) are not determined by the distribution of $\mathbf{X}$: we may replace the pair $(\mathbf{A}, \mathbf{Y})$ in (9.3) by $(c\mathbf{A}, c^{-1}\mathbf{Y})$ for any constant $c \in (0, \infty)$. A matrix Σ satisfying $\Sigma = \mathbf{AA}^\mathsf{T}$ is called a dispersion matrix of the elliptically distributed vector $\mathbf{X}$. If the covariance matrix $\mathrm{Cov}(\mathbf{X})$ exists finitely, then $\mathrm{Cov}(\mathbf{X}) = c\Sigma$ for some constant $c \in (0, \infty)$. To verify this claim, we note that, by (9.3) and Proposition 9.2,

$$\mathrm{Cov}(\mathbf{X}) = \mathrm{E}[(\mathbf{X} - \mu)(\mathbf{X} - \mu)^\mathsf{T}] = \mathrm{E}[R^2]\mathbf{A}\,\mathrm{E}[\mathbf{U}\mathbf{U}^\mathsf{T}]\mathbf{A}^\mathsf{T} = \frac{\mathrm{E}[R^2]}{d}\Sigma.$$

The last equality above can be proven as follows. Consider a standard normally distributed vector $\mathbf{Z}$ and recall that $\mathbf{Z}/|\mathbf{Z}|$ is uniformly distributed on the unit sphere and $\mathrm{E}[|\mathbf{Z}|^2] = d$. Therefore,

$$\mathbf{I} = \mathrm{Cov}(\mathbf{Z}) = \mathrm{E}[|\mathbf{Z}|^2]\,\mathrm{E}[\mathbf{U}\mathbf{U}^\mathsf{T}] = d\,\mathrm{E}[\mathbf{U}\mathbf{U}^\mathsf{T}].$$

For a dispersion matrix Σ with nonzero diagonal entries we define the linear correlation parameter $\rho_{ij} = \Sigma_{ij} / (\Sigma_{ii} \Sigma_{jj})^{1/2}$. If Cov($\mathbf{X}$) exists finitely, then $\rho_{ij} =$ Cor(X_i, X_j), i.e., the linear correlation parameter coincides with the ordinary linear correlation coefficient.

The normal variance mixture distributions are the distributions of random vectors with stochastic representation

$$\mathbf{X} \overset{\mathrm{d}}{=} \mu + W\mathbf{A}\mathbf{Z}, \tag{9.4}$$

where $\mathbf{A}$ and $\mathbf{Z}$ are the same as in (9.2) and W is a nonnegative random variable independent of $\mathbf{Z}$. From Example 9.2 it follows that a normal variance mixture distribution is an elliptical distribution. By conditioning on $W = w$, we see that $\mathbf{X}|W = w$ is $N_d(\mu, w^2\Sigma)$-distributed, which explains the name normal variance mixture. If $E[W^2] < \infty$, then $\mathbf{X}$ has a well-defined mean vector $\mu = E[\mathbf{X}]$ and covariance matrix

$$\text{Cov}(\mathbf{X}) = E[(\mathbf{X} - \mu)(\mathbf{X} - \mu)^{\mathrm{T}}] = E[W^2]\mathbf{A}\,E[\mathbf{Z}\mathbf{Z}^{\mathrm{T}}]\mathbf{A}^{\mathrm{T}} = E[W^2]\Sigma.$$

Example 9.4 (Multivariate Student's t). If we take $W^2 \overset{\mathrm{d}}{=} \nu/S_\nu$, where S_ν has a Chi-square distribution with ν degrees of freedom, then the resulting distribution of $\mathbf{X} = \mu + W\mathbf{A}\mathbf{Z}$ is called a multivariate Student's t distribution with ν degrees of freedom, written $t_d(\mu, \Sigma, \nu)$. Note that Σ is not the covariance matrix of $\mathbf{X}$. Since $E[W^2] = \nu/(\nu - 2)$ if $\nu > 2$, it follows that Cov($\mathbf{X}$) = $(\nu/(\nu - 2))\Sigma$.

For a normally distributed random vector $\mathbf{X} \overset{\mathrm{d}}{=} \mu + \mathbf{A}\mathbf{Z}$, where $\mathbf{A}\mathbf{A}^{\mathrm{T}} = \Sigma$, any linear combination of the components of $\mathbf{X}$ is again normally distributed. That is, for any nonrandom vector $\mathbf{w}$ of the same dimension,

$$\mathbf{w}^{\mathrm{T}}\mathbf{X} \overset{\mathrm{d}}{=} \mathbf{w}^{\mathrm{T}}\mu + \mathbf{w}^{\mathrm{T}}\mathbf{A}\mathbf{Z}$$
$$= \mathbf{w}^{\mathrm{T}}\mu + (\mathbf{A}^{\mathrm{T}}\mathbf{w})^{\mathrm{T}}\mathbf{Z}$$
$$\overset{\mathrm{d}}{=} \mathbf{w}^{\mathrm{T}}\mu + (\mathbf{w}^{\mathrm{T}}\Sigma\mathbf{w})^{1/2}Z_1.$$

A similar property holds for arbitrary elliptical distributions.

Proposition 9.3. *If $\mathbf{X}$ has an elliptical distribution with stochastic representation $\mathbf{X} \overset{\mathrm{d}}{=} \mu + \mathbf{A}\mathbf{Y}$, where $\mathbf{Y}$ is spherically distributed, then for any vector $\mathbf{a}$ of the same dimension $\mathbf{a}^{\mathrm{T}}\mathbf{X} \overset{\mathrm{d}}{=} \mathbf{a}^{\mathrm{T}}\mu + (\mathbf{a}^{\mathrm{T}}\Sigma\mathbf{a})^{1/2}Y_1$, where $\Sigma = \mathbf{A}\mathbf{A}^{\mathrm{T}}$.*

The proof is omitted since the result follows immediately from Proposition 9.1 and the defining property (9.3) of elliptical distributions.

As was previously mentioned, normal variance mixture distributions and, more generally, elliptical distributions share many of the attractive properties of normal distributions. However, there are important exceptions. Recall that the components of the $N_d(\mu, \Sigma)$-distributed vector $\mu + \mathbf{A}\mathbf{Z}$ are independent if and only if $\mathbf{A}\mathbf{A}^{\mathrm{T}} = \Sigma$ is a diagonal matrix, that is, if the components are uncorrelated. This property does not hold for arbitrary normal variance mixture distributions. If $\mathbf{X} \overset{\mathrm{d}}{=} \mu + W\mathbf{A}\mathbf{Z}$

with $\mathbf{A}\mathbf{A}^T = \Sigma$ a diagonal matrix, then the components of $\mathbf{X}$ are uncorrelated. If Σ is a diagonal matrix with strictly positive diagonal entries, then $(X_k, X_l) \overset{d}{=} (\mu_k + WA_{k,k}Z_k, \mu_l + WA_{l,l}Z_l)$, where $A_{k,k}, A_{l,l} > 0$. Clearly, X_k and X_l are not independent unless W is a constant.

The sum of independent elliptically distributed random vectors with the same (up to a constant factor) dispersion matrix is elliptically distributed.

Proposition 9.4. *If the random vectors $\mathbf{X}_1$ and $\mathbf{X}_2$ in $\mathbb{R}^d$ are independent and elliptically distributed with common dispersion matrix Σ, then $\mathbf{X}_1 + \mathbf{X}_2$ is elliptically distributed.*

Proof. For a matrix $\mathbf{A}$ such that $\mathbf{A}\mathbf{A}^T = \Sigma$ we may write $\mathbf{X}_1 + \mathbf{X}_2 \overset{d}{=} \mu_1 + \mu_2 + \mathbf{A}(\mathbf{Y}_1 + \mathbf{Y}_2)$ for some independent spherically distributed vectors $\mathbf{Y}_1$ and $\mathbf{Y}_2$. It remains to show that $\mathbf{Y}_1 + \mathbf{Y}_2$ is spherically distributed. For every orthogonal matrix $\mathbf{O}$ and $\mathbf{y}$ in $\mathbb{R}^d$,

$$P(\mathbf{O}(\mathbf{Y}_1 + \mathbf{Y}_2) \le \mathbf{y}) = \int P(\mathbf{O}\mathbf{Y}_1 + \mathbf{z} \le \mathbf{y} \mid \mathbf{O}\mathbf{Y}_2 = \mathbf{z}) dF_{\mathbf{O}\mathbf{Y}_2}(\mathbf{z})$$

$$= \int P(\mathbf{Y}_1 + \mathbf{z} \le \mathbf{y}) dF_{\mathbf{Y}_2}(\mathbf{z})$$

$$= P(\mathbf{Y}_1 + \mathbf{Y}_2 \le \mathbf{y}),$$

i.e., $\mathbf{O}(\mathbf{Y}_1 + \mathbf{Y}_2) \overset{d}{=} \mathbf{Y}_1 + \mathbf{Y}_2$, from which the conclusion follows. $\square$

Example 9.5 (Summation of log returns). Consider a set of identically distributed and uncorrelated random variables $X_1, \ldots, X_n$ that represent future daily log returns for some asset. Suppose that each log return has a finite mean μ and standard deviation σ. If the log returns are independent, then by the central limit theorem, $X_1 + \cdots + X_n$ is approximately $N(n\mu, n\sigma^2)$-distributed for n large. If the vector $\mathbf{X} = (X_1, \ldots, X_n)^T$ of log returns has an elliptical distribution, then Proposition 9.3 implies that

$$X_1 + \cdots + X_n = \mathbf{1}^T\mathbf{X} \overset{d}{=} n\mu + n^{1/2}(X_1 - \mu).$$

We see that the n-day log return and the 1-day log return belong to the same location-scale family of distributions. For instance, if the 1-day log return has a heavy-tailed Student's t distribution with a low-degree-of-freedom parameter, then so does the n-day log return.

9.2.1 Goodness of Fit of an Elliptical Model

Consider a random vector $\mathbf{X}$ with an elliptical distribution with representation $\mathbf{X} = \mu + \mathbf{A}\mathbf{Y}$, where $\mathbf{Y}$ has a spherical distribution and $\Sigma = \mathbf{A}\mathbf{A}^T$ is invertible. By Proposition 9.3,

$$\mathbf{w}^T\mathbf{X} \overset{d}{=} \mathbf{w}^T\mu + (\mathbf{w}^T\Sigma\mathbf{w})^{1/2}Y_1 \quad \text{for all nonrandom vectors } \mathbf{w} \ne \mathbf{0}$$

or, equivalently,

$$\frac{\mathbf{w}^T\mathbf{X} - \mathbf{w}^T\boldsymbol{\mu}}{(\mathbf{w}^T\boldsymbol{\Sigma}\mathbf{w})^{1/2}} \overset{\text{d}}{=} Y_1 \quad \text{for all nonrandom vectors } \mathbf{w} \neq \mathbf{0}. \tag{9.5}$$

The property (9.5) can be used to investigate whether or not a multivariate sample is likely to come from an elliptical distribution. Let us illustrate the procedure by an example.

Example 9.6 (Estimation and fit of an elliptical model). Consider a sample of size 500 of pairs of daily log returns for the Dow Jones Industrial Average (DJIA) and Nasdaq Composite indices (index values from November 11, 2008 until November 4, 2010). The scatter plot of the pairs of log returns is shown in the upper left plot in Fig. 9.2. The log-return sample is denoted $\{\mathbf{x}_1, \ldots, \mathbf{x}_{500}\}$. We assume initially that the sample can be seen as outcomes from an elliptically distributed vector $\mathbf{X}$ and investigate whether this assumption can be rejected or not. If it is not rejected, then we also want to determine the elliptical distribution of $\mathbf{X}$. We assume that the location parameter $\boldsymbol{\mu}$ and a scalar multiple of the shape parameter $\mathbf{C} = c\boldsymbol{\Sigma}$, which is assumed invertible, can be estimated. Note that (9.5) can be expressed as

$$\frac{\mathbf{w}^T\mathbf{X} - \mathbf{w}^T\boldsymbol{\mu}}{(\mathbf{w}^T\mathbf{C}\mathbf{w})^{1/2}} \overset{\text{d}}{=} c^{-1/2}Y_1 \quad \text{for all nonrandom vectors } \mathbf{w} \neq \mathbf{0}.$$

If the covariance matrix $\text{Cov}(\mathbf{X})$ exists finitely, then $\boldsymbol{\mu} = \text{E}[\mathbf{X}]$, and we may take $\mathbf{C} = \text{Cov}(\mathbf{X})$. Here we estimate $\boldsymbol{\mu}$ and $\mathbf{C}$ by the sample mean and sample covariance, respectively. The estimates are denoted $\widehat{\boldsymbol{\mu}}$ and $\widehat{\mathbf{C}}$. Consider a large set of vectors $\{\mathbf{w}_1, \ldots, \mathbf{w}_n\}$ of unit length. For each $\mathbf{w}_k$ we construct the sample $\{y_{k,1}, \ldots, y_{k,500}\}$ by

$$y_{k,l} = \frac{\mathbf{w}_k^T\mathbf{x}_l - \mathbf{w}_k^T\widehat{\boldsymbol{\mu}}}{(\mathbf{w}_k^T\widehat{\mathbf{C}}\mathbf{w}_k)^{1/2}} \quad \text{for } k = 1, \ldots, n, \quad l = 1, \ldots, 500.$$

Each such sample can be viewed as a sample from $c^{-1/2}Y_1$. If the data were generated by the elliptical distribution of $\mathbf{X}$, then all the n constructed samples must come from the same distribution, the distribution of $c^{-1/2}Y_1$. By overlaying the n q–q plots of the empirical quantiles for the n samples against the quantiles of a chosen reference distribution, we can check graphically whether the data appear to be consistent with an elliptical distribution or not. Moreover, the distribution of $c^{-1/2}Y_1$ can be estimated from the q–q plots.

Here we take $n = 100$ and sample the $\mathbf{w}_k$ randomly from the uniform distribution of the unit sphere by setting $\mathbf{w}_k = \mathbf{z}_k/|\mathbf{z}_k|$, where the $\mathbf{z}_k$ are outcomes of independent $N_2(\mathbf{0}, \mathbf{I})$-distributed random vectors. The upper left plot in Fig. 9.2 is a scatter plot of the sample $\{\mathbf{x}_1, \ldots, \mathbf{x}_{500}\}$. The upper right plot in Fig. 9.2 shows the $n = 100$ q–q plots of the empirical quantiles of the samples $\{y_{k,1}, \ldots, y_{k,500}\}$ (y-axis) against the quantiles of the standard normal distribution (x-axis). The q–q plots indicate a reasonable fit to a common distribution with heavier tails than the normal distribution.

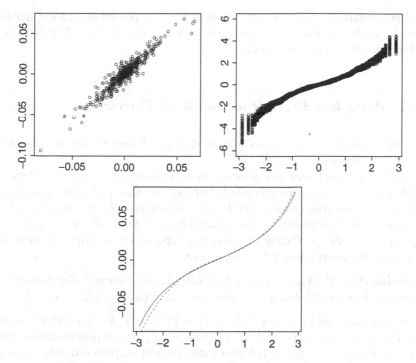

Fig. 9.2 *Scatter plot* showing pairs (x_D, x_N) of DJIA and Nasdaq log returns. *Upper right plot*: 100 overlaid q–q plots for empirical quantiles for each of 100 samples $\{y_{k,1}, \ldots, y_{k,500}\}$ (*y*-axis) against standard normal quantiles (*x*-axis). The *solid curve* in the *lower plot* shows the quantiles of the model for the Nasdaq log returns based on the fitted bivariate Student's t model (*y*-axis) against standard normal quantiles (*x*-axis). The *dashed curve* in the *lower plot* shows the polynomial normal quantiles (Example 8.11) fitted to the Nasdaq log returns (*y*-axis) against the standard normal quantiles (*x*-axis)

Under the assumption that $\{\mathbf{x}_1, \ldots, \mathbf{x}_{500}\}$ is a sample from the bivariate Student's t_ν distribution with $\nu > 2$ (otherwise it does not make sense to use the sample covariance matrix) and under the assumption that $\widehat{\mu} = \mu$ and $\widehat{\mathbf{C}} = \text{Cov}(\mathbf{X})$, it holds that all the samples $\{y_{k,1}, \ldots, y_{k,500}\}$ are samples from the distribution of $((\nu-2)/\nu)^{1/2} Z$, where Z is standard t_ν-distributed. Least-squares estimation based on all 100 univariate samples gives the estimate $\widehat{\nu} \approx 4.09$. The selected model for the sample $\{\mathbf{x}_1, \ldots, \mathbf{x}_{500}\}$ is the distribution $t_2(\widehat{\mu}, ((\widehat{\nu} - 2)/\widehat{\nu})\widehat{\mathbf{C}}, \widehat{\nu})$.

The second marginal distribution (a univariate Student's t distribution) of the bivariate Student's t distribution for the pair of DJIA and Nasdaq log returns provides a model for the Nasdaq log returns. In the lower plot in Fig. 9.2, we compare this model to the polynomial normal model in Example 8.11. The solid curve in the lower plot is a q–q plot of the quantiles for the model for the Nasdaq log returns (*y*-axis) against standard normal quantiles (*x*-axis). The quantiles of the fitted polynomial normal model in Example 8.11 are plotted against the standard normal quantiles as the dashed curve in the lower plot in Fig. 9.2. It is hard to distinguish between the two models.

9.2.2   Asymptotic Dependence and Rank Correlation

We now introduce general notions of dependence and study them in the context of elliptical distributions.

The first notion of dependence measures the dependence of extreme values and is called tail dependence or asymptotic dependence. Consider a pair (X_1, X_2) of random variables with equally distributed components. We say that X_1 and X_2 are asymptotically dependent in the lower left tail if the limit $\lim_{x \to -\infty} P(X_2 \leq x \mid X_1 \leq x)$, the coefficient of lower tail dependence, is strictly positive and asymptotically independent if the limit is zero.

Proposition 9.5. *If (X_1, X_2) has a bivariate standard normal distribution with linear correlation coefficient $\rho < 1$, then $\lim_{x \to -\infty} P(X_2 \leq x \mid X_1 \leq x) = 0$.*

Proof. First note that $P(X_2 \leq x \mid X_1 \leq x) = P(X_1 \leq x, X_2 \leq x)/\Phi(x)$ and that $(X_1, X_2) \stackrel{d}{=} (Z_1, \rho Z_1 + (1 - \rho^2)^{1/2} Z_2)$, where Z_1, Z_2 are independent and standard normally distributed. If $\rho = -1$, then the statement of the proposition holds, so we may without loss of generality assume that $|\rho| < 1$. We may write

$$P(X_1 \leq x, X_2 \leq x) = \int_{-\infty}^{\infty} P\left(Z_1 \leq x, \rho Z_1 + (1 - \rho^2)^{1/2} t \leq x\right) \phi(t) dt$$

$$= \int_{-\infty}^{a(x)} \Phi(x) \phi(t) dt + \int_{a(x)}^{\infty} \Phi((x - (1 - \rho^2)^{1/2} t)/\rho) \phi(t) dt,$$

where $a(x) = ((1 - \rho)/(1 + \rho))^{1/2} x$. Therefore,

$$\lim_{x \to -\infty} P(X_2 \leq x \mid X_1 \leq x) = \lim_{x \to -\infty} \left(\Phi(a(x)) + \frac{\int_{a(x)}^{\infty} \Phi((x - (1 - \rho^2)^{1/2} t)/\rho) \phi(t) dt}{\Phi(x)} \right)$$

$$= \lim_{x \to -\infty} \frac{\int_{a(x)}^{\infty} \Phi((x - (1 - \rho^2)^{1/2} t)/\rho) \phi(t) dt}{\Phi(x)}.$$

Applying l'Hôpital's rule gives

$$\lim_{x \to -\infty} P(X_2 \leq x \mid X_1 \leq x)$$

$$= - \lim_{x \to -\infty} \frac{\Phi(x)}{\phi(x)} \left(\frac{1 - \rho}{1 + \rho} \right)^{1/2} + \lim_{x \to -\infty} \frac{1}{\rho \phi(x)} \int_{a(x)}^{\infty} \phi((x - (1 - \rho^2)^{1/2} t)/\rho) \phi(t) dt.$$

We saw in Example 8.1 that $\Phi(x) \sim -\phi(x)/x$ as $x \to -\infty$, so we only need to compute the last limit given above. By writing up explicitly the standard normal densities and making a substitution of integration variable, we arrive at

$$\frac{1}{\rho\phi(x)} \int_{a(x)}^{\infty} \phi((x-(1-\rho^2)^{1/2}t)/\rho)\phi(t)dt = \int_{-\infty}^{a(x)} \phi(u)du = \Phi(a(x)),$$

which tends to 0 as $x \to -\infty$. $\qquad\square$

Unlike the components of a normally distributed random vector, the components of a vector with a bivariate Student's t distribution are asymptotically dependent. We omit the proof of the following proposition and refer the reader to Sect. 9.6 for further details.

Proposition 9.6. *Let (X_1, X_2) have an elliptical distribution with linear correlation parameter ρ. If X_1 and X_2 are equally distributed, and if $P(X_1 \leq x)$ is regularly varying at $-\infty$ with index $-\alpha$, then*

$$\lim_{x\to-\infty} P(X_2 \leq x \mid X_1 \leq x) = \frac{\int_{(\pi/2-\arcsin\rho)/2}^{\pi/2}(\cos t)^\alpha dt}{\int_0^{\pi/2}(\cos t)^\alpha dt}.$$

Zero correlation does not imply asymptotic independence, and covariances and correlations do not provide sufficient information to assess dependence between extreme values. For example, a quadratic hedge—based on a covariance structure—may perform poorly when it matters the most if the liability and the hedging instruments are asymptotically dependent. There are many examples from financial markets of simultaneous extreme price movements for assets whose log returns are weakly correlated between the assets.

Consider an elliptically distributed random vector (X_1, X_2) with a dispersion matrix Σ. Recall that any matrix $\Sigma_c = c\Sigma$ is a dispersion matrix for (X_1, X_2). However, the linear correlation parameter $\rho = \Sigma_{1,2}/(\Sigma_{1,1}\Sigma_{2,2})^{1/2}$ is uniquely determined by the elliptical distribution. Since $\rho = \mathrm{Cor}(X_1, X_2)$, whenever $\mathrm{Cor}(X_1, X_2)$ exists [the variances $\mathrm{Var}(X_1)$ and $\mathrm{Var}(X_2)$ are nonzero and finite], we may estimate ρ as the sample correlation coefficient. However, for heavy-tailed data (corresponding to distributions with finite variances) the sample correlation coefficient is an estimator of ρ with a large—or infinite—variance. An alternative approach to estimating the linear correlation parameter ρ is based on estimating another (rank) correlation coefficient called Kendall's tau, whose value for an elliptical distribution can be expressed in terms of the linear correlation parameter ρ. This approach allows for estimation of ρ also for elliptical distributions whose marginal distributions have infinite variances.

Kendall's tau for the random vector (X_1, X_2) is defined as

$$\tau(X_1, X_2) = P((X_1 - X_1')(X_2 - X_2') > 0) - P((X_1 - X_1')(X_2 - X_2') < 0), \quad (9.6)$$

where (X_1', X_2') is an independent copy of (X_1, X_2).

Proposition 9.7. *Let (X_1, X_2) have an elliptical distribution with location parameter (μ_1, μ_2) and linear correlation parameter ρ. If $P(X_1 = \mu_1) = P(X_2 = \mu_2) = 0$, then*

$$\tau(X_1, X_2) = \frac{2}{\pi} \arcsin \rho. \tag{9.7}$$

Proof. Without loss of generality we may consider the case $|\rho| < 1$. Since $P((X_1 - X_1')(X_2 - X_2') = 0) = 0$, we find that

$$\tau(X_1, X_2) = 2\,P((X_1 - X_1')(X_2 - X_2') > 0) - 1.$$

The independence of $\mathbf{X} = (X_1, X_2)^{\mathrm{T}}$ and $\mathbf{X}' = (X_1', X_2')^{\mathrm{T}}$ and representation (9.3) imply that

$$(\mathbf{X}, \mathbf{X}') \stackrel{\mathrm{d}}{=} (\boldsymbol{\mu}, \boldsymbol{\mu}) + \mathbf{A}(R\mathbf{U}, R'\mathbf{U}'),$$

where $R, R', \mathbf{U}, \mathbf{U}'$ are independent. From Proposition 9.4 we know that $\mathbf{X} - \mathbf{X}' \stackrel{\mathrm{d}}{=} \mathbf{A}R^*\mathbf{U}^*$, where R^* and $\mathbf{U}^*$ are independent, and the assumption $P(X_1 = \mu_1) = P(X_2 = \mu_2) = 0$ implies that $P(R^* = 0) = 0$. With $\mathbf{W} = \mathbf{A}\mathbf{U}^*$ we have found that

$$\tau(X_1, X_2) = 2\,P(R^* W_1 W_2 > 0) - 1 = 2\,P(W_1 W_2 > 0) - 1.$$

Write

$$\boldsymbol{\Sigma} = \begin{pmatrix} \sigma_1^2 & \sigma_1\sigma_2\rho \\ \sigma_1\sigma_2\rho & \sigma_2^2 \end{pmatrix}, \quad \mathbf{A} = \begin{pmatrix} \sigma_1(1-\rho^2)^{1/2} & \sigma_1\rho \\ 0 & \sigma_2 \end{pmatrix}, \quad \mathbf{U}^* \stackrel{\mathrm{d}}{=} \begin{pmatrix} \cos U \\ \sin U \end{pmatrix},$$

where U is uniformly distributed on $[-\pi, \pi)$. Then

$$
\begin{aligned}
P(W_1 W_2 > 0) &= 2\,P(W_1 > 0, W_2 > 0) \\
&= 2\,P(\sigma_1(1-\rho^2)^{1/2}\cos U + \sigma_1\rho\sin U > 0, \sigma_2\sin U > 0) \\
&= 2\,P((1-\rho^2)^{1/2}\cos U + \rho\sin U > 0, \sin U > 0) \\
&= 2\,P(\cos y \cos U + \sin y \sin U > 0, \sin U > 0),
\end{aligned}
$$

where $y = \arcsin \rho \in [-\pi/2, \pi/2]$. Clearly, $\sin U > 0$ is here equivalent to $U \in (0, \pi)$. Since $\cos y \cos U + \sin y \sin U = \cos(U - y)$ and $\cos(U - y) > 0$ is here equivalent to $U \in (y - \pi/2, y + \pi/2)$, we find that

$$P(\cos y \cos U + \sin y \sin U > 0, \sin U > 0) = P(U \in (y - \pi/2, y + \pi/2) \cap (0, \pi))$$

$$= P(U \in (0, y + \pi/2)).$$

Putting the pieces together gives

$$\tau(X_1, X_2) = 4\frac{\arcsin \rho + \pi/2}{2\pi} - 1 = \frac{2}{\pi}\arcsin \rho. \qquad \square$$

Consider the function sign(x) with value 0 for $x = 0$ and the value $x/|x|$ otherwise. Kendall's tau in (9.6) can be written as

$$\tau(X_1, X_2) = \mathrm{E}\left[\mathrm{sign}\left((X_1 - X_1')(X_2 - X_2')\right)\right]. \tag{9.8}$$

Given a sample $\{\mathbf{X}_1, \ldots, \mathbf{X}_n\}$ of identically distributed vectors $\mathbf{X}_k = (X_{k,1}, X_{k,2})^\mathrm{T}$, we estimate (9.8) by the number of index pairs (j, k), where $j < k$ such that $(X_{j,1} - X_{k,1})(X_{j,2} - X_{k,2}) > 0$ minus the number of index pairs such that $(X_{j,1} - X_{k,1})(X_{j,2} - X_{k,2}) < 0$ divided by the total number of index pairs:

$$\widehat{\tau} = \binom{n}{2}^{-1} \sum_{j<k} \mathrm{sign}\left((X_{j,1} - X_{k,1})(X_{j,2} - X_{k,2})\right).$$

Finally, if the $\mathbf{X}_k$ are elliptically distributed such that the condition in Proposition 9.7 holds, then the estimator of the linear correlation parameter ρ is chosen as

$$\widehat{\rho} = \sin\left(\frac{\pi}{2}\widehat{\tau}\right). \tag{9.9}$$

To assess the accuracy of the estimator in (9.9) and compare it to the sample correlation coefficient, we consider a simulation study that is summarized in Fig. 9.3. For samples from a bivariate normal distribution the two estimators perform similarly. For samples from a bivariate Student's t distribution with three degrees of freedom we find that the estimator in (9.9), a nonlinear transformation of Kendall's tau estimator, performs much better than the sample correlation coefficient and similarly to its performance on data from a bivariate normal distribution.

9.2.3 Linearization and Elliptical Distributions

Suppose that the future value of a financial portfolio can be expressed as $g(\mathbf{X})$, where the function g is a known function and $\mathbf{X}$ is a random vector whose components represent, e.g., log returns for a given set of assets over a given future time period. If the time period is rather short and if $\mathbf{X}$ is likely to take a value that is not too far from its expected value $\boldsymbol{\mu} = \mathrm{E}[\mathbf{X}]$, then the first-order approximation

$$g(\mathbf{X}) \approx g(\boldsymbol{\mu}) + \nabla g^\mathrm{T}(\boldsymbol{\mu})(\mathbf{X} - \boldsymbol{\mu}) = g(\boldsymbol{\mu}) + \sum_{k=1}^{d} \frac{\partial g}{\partial x_k}(\boldsymbol{\mu})(X_k - \mu_k)$$

can be assumed to be accurate. The approximation replaces the nonlinear expression in the components of $\mathbf{X}$ by a weighted sum of the components translated by a constant. However, it is typically hard to determine the probability distribution of a

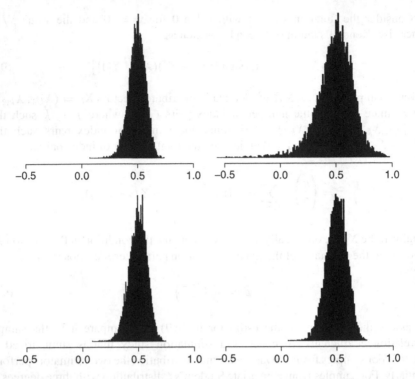

Fig. 9.3 Histograms based on 10,000 estimates of linear correlation parameter, where each estimate is based on a sample of size 100 from a bivariate elliptical distribution with linear correlation parameter 0.5. *Plots* to *left* show estimates based on samples from a bivariate normal distribution. *Plots* to *right* show estimates based on sample from a bivariate Student's *t* distribution with three degrees of freedom. The estimates in the *upper plots* are ordinary sample correlations. The estimates in the *lower plots* are transformations of Kendall's tau estimates as in (9.9)

sum of dependent random variables. An important exception is when $\mathbf{X}$ is elliptically distributed. In this case, $\mathbf{X}$ has the stochastic representation $\mathbf{X} \overset{d}{=} \boldsymbol{\mu} + \mathbf{A}\mathbf{Y}$, where $\mathbf{Y}$ has a spherical distribution, so Proposition 9.3 gives

$$g(\mathbf{X}) \approx g(\boldsymbol{\mu}) + \nabla g^{\mathrm{T}}(\boldsymbol{\mu})(\mathbf{X} - \boldsymbol{\mu}) \overset{d}{=} g(\boldsymbol{\mu}) + \left(\nabla g^{\mathrm{T}}(\boldsymbol{\mu}) \boldsymbol{\Sigma} \nabla g(\boldsymbol{\mu})\right)^{1/2} Y_1, \quad (9.10)$$

where $\boldsymbol{\Sigma} = \mathbf{A}\mathbf{A}^{\mathrm{T}}$ or, more explicitly,

$$g(\mathbf{X}) \overset{d}{\approx} g(\boldsymbol{\mu}) + \left(\sum_{j,k=1}^{d} \frac{\partial g}{\partial x_j}(\boldsymbol{\mu}) \frac{\partial g}{\partial x_k}(\boldsymbol{\mu}) \boldsymbol{\Sigma}_{j,k}\right)^{1/2} Y_1.$$

The accuracy of this approximation clearly depends strongly on how concentrated the probability mass of $\mathbf{X}$ is around its expected value $\boldsymbol{\mu}$. We illustrate the accuracy of the linearization with an example for one-dimensional elliptical distributions and a specific function g.

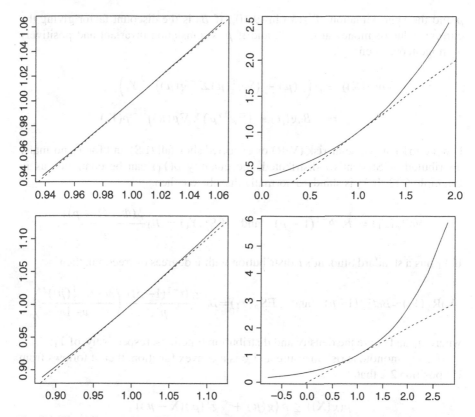

Fig. 9.4 These four q–q plots illustrate the approximation error from linearization. The *plots* show the quantiles of e^X (*y*-axis) against the quantiles of $1 + X$ (*x*-axis). The *upper plots* correspond to X's being N$(0, 0.02^2)$-distributed (*left*) and N$(0, 0.3^2)$-distributed (*right*). The *lower plots* correspond to X's having a Student's t distribution with three degrees of freedom and standard deviation 0.02 (*left*) and 0.3 (*right*)

Example 9.7 (Linearization). Let $g(x) = e^x$ and consider a random variable X with a spherical distribution with distribution function F. The quantile function of $g(X)$ is $g(F^{-1}(p))$, whereas the that of $g(0) + g'(0)X = 1 + X$ is $1 + F^{-1}(p)$. Figure 9.4 plots the quantiles of e^X (*y*-axis) against the quantiles of $1 + X$ (*x*-axis) together with the dashed straight line corresponding to a perfect fit. The upper plots correspond to X's being normally distributed with standard deviation 0.02 (left) and 0.3 (right). The lower plots correspond to X's having a Student's t distribution with three degrees of freedom and standard deviation 0.02 (left) and 0.3 (right). We see that the smaller the standard deviation is and the lighter the tails are, the more accurate is the linear approximation.

Example 9.8 (Linearization and risk measures). Suppose that $g(\mathbf{X})$ represents the value at time T of a portfolio of financial assets, where $\mathbf{X}$ has an elliptical distribution with stochastic representation $\mathbf{X} \stackrel{\mathrm{d}}{=} \boldsymbol{\mu} + \mathbf{A}\mathbf{Y}$. Consider a risk measure

ρ and the approximation of $g(\mathbf{X})$ in (9.10). If B_0 is the discount factor giving the current value of money at time T, and if ρ is translation invariant and positively homogeneous, then

$$\rho(g(\mathbf{X})) \approx \rho\left(g(\boldsymbol{\mu}) + \left(\nabla g^{\mathrm{T}}(\boldsymbol{\mu})\boldsymbol{\Sigma}\nabla g(\boldsymbol{\mu})\right)^{1/2} Y_1\right)$$

$$= -B_0 g(\boldsymbol{\mu}) + \left(\nabla g^{\mathrm{T}}(\boldsymbol{\mu})\boldsymbol{\Sigma}\nabla g(\boldsymbol{\mu})\right)^{1/2} \rho(Y_1).$$

For ρ chosen as value-at-risk (VaR) or expected shortfall (ES) and for Y_1 normally distributed or Student's t-distributed, the quantity $\rho(Y_1)$ can be computed as in Example 6.13. If Y_1 is standard normally distributed, then

$$\mathrm{VaR}_p(Y_1) = B_0\Phi^{-1}(1-p) \quad \text{and} \quad \mathrm{ES}_p(Y_1) = B_0\frac{\phi(\Phi^{-1}(1-p))}{p}.$$

If Y_1 has a standard Student's t distribution with ν degrees of freedom, then

$$\mathrm{VaR}_p(Y_1) = B_0 t_\nu^{-1}(1-p) \quad \text{and} \quad \mathrm{ES}_p(Y_1) = B_0\frac{g_\nu(t_\nu^{-1}(1-p))}{p}\left(\frac{\nu+(t_\nu^{-1}(p))^2}{\nu-1}\right),$$

where g_ν and t_ν are the density and distribution functions, respectively, of Y_1.

If ρ is a monotone risk measure and g is a convex function, then it follows from Proposition 2.2 that

$$\rho(g(\mathbf{X})) \leq \rho\left(g(\boldsymbol{\mu}) + \nabla g^{\mathrm{T}}(\boldsymbol{\mu})(\mathbf{X} - \boldsymbol{\mu})\right),$$

i.e., linearization overestimates the risk. If ρ is also translation invariant and positively homogeneous, then

$$\rho(g(\mathbf{X})) \leq \rho\left(g(\boldsymbol{\mu}) + \nabla g^{\mathrm{T}}(\boldsymbol{\mu})(\mathbf{X} - \boldsymbol{\mu})\right)$$

$$= -B_0 g(\boldsymbol{\mu}) + \left(\nabla g^{\mathrm{T}}(\boldsymbol{\mu})\boldsymbol{\Sigma}\nabla g(\boldsymbol{\mu})\right)^{1/2} \rho(Y_1).$$

As an illustration, let $\mathbf{X}$ be a vector of log returns of d assets and consider a linear portfolio consisting of a long position of current value $w_k \geq 0$ in the kth asset, for every k. Then the future portfolio value is $g(\mathbf{X}) = w_1 e^{X_1} + \cdots + w_d e^{X_d}$ and g is convex.

Example 9.8 illustrates how linearization and an elliptical approximation can be used to construct explicit approximation formulas for risk measures. This approach must be used with caution. The accuracy of the first-order approximation of g around $\boldsymbol{\mu}$ evaluated at $\mathbf{X}$ is best around $\boldsymbol{\mu}$. However, risk measures of $g(\mathbf{X})$, such as VaR and ES, typically depend on the behavior of $\mathbf{X}$ far from $\boldsymbol{\mu}$.

Example 9.9 (Linearization over a short time horizon). Considering a portfolio of shares of two stocks. The portfolio contains h_1 and h_2 shares of the two stocks.

The spot prices at time t are given by S_t^1 and S_t^2, respectively. Suppose that we want to compute $\mathrm{VaR}_p(V_T - V_0/B_0)$, where $V_T - V_0/B_0$ is the change in portfolio value from now until time T, measured in money at time T. We have

$$
\begin{aligned}
V_T - V_0/B_0 &= h_1(S_T^1 - S_0^1/B_0) + h_2(S_T^2 - S_0^2/B_0) \\
&= h_1 S_0^1(e^{X_1} - 1/B_0) + h_2 S_0^2(e^{X_2} - 1/B_0) \\
&= g(X_1, X_2),
\end{aligned}
$$

where $(X_1, X_2) = (\log(S_T^1/S_0^1), \log(S_T^2/S_0^2))$ is the log-return pair from now until time T. If T is small (a couple of days, say), then it may be reasonable to set $\mu_1 = \mu_2 = 0$ and $B_0 = 1$, which yields

$$
\begin{aligned}
g(X_1, X_2) &\approx g(\mu_1, \mu_2) + \sum_{k=1}^{2} \frac{\partial g}{\partial x_k}(\mu_1, \mu_2)(X_k - \mu_k) \\
&= \sum_{k=1}^{2} h_k S_0^k(e^{\mu_k} - 1/B_0) + \sum_{k=1}^{2} h_k S_0^k e^{\mu_k}(X_k - \mu_k) \\
&= h_1 S_0^1 X_1 + h_2 S_0^2 X_2.
\end{aligned}
$$

If $\mathbf{X} = (X_1, X_2)^{\mathrm{T}}$ has an elliptical distribution with representation $\mathbf{X} = \mathbf{AY}$, where $\mathbf{AA}^{\mathrm{T}} = \boldsymbol{\Sigma}$, then

$$
\begin{aligned}
\mathrm{VaR}_p(V_1 - V_0/B_0) &\approx V_0 + \mathrm{VaR}_p((\mathbf{w}^{\mathrm{T}} \boldsymbol{\Sigma} \mathbf{w})^{1/2} Y_1) \\
&= V_0 + (\mathbf{w}^{\mathrm{T}} \boldsymbol{\Sigma} \mathbf{w})^{1/2} F_{Y_1}^{-1}(1 - p),
\end{aligned}
$$

where $\mathbf{w}^{\mathrm{T}} = (h_1 S_0^1, h_2 S_0^2)$.

Example 9.10 (Linearization over a long time horizon). Suppose that we want to compute $\mathrm{VaR}_p(V_T - V_0/B_0)$ for a portfolio over a T-day period. Suppose further that V_T can be expressed as a function g of the T-day log returns and that the vectors $\mathbf{X}_1, \ldots, \mathbf{X}_T$ of 1-day log returns are independent and identically elliptically distributed with mean $\boldsymbol{\mu} = \mathrm{E}[\mathbf{X}_1]$ and covariance matrix $\boldsymbol{\Sigma} = \mathrm{Cov}(\mathbf{X}_1)$. Set $\mathbf{W} = \mathbf{X}_1 + \cdots + \mathbf{X}_T$ and note that $\mathbf{W}$, with $\mathrm{E}[\mathbf{W}] = T\boldsymbol{\mu}$ and $\mathrm{Cov}(\mathbf{W}) = T\boldsymbol{\Sigma}$, is the vector of log returns for the entire T-day period. From Proposition 9.4 we know that $\mathbf{W}$ is elliptically distributed, however, in general (unless $\mathbf{W}$ is normally distributed) of a different type than $\mathbf{X}_1$. The elliptical distribution of $\mathbf{W}$ is not easily inferred from the distribution of $\mathbf{X}_1$. However, if T is sufficiently large, then it may be reasonable (based on the central limit theorem) to assume that $\mathbf{W}$ is approximately normally distributed. However, one should be aware that the convergence in distribution to the normal distribution in the central limit theorem is slow in the tail regions.

Linearization, together with the normal approximation, gives

$$g(\mathbf{W}) = \sum_{k=1}^{d} h_k S_0^k (e^{W_k} - 1/B_0)$$

$$\approx \sum_{k=1}^{d} h_k S_0^k (e^{T\mu_k} - 1/B_0) + \sum_{k=1}^{d} h_k S_0^k e^{T\mu_k} (W_k - T\mu_k)$$

$$\stackrel{d}{\approx} \sum_{k=1}^{d} h_k S_0^k (e^{T\mu_k} - 1/B_0) + T^{1/2} \left(\sum_{j,k=1}^{d} h_j h_k S_0^j S_0^k e^{T(\mu_j + \mu_k)} \Sigma_{j,k} \right)^{1/2} Z,$$

where Z is standard normally distributed. In particular,

$$\text{VaR}_p(V_T - V_0/B_0) \approx \sum_{k=1}^{d} h_k S_0^k \left(1 - B_0 e^{T\mu_k}\right)$$

$$+ T^{1/2} B_0 \left(\sum_{j,k=1}^{d} h_j h_k S_0^j S_0^k e^{T(\mu_j + \mu_k)} \Sigma_{j,k} \right)^{1/2} \Phi^{-1}(1-p).$$

If $B_0 e^{T\mu_k} \approx 1$ for all k, then the estimate of $\text{VaR}_p(V_T - V_0/B_0)$ is approximately proportional to the square root of the length T of the time period.

As an illustration, we consider the situation where $\mathbf{X}_1$ has a ten-dimensional Student's t distribution with three degrees of freedom, with zero mean and standard deviations 0.01 and pairwise linear correlation coefficients of 0.4. Moreover, we assume that we hold one share of each stock ($h_k = 1$), that the current share price is 10 for each stock ($S_0^k = 10$), and that interest rates can be ignored ($B_0 = 1$). This gives

$$\text{VaR}_p(V_T - V_0/B_0) \approx T^{1/2}(d(1 + 0.4(d-1)))^{1/2} \Phi^{-1}(1-p)$$

$$= (27T)^{1/2} \Phi^{-1}(1-p).$$

We now compare this estimate to the empirical estimate based on a large simulated sample of independent copies of $V_T - V_0/B_0$. The results are shown in Fig. 9.5. It is interesting to note that for T small, the underestimation of $\text{VaR}_p(V_T - V_0/B_0)$ for p small due to the lighter tails of the normal distribution is offset by the overestimation of $\text{VaR}_p(V_T - V_0/B_0)$ due to linearization.

Fig. 9.5 Illustration of accuracy of estimates of $\mathrm{VaR}_{0.05}(V_T - V_0)$ and $\mathrm{VaR}_{0.01}(V_T - V_0)$ based on linearization and a normal approximation, as functions of $T \in \{1, \ldots, 100\}$ (*dashed curves*). The *solid curves* show the empirical VaR estimates based on simulated samples of size 10^5

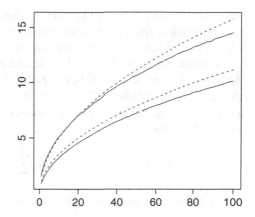

9.3 Applications of Elliptical Distributions in Risk Management

In this section, we consider five applications of elliptical distributions in risk management. In the first application, we derive a risk-aggregation formula that relates the risk, in terms of a translation-invariant and positively homogeneous risk measure, for a sum of jointly elliptically distributed random values to the risk of the terms in the sum. The second application shows how linearization and a normal approximation can be used to approximate the risk measure $\mathrm{VaR}_{0.005}(A - L)$ used to determine the solvency of an insurance company. This application presents the idea behind the so-called standard formula that is used in the measurement of risk in the insurance industry. The third application suggests a hedging approach to European call options that is more appropriate than delta hedging if the joint distribution for the log return of the underlying asset value and the change in the implied volatility can be assumed to be elliptical. The fourth application presents how a trader might design a bet on changes in implied volatility for two maturity times and considers ways to investigate the risk of such a bet. The fifth application illustrates that if the vector of returns on a set of risky assets can be assumed to be elliptically distributed, then portfolio investment problems can often be reduced to the trade-off investment problem (4.7).

9.3.1 Risk Aggregation with Elliptical Distributions

Consider a company divided into n business units with future net values of assets and liabilities given by $X_1, \ldots, X_n$. Suppose that each business unit is able to accurately estimate $\mathrm{E}[X_k]$ and $\rho(X_k)$, where ρ is some translation-invariant and positively homogeneous risk measure. The company wants to compute $\rho(X_1 + \cdots + X_n)$ to get a measurement on the aggregate risk for the whole company. There is no straightforward way to combine the individual risk estimates $\rho(X_k)$

and the expected values $E[X_k]$ into an aggregate risk estimate. However, there is a convenient risk-aggregation formula that is valid under the assumption that $(X_1, \ldots, X_n)^T$ has an elliptical distribution.

Suppose that $\mathbf{X} = (X_1, \ldots, X_n)^T$ has an elliptical distribution so that $\mathbf{X} \overset{d}{=} \boldsymbol{\mu} + \mathbf{AY}$, with $\mathbf{AA}^T = \boldsymbol{\Sigma}$, and $\mathbf{Y}$ has a spherical distribution. Matrix $\boldsymbol{\Sigma}$ can always be expressed as the product $\mathbf{DCD}$, where $\mathbf{D}$ is a diagonal matrix with diagonal entries $D_{k,k} = \Sigma_{k,k}^{1/2}$ and $\mathbf{C}$ is a correlation matrix (the linear correlation matrix of $\mathbf{X}$ if it exists). Note that

$$\rho(X_1 + \cdots + X_n) = -B_0 \sum_{k=1}^{n} \mu_k + \rho\left(\sum_{k=1}^{n}(X_k - \mu_k)\right),$$

where B_0 is the discount factor between now and the considered future time, and

$$\sum_{k=1}^{n}(X_k - \mu_k) = \mathbf{1}^T \mathbf{AY} \overset{d}{=} (\mathbf{1}^T \boldsymbol{\Sigma} \mathbf{1})^{1/2} Y_1.$$

Since $\mathbf{1}^T \boldsymbol{\Sigma} \mathbf{1} = \Sigma_{1,1} + \Sigma_{1,2} + \cdots + \Sigma_{n,n}$ and $\Sigma_{j,k} = D_{j,j} C_{j,k} D_{k,k}$, it holds that

$$\rho\left(\sum_{k=1}^{n}(X_k - \mu_k)\right) = \rho\left(\left(\sum_{j,k} \Sigma_{j,k}\right)^{1/2} Y_1\right)$$

$$= \left(\sum_{j,k} C_{j,k} D_{j,j} D_{k,k}\right)^{1/2} \rho(Y_1)$$

$$= \left(\sum_{j,k} C_{k,l} D_{j,j} D_{k,k} \rho(Y_1)^2\right)^{1/2}$$

$$= \left(\sum_{j,k} C_{j,k} \rho(D_{j,j} Y_j) \rho(D_{k,k} Y_k)\right)^{1/2}$$

$$= \left(\sum_{j,k} C_{j,k} \rho(X_j - \mu_j) \rho(X_k - \mu_k)\right)^{1/2}.$$

We have found that if $\mathbf{X} = (X_1, \ldots, X_n)^T$ has an elliptical distribution and if ρ is a translation-invariant and positively homogeneous risk measure, then

$$\rho(X_1 + \cdots + X_n) = \left(\sum_{j,k} C_{j,k}\{B_0\mu_j + \rho(X_j)\}\{B_0\mu_k + \rho(X_k)\}\right)^{1/2} - B_0 \sum_{k} \mu_k.$$

The only additional input needed, besides the individual risk estimates $\rho(X_k)$ and the means μ_k, are the linear correlation coefficients $C_{j,k}$.

9.3.2 Solvency of an Insurance Company

In this section, we present another example of linearization and normal approximation in the context of the solvency of an insurance company.

Consider an insurance company with assets and liabilities. Let A and L denote the time 1 (1 year from now) values of the assets and liabilities, respectively. We consider the insurance company to be solvent if

$$\text{VaR}_{0.005}(A - L) \leq 0.$$

If r_1 is the current risk-free, 1-year zero rate, then we may write

$$\text{VaR}_{0.005}(A - L) = F^{-1}_{e^{-r_1}(L-A)}(0.995).$$

We consider a stylized model for the assets and liabilities and assume that the liabilities correspond to the stochastic cash flow $(C_1, \ldots, C_n)$, where C_k is the amount the insurer has to pay at the end of year k due to the occurrence of claims before the end of year 1. Each written contract offers a protection for the insured over a 1-year period. Operating expenses for the insurer could be included in the C_k or dealt with in other ways. The expectation $E[C_k]$ is the expected claim amount to be paid at time k, and $e^{-r_k} E[C_k]$ is the present value of this amount. The expected claim amount $E[C_k]$ could be determined by some stochastic claim-reserving method, such as the chain ladder method presented in Sect. 7.6.1. The best estimate, at time 0, of the present value of the liabilities is

$$L_0 = \sum_{k=1}^{n} E[C_k]e^{-r_k k}.$$

At time 1 we observe C_1 and receive new information about the future payments C_k. If $\mathbf{I}_1$ denotes the information available at time 1, then $E[C_k \mid \mathbf{I}_1]$ is the updated prediction of the payment due at time k. The time 1 value of the liabilities is therefore given by

$$L = \sum_{k=1}^{n} E[C_k \mid \mathbf{I}_1]e^{-(r_{k-1}+\Delta r_{k-1})(k-1)},$$

where $\Delta \mathbf{r}$ is the vector of zero rate changes from time 0 to 1. Suppose for simplicity that the assets of the insurer consist of a bond portfolio designed to match future claim payments and K units of cash on a bank account. The time 0 value A_0 of the assets and the time 1 value A of the assets are given by

$$A_0 = \sum_{k=1}^{n} E[C_k]e^{-r_k k} + K,$$

$$A = \sum_{k=1}^{n} E[C_k]e^{-(r_{k-1}+\Delta r_{k-1})(k-1)} + Ke^{r_1}.$$

The time 0 value of the bond portfolio precisely matches the time 0 value of the liability, $A_0 - K = L_0$. Moreover,

$$e^{-r_1}(L - A) = e^{-r_1} \sum_{k=1}^{n} (E[C_k \mid I_1] - E[C_k]) e^{-(r_{k-1} + \Delta r_{k-1})(k-1)} - K$$

$$= \sum_{k=1}^{n} e^{-r_k k} E[C_k] Y_k e^{X_k} - K$$

$$= g(X_1, \ldots, X_n, Y_1, \ldots, Y_n),$$

where $X_k = -r_1 - (r_{k-1} + \Delta r_{k-1})(k-1) + r_k k$, $Y_k = (E[C_k \mid I_1] - E[C_k])/E[C_k]$ for $k = 1, \ldots, n$, and

$$g(\mathbf{x}, \mathbf{y}) = \sum_{k=1}^{n} e^{-r_k k} E[C_k] y_k e^{x_k} - K.$$

The quantity $Y_1 = (C_1 - E[C_1])/E[C_1]$ measures the relative deviation of the actual amount paid at the end of the year from the current prediction. For $k \geq 2$, $Y_k = (E[C_k \mid I_1] - E[C_k])/E[C_k]$ measures the relative deviation of the updated prediction at the end of the year of the claim payments at time k, for claims incurred before the end of the year, from the current prediction.

Since g is a nonlinear function of the risk factors $(X_1, \ldots, X_n, Y_1, \ldots, Y_n)$, the computation of VaR is simplified substantially by linearization. Let $\mu_k = E[X_k]$, and note that $E[Y_k] = 0$. Therefore, it makes sense to consider the first-order approximation of g around $(\mu_1, \ldots, \mu_n, 0, \ldots, 0)$, which gives

$$g(X_1, \ldots X_n, Y_1, \ldots, Y_n) \approx g(\mu_1, \ldots, \mu_k, 0, \ldots, 0) + \sum_{k=1}^{n} e^{-r_k k} E[C_k] Y_k e^{\mu_k}$$

$$= -K + \sum_{k=1}^{n} e^{-r_k k} E[C_k] Y_k e^{\mu_k}$$

$$= -K + \mathbf{w}^{\mathrm{T}} \mathbf{Y},$$

where $w_k = e^{-r_k k} E[C_k] e^{\mu_k}$. Because of the linearization, the effect of the X_k vanishes. The contributions to the risk coming from changes in the zero rates are second-order effects and do not show up in the linearized version of g. Although ignoring second-order effects is convenient for explicit computations, it leads to a crude approximation.

If $\mathbf{Y}$ is $N(\mathbf{0}, \Sigma)$-distributed, then we find that

$$\mathrm{VaR}_{0.005}(A - L) \approx F^{-1}_{-K + \mathbf{w}^{\mathrm{T}} \mathbf{Y}}(0.995) = -K + (\mathbf{w}^{\mathrm{T}} \Sigma \mathbf{w})^{1/2} \Phi^{-1}(0.995).$$

Taking this approximation as an equality we find that the solvency condition $\text{VaR}_{0.005}(A - L) \leq 0$ is equivalent to

$$K \geq (\mathbf{w}^{\mathsf{T}} \boldsymbol{\Sigma} \mathbf{w})^{1/2} \Phi^{-1}(0.995).$$

The outlined procedure is the basic idea behind the standard formula in the Solvency II framework for the computation of sufficient buffer capital for an insurance company. Of course, in practice, many more risk factors need to be included and the insurer's asset portfolio is more complex. Nevertheless, the linearization approach and the normal approximation is at the heart of the standard formula. To compensate for the inaccuracies of linearization and the normal approximation, the covariance matrix $\boldsymbol{\Sigma}$ is not estimated from data but given exogenously by the regulators.

9.3.3 Hedging of a Call Option When the Volatility Is Stochastic

Suppose that now at time 0 we have issued a European call option with strike price K on the value S_T of a stock market index at time T. Suppose also that we want to hedge against changes in the option price from now until time $t < T$ by taking a position in the underlying index and deposit cash to minimize

$$\mathrm{E}[(h_0 + h_1 S_t - C_t)^2],$$

where C_t is the call option price at time t. If t is small, then the delta-hedging approach in Sect. 3.5 gives an approximative solution to the quadratic hedging problem. Suppose that the option price is expressed in terms of the Black–Scholes formula (1.7) as a function $C_t = C(S_t, \sigma_t, r_t, t, T - t)$, where the arguments correspond to the value of the underlying index at time t, the option's implied volatility at time t, interest rate prevailing between time t and the maturity time T of the option, and the remaining time to maturity. The delta-hedging approach relies on the first-order approximation

$$C_t \approx C_0 + \frac{\partial C_0}{\partial S_0}(S_t - S_0),$$

which gives the delta-hedge position $(h_0^\delta, h_1^\delta) \approx (h_0, h_1)$, where

$$h_1^\delta = \frac{\partial C_0}{\partial S_0} \quad \text{and} \quad h_0^\delta = C_0 - \frac{\partial C_0}{\partial S_0} S_0.$$

The Black–Scholes formula reads

$$C_t = S_t \Phi(d_1) - K e^{-r_t(T-t)} \Phi(d_2),$$

$$d_1 = \frac{\log(S_t/K) + (r_t + \sigma_t^2/2)(T - t)}{\sigma_t \sqrt{T - t}} \quad \text{and} \quad d_2 = d_1 - \sigma_t \sqrt{T - t}$$

and gives

$$\frac{\partial C_0}{\partial S_0} = \Phi(d_1), \quad d_1 = \frac{\log(S_0/K) + (r_0 + \sigma_0^2/2)T}{\sigma_0\sqrt{T}}.$$

The hedging error at time t is

$$h_0^\delta + h_1^\delta S_t - C_t = C_0 - C_t + \Phi(d_1)(S_t - S_0).$$

The change in the interest rate from r_0 to r_t typically does not contribute much to the hedging error, and therefore we may approximate $r_t \approx r_0$. We may therefore view the hedging error as a function of the changes in the index value and in the implied volatility or, equivalently, as a function $g(\mathbf{z})$ evaluated at $\mathbf{Z} = (Z_1, Z_2)$, where $Z_1 = \log(S_t/S_0)$ and $Z_2 = \sigma_t - \sigma_0$. Therefore, a model for (Z_1, Z_2) implies a model for the hedging error and the latter model can be analyzed by, e.g., simulation from (Z_1, Z_2). The sample from (Z_1, Z_2) can then be converted to a sample from the distribution of the hedging error whose empirical distribution can be studied. Alternatively, we could linearize the nonlinear function $g(\mathbf{z})$ and evaluate the linear approximation at $\mathbf{Z} = (Z_1, Z_2)$. The linearization approach may give an approximation of the distribution for the hedging error that can be analyzed analytically, without simulation. Consider the first-order approximation

$$C_t \approx g(\mathbf{0}) + \frac{\partial g}{\partial z_1}(\mathbf{0})Z_1 + \frac{\partial g}{\partial z_2}(\mathbf{0})Z_2,$$

where $g(\mathbf{z}) = g_1(g_2(z_1), g_3(z_2))$ with $g_2(z_1) = S_0 e^{z_1}$, $g_3(z_2) = z_2 + \sigma_0$, and

$$g_1(s, \sigma) = s\Phi(d_1) - Ke^{-r_0(T-t)}\Phi(d_2),$$

$$d_1 = \frac{\log(s/K) + (r_0 + \sigma^2/2)(T - t)}{\sigma\sqrt{T - t}} \quad \text{and} \quad d_2 = d_1 - \sigma\sqrt{T - t}.$$

The chain rule, together with the expressions for the partial derivatives of the Black–Scholes formula (Sect. 1.2.2), gives

$$\frac{\partial g}{\partial z_1}(\mathbf{0}) = \frac{\partial g_1}{\partial s}(S_0, \sigma_0)\frac{dg_2}{dz_1}(0) = \Phi(d_1)S_0,$$

$$\frac{\partial g}{\partial z_2}(\mathbf{0}) = \frac{\partial g_1}{\partial \sigma}(S_0, \sigma_0)\frac{dg_3}{dz_2}(0) = \phi(d_1)S_0\sqrt{T - t}.$$

Summing up, we arrive at the following approximation of the hedging error:

$$h_0^\delta + h_1^\delta S_t - C_t = C_0 - C_t + \Phi(d_1)(S_t - S_0)$$

$$\approx C_0 - C_0 - \Phi(d_1)S_0 Z_1 - \phi(d_1)S_0\sqrt{T - t}\,Z_2$$

$$+\Phi(d_1)(S_0(1 + Z_1) - S_0)$$

$$= -\phi(d_1)S_0\sqrt{T - t}\,(\sigma_t - \sigma_0).$$

We see that the position, the delta hedge and the issued call option, is immune against changes in the index value (approximately, over a short time period) and that the hedging error is due to changes in the implied volatility. We also find that the variance of the hedging error is

$$\mathrm{Var}(h_0^\delta + h_1^\delta S_t - C_t) \approx \phi(d_1)^2 S_0^2 (T - t)\, \mathrm{Var}(\sigma_t).$$

We now want to reduce the hedging error by replacing the delta hedge by a similar hedge that also takes changes in the implied volatility into account. The position in the underlying index and in cash for the optimal quadratic hedge is

$$h_1 = \frac{\mathrm{Cov}(S_t, C_t)}{\mathrm{Var}(S_t)} \quad \text{and} \quad h_0 = \mathrm{E}[C_t] - h_1\, \mathrm{E}[S_t].$$

Here we approximate

$$\mathrm{Cov}(S_t, C_t) \approx \mathrm{Cov}(S_0 Z_1, \Phi(d_1) S_0 Z_1 + \phi(d_1) S_0 \sqrt{T - t} Z_2)$$
$$= S_0^2 \Phi(d_1)\, \mathrm{Var}(Z_1) + S_0^2 \phi(d_1) \sqrt{T - t}\, \mathrm{Cov}(Z_1, Z_2),$$
$$\mathrm{Var}(S_t) \approx S_0^2\, \mathrm{Var}(Z_1),$$
$$\mathrm{E}[C_t] \approx C_0,$$
$$\mathrm{E}[S_t] \approx S_0.$$

This gives the hedge $(h_0^*, h_1^*) \approx (h_0, h_1)$, where

$$h_1^* = \Phi(d_1) + \phi(d_1)\sqrt{T - t}\frac{\mathrm{Cov}(Z_1, Z_2)}{\mathrm{Var}(Z_1)}$$
$$= \Phi(d_1) + \phi(d_1)\sqrt{T - t}\frac{\sigma_{Z_2}}{\sigma_{Z_1}}\rho,$$
$$h_0^* = C_0 - h_1^* S_0,$$

where $\sigma_{Z_k} = \mathrm{Var}(Z_k)^{1/2}$ and $\rho = \mathrm{Cor}(Z_1, Z_2)$. We observe that the position h_1^* in the underlying index corresponds to the delta-hedge position h_1^δ plus a correction term. We get the following approximation of the hedging error:

$$h_0^* + h_1^* S_t - C_t \approx C_0 + \left(\Phi(d_1) + \phi(d_1)\sqrt{T - t}\frac{\sigma_{Z_2}}{\sigma_{Z_1}}\rho \right) S_0 Z_1$$
$$-C_0 - \Phi(d_1) S_0 Z_1 - \phi(d_1) S_0 \sqrt{T - t} Z_2$$
$$= \phi(d_1) S_0 \sqrt{T - t} \left(\frac{\sigma_{Z_2}}{\sigma_{Z_1}}\rho Z_1 - Z_2 \right).$$

In particular, the variance of the hedging error is approximately

$$\text{Var}(h_0^* + h_1^* S_t - C_t) \approx \text{Var}\left(\phi(d_1) S_0 \sqrt{T-t}\left(\frac{\sigma_{Z_2}}{\sigma_{Z_1}}\rho Z_1 - Z_2\right)\right)$$

$$= \phi(d_1)^2 S_0^2 (T-t)\,\text{Var}(\sigma_t)(1-\rho^2),$$

where the last equality can be verified by straightforward computations of the variance of the sum of two correlated terms. Notice that taking changes in implied volatility into account when computing the approximation of the quadratic hedge makes the variance of the hedging error smaller by a factor of $(1-\rho^2)$.

9.3.4 Betting on Changes in Volatility

Suppose that a trader is betting on changes in implied volatility from time 0 today until time $t > 0$ in the future for two future maturity times and that we want to analyze the riskiness of this volatility bet. Consider two call options on the values of an index at two future times $0 < T_1 < T_2$. The trader believes that over a short period of time the change in implied volatility $\sigma_t^1 - \sigma_0^1$ for the nearer maturity time T_1 will be greater than that for the more distant maturity time T_2, $\sigma_t^2 - \sigma_0^2$. The trader wants to capitalize on this belief but at the same time not bet on other potential movements of the underlying index value. We first determine the particular portfolio corresponding to the volatility bet.

Consider a long position of size h_2 in a call option with strike K_1 maturing at time T_1 and a short position of size h_3 in a call option with strike K_2 maturing at time T_2. The future value of this position is, to a first-order approximation and with the expressions for the partial derivatives of the Black–Scholes formula,

$$h_2 C_t^1 - h_3 C_t^2 \approx h_2 C_0^1 - h_3 C_0^2$$

$$+ h_2\left(\Phi(d_1^1)(S_t - S_0) + \phi(d_1^1) S_0 \sqrt{T_1}(\sigma_t^1 - \sigma_0^1)\right)$$

$$- h_3\left(\Phi(d_1^2)(S_t - S_0) + \phi(d_1^2) S_0 \sqrt{T_2}(\sigma_t^2 - \sigma_0^2)\right),$$

where

$$d_1^j = \frac{\log(S_0/K_j) + (r_j + (\sigma_0^j)^2/2)T_j}{\sigma_0^j \sqrt{T_j}} \quad \text{for } j = 1, 2.$$

With $Z_1 = \log(S_t/S_0)$, $Z_2 = \sigma_t^1 - \sigma_0^1$, and $Z_3 = \sigma_t^2 - \sigma_0^2$, and the approximation $S_t - S_0 \approx S_0 Z_1$, we get

$$h_2 C_t^1 - h_3 C_t^2 \approx h_2 C_0^1 - h_3 C_0^2 + (h_2 \Phi(d_1^1) - h_3 \Phi(d_1^2)) S_0 Z_1$$

$$+ h_2 S_0 \phi(d_1^1) \sqrt{T_1} Z_2 - h_3 S_0 \phi(d_1^2) \sqrt{T_2} Z_3.$$

The volatility bet is a bet on the occurrence of the event $Z_2 > Z_3$, and on nothing else. Therefore, the trader chooses h_2 and h_3 so that

$$h_2\phi(d_1^1)\sqrt{T_1} - h_3\phi(d_1^2)\sqrt{T_2} = 0,$$

meaning that the impact of a parallel shift in the implied volatility should be approximately zero. Moreover, the trader wants the bet to be immune to changes in the value of the underlying index. Therefore, the trader takes the position

$$h_1 = -(h_2\Phi(d_1^1) - h_3\Phi(d_1^2))$$

in the index and a position

$$h_0 = -h_1 S_0 - h_2 C_0^1 + h_3 C_0^2$$

in cash. Summing up, we find that the volatility bet corresponds to the portfolio weights h_0, h_1, h_2, h_3 and the future portfolio value

$$h_0 + h_1 S_t + h_2 C_t^1 - h_3 C_t^2 \approx h_2 S_0 \phi(d_1^1)\sqrt{T_1} Z_2 - h_2 \frac{\phi(d_1^1)\sqrt{T_1}}{\phi(d_1^2)\sqrt{T_2}} S_0 \phi(d_1^2)\sqrt{T_2} Z_3$$

$$= h_2 S_0 \phi(d_1^1)\sqrt{T_1}(Z_2 - Z_3).$$

To estimate the risk of holding this portfolio until time t, we could now assign a bivariate elliptical distribution to (Z_2, Z_3), determine the corresponding univariate elliptical distribution of $Z_2 - Z_3$, and finally compute $\rho(h_2 S_0 \phi(d_1^1)\sqrt{T_1}(Z_2 - Z_3))$ for a suitable choice of risk measure ρ. However, this apparent straightforward approach to measuring the riskiness of the volatility bet is not unproblematic. Assigning a bivariate model to (Z_2, Z_3) can at best be guided by historical data on implied volatility changes but will to a large extent be based on subjective beliefs. Moreover, if the sizes of the option positions are large, then it may be unrealistic to assume that the positions can be closed at time t if t is small. In that case, we need a longer time period for the risk modeling, and this makes the whole linearization approach questionable.

9.3.5 Portfolio Optimization with Elliptical Distributions

Suppose vector $\mathbf{R}$ of returns on a collection of risky assets can be modeled by a normal variance mixture distribution so that $\mathbf{R} \overset{d}{=} \mu + W\mathbf{AZ}$, where $\mathbf{Z}$ is $N_d(\mathbf{0}, \mathbf{I})$-distributed and independent of $W \geq 0$, and $\mathbf{AA}^\mathsf{T} = \Sigma$. If R_0 is the return on a risk-free asset, then the future value of a portfolio with monetary portfolio weights $\mathbf{w}$ in the risky assets and w_0 in the risk-free asset can be expressed as

$$V_1 = w_0 R_0 + \mathbf{w}^\mathsf{T}\mathbf{R}$$

$$\overset{d}{=} w_0 R_0 + \mathbf{w}^\mathsf{T}\mu + (\mathbf{w}^\mathsf{T}\Sigma\mathbf{w})^{1/2} W Z_1. \tag{9.11}$$

Suppose the variance $\text{Var}(V_1) = \sigma^2 \mathbf{w}^T \boldsymbol{\Sigma} \mathbf{w}$, where $\sigma^2 = \text{Var}(WZ_1)$, exists. Then the solution to the investment problem

$$\text{maximize} \quad w_0 R_0 + \mathbf{w}^T \boldsymbol{\mu} - \tfrac{c}{2V_0}\sigma^2 \mathbf{w}^T \boldsymbol{\Sigma} \mathbf{w}$$
$$\text{subject to} \quad w_0 + \mathbf{w}^T \mathbf{1} \leq V_0$$

follows from the solution to the trade-off investment problem (4.7) by replacing $\boldsymbol{\Sigma}$ in (4.7) by $\sigma^2 \boldsymbol{\Sigma}$ and is given by

$$\mathbf{w} = \frac{V_0}{c}(\sigma^2 \boldsymbol{\Sigma})^{-1}(\boldsymbol{\mu} - R_0 \mathbf{1}) \quad \text{and} \quad w_0 = V_0 - \mathbf{w}^T \mathbf{1}.$$

A convenient feature of having an elliptical distribution for vector $\mathbf{R}$ of returns is that portfolio optimization problems often reduce to the trade-off investment problem (4.7). Consider the problem of portfolio optimization in the context of a spectral risk measure.

Example 9.11 (Spectral risk measures). Portfolio optimization with respect to a spectral risk measure (Sect. 6.5) amounts to minimizing a spectral risk measure $\rho_\phi(X)$, where X denotes a future portfolio value, under a budget constraint (and possibly additional constraints). By the stochastic representation (9.11), we can express the quantile function of V_1 as

$$F_{V_1}^{-1}(p) = w_0 R_0 + \mathbf{w}^T \boldsymbol{\mu} + (\mathbf{w}^T \boldsymbol{\Sigma} \mathbf{w})^{1/2} F_{WZ_1}^{-1}(p).$$

Therefore, the spectral risk measure

$$\rho_\phi(X) = -\int_0^1 \phi(p) F_{X/R_0}^{-1}(p)dp,$$

applied to $X = V_1 - V_0 R_0$, can be expressed as

$$\rho_\phi(V_1 - V_0 R_0) = -\int_0^1 \phi(p) F_{V_1/R_0}^{-1}(p)dp + V_0$$

$$= \frac{1}{R_0}\left(-w_0 R_0 - \mathbf{w}^T \boldsymbol{\mu} - (\mathbf{w}^T \boldsymbol{\Sigma} \mathbf{w})^{1/2} \int_0^1 \phi(p) F_{WZ_1}^{-1}(p)dp\right) + V_0.$$

In particular, we can formulate the portfolio optimization problem

$$\text{minimize} \quad \rho_\phi(w_0 R_0 + \mathbf{w}^T \mathbf{R} - V_0 R_0)$$
$$\text{subject to} \quad w_0 + \mathbf{w}^T \mathbf{1} \leq V_0$$

as the trade-off problem

$$\text{maximize} \quad w_0 R_0 + \mathbf{w}^T \boldsymbol{\mu} - \tfrac{c}{2V_0}(\mathbf{w}^T \boldsymbol{\Sigma} \mathbf{w})^{1/2}$$
$$\text{subject to} \quad w_0 + \mathbf{w}^T \mathbf{1} \leq V_0,$$

where

$$c = -2V_0 \int_0^1 \phi(p) F_{WZ_1}^{-1}(p) dp.$$

We conclude that, for an elliptical model for vector $\mathbf{R}$ of returns, minimizing the spectral risk measure of the future portfolio value subject to a budget constraint is equivalent to solving a trade-off problem with the trade-off parameter given above.

9.4 Copulas

A rather common situation arises when we search for a multivariate model for a set of random variables $Y_1, \ldots, Y_d$ whose univariate distributions are rather well understood but whose joint distribution is only partially understood. A useful approach to the construction of a multivariate distribution for $\mathbf{Y} = (Y_1, \ldots, Y_d)$ with specified univariate marginal distribution functions $G_1, \ldots, G_d$, the distribution functions of the vector's components, is obtained by combining the so-called probability and quantile transforms. The probability transform says that if X is a random variable with a continuous distribution function F, then $F(X)$ is uniformly distributed on the interval $(0, 1)$. The quantile transform says that if U is uniformly distributed and if G is any distribution function, then $G^{-1}(U)$ has distribution function G. This implies that for any random vector $\mathbf{X} = (X_1, \ldots, X_d)$ whose components have continuous distribution functions $F_1, \ldots, F_d$, the random vector $\mathbf{Y} = (G_1^{-1}(F_1(X_1)), \ldots, G_d^{-1}(F_d(X_d)))$ corresponds to a multivariate model with prespecified univariate marginal distributions. If all F_k and G_k are both continuous and strictly increasing, then the preceding statement is actually straightforward to verify:

$$P(G_k^{-1}(F_k(X_k)) \le y) = P(F_k(X_k) \le G_k(y)) = G_k(y),$$

which shows that Y_k has distribution function G_k. The difficulty when it comes to constructing a good multivariate model for $\mathbf{Y}$ using this approach clearly lies in the choice of the distribution for vector $\mathbf{X}$ since the dependence between the X_k will be inherited by the Y_k.

Example 9.12. Consider the two scatter plots in Fig. 9.6. The left scatter plot shows a sample of size 2,000 from a bivariate standard normal distribution with linear correlation 0.5. The right scatter plot shows a sample of size 2,000 from a bivariate distribution with standard normal marginal distributions and a dependence structure inherited from a bivariate Student's t distribution with one degree of freedom. The points of the right scatter plot were obtained from the points of the left scatter plot as follows. Write $\mathbf{Z}_1, \ldots, \mathbf{Z}_{2000}$ for the independent bivariate normal random vectors whose outcomes are shown in the left plot. Let $S_1, \ldots, S_{2000}$ be independent χ_1^2-distributed random variables independent of the sample from the bivariate normal distribution. A sample of independent bivariate Student's t_1-distributed vectors was obtained by setting $\mathbf{X}_k = S_k^{-1/2} \mathbf{Z}_k$ for $k = 1, \ldots, 2000$. Finally, the random vectors whose outcomes are shown in the plot to the right were constructed as $\mathbf{Y}_k = (\Phi^{-1}(t_1(X_{k,1})), \Phi^{-1}(t_1(X_{k,2})))^{\mathrm{T}}$ for $k = 1, \ldots, 2000$.

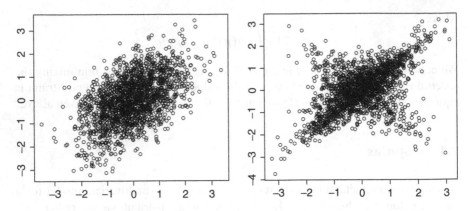

Fig. 9.6 Samples of size 2,000 from two bivariate distributions with standard normal marginal distributions. *Left plot*: sample from a bivariate standard normal with linear correlation 0.5. *Right plot*: sample from a bivariate standard Student's t distribution with one degree of freedom, with marginal distributions transformed to a standard normal

Suppose that we want to build a multivariate model corresponding to a random vector $\mathbf{X} = (X_1, \ldots, X_d)$ with a nontrivial dependence between its components and certain marginal distribution functions $F_1, \ldots, F_d$. Then the quantile transform says that we may start with a suitable vector $\mathbf{U} = (U_1, \ldots, U_d)$ whose components are uniformly distributed on $(0, 1)$ and specify $\mathbf{X}$ as

$$\mathbf{X} = (F_1^{-1}(U_1), \ldots, F_d^{-1}(U_d)).$$

The random vector $\mathbf{X}$ inherits the dependence among its components from vector $\mathbf{U}$. The distribution function C of a random vector $\mathbf{U}$ whose components U_k are uniformly distributed on $(0, 1)$ is called a copula, i.e.,

$$C(u_1, \ldots, u_d) = \mathrm{P}(U_1 \leq u_1, \ldots, U_d \leq u_d), \quad (u_1, \ldots, u_d) \in (0, 1)^d.$$

Let $(X_1, \ldots, X_d)$ be a random vector with distribution function $F(x_1, \ldots, x_d) = \mathrm{P}(X_1 \leq x_1, \ldots, X_d \leq x_d)$ and suppose that $F_k(x) = \mathrm{P}(X_k \leq x)$ is a continuous function for every k. The probability transform, statement (iv) of Proposition 6.1, implies that the components of the vector $\mathbf{U} = (U_1, \ldots, U_d) = (F_1(X_1), \ldots, F_d(X_d))$ are uniformly distributed on $(0, 1)$. In particular, the distribution function C of $\mathbf{U}$ is a copula and we call it the copula of $\mathbf{X}$. Using statement (i) of Proposition 6.1 we find that

$$C(F_1(x_1), \ldots, F_d(x_d)) = \mathrm{P}(U_1 \leq F_1(x_1), \ldots, U_d \leq F_d(x_d))$$
$$= \mathrm{P}(F_1^{-1}(U_1) \leq x_1, \ldots, F_d^{-1}(U_d) \leq x_d)$$
$$= F(x_1, \ldots, x_d).$$

This representation of the joint distribution function F in terms of the copula C and the marginal distribution functions $F_1, \ldots, F_d$ explains the name "copula": a function that "couples" the joint distribution function to its univariate marginal distribution functions.

Example 9.13 (Gaussian and Student's t copulas). The copula C_R^{Ga} of a d-dimensional standard normal distribution, with linear correlation matrix R, is the distribution function of the random vector $(\Phi(X_1), \ldots, \Phi(X_d))$, where Φ is the univariate standard normal distribution function and X is $N_d(0, R)$-distributed. Hence,

$$C_R^{\mathrm{Ga}}(\mathbf{u}) = P(\Phi(X_1) \le u_1, \ldots, \Phi(X_d) \le u_d) = \Phi_R^d(\Phi^{-1}(u_1), \ldots, \Phi^{-1}(u_d)),$$

where Φ_R^d is the distribution function of X. Copulas of the preceding form are called Gaussian copulas.

The copula $C_{\nu,R}^t$ of a d-dimensional standard Student's t distribution with $\nu > 0$ degrees of freedom and linear correlation matrix R is the distribution of the random vector $(t_\nu(X_1), \ldots, t_\nu(X_d))$, where X has a $t_d(0, R, \nu)$ distribution and t_ν is the univariate standard Student's t_ν distribution function. Hence,

$$C_{\nu,R}^t(\mathbf{u}) = P(t_\nu(X_1) \le u_1, \ldots, t_\nu(X_d) \le u_d) = t_{\nu,R}^d(t_\nu^{-1}(u_1), \ldots, t_\nu^{-1}(u_d)),$$

where $t_{\nu,R}^d$ the distribution function of X. Copulas of the preceding form are called Student's t copulas.

Consider a random vector (Y_1, Y_2) with continuous strictly increasing marginal distribution functions G_1 and G_2 and the copula of a Student's t distribution with linear correlation parameter ρ. We consider here the question of how ρ can be estimated from a sample from the distribution of (Y_1, Y_2). We may write $(Y_1, Y_2) = (G_1^{-1}(F_1(X_1)), G_2^{-1}(F_2(X_2)))$, where (X_1, X_2) has a Student's t distribution with linear correlation parameter ρ. In particular, the functions T_1 and T_2 given by $T_k(x) = G_k^{-1}(F_k(x))$ are continuous and strictly increasing, so for an independent copy (X_1', X_2') of (X_1, X_2) it holds that

$$
\begin{aligned}
\tau(Y_1, Y_2) &= \tau(T_1(X_1), T_2(X_2)) \\
&= 2\, P((T_1(X_1) - T_1(X_1'))(T_2(X_2) - T_2(X_2')) > 0) - 1 \\
&= 2\, P((X_1 - X_1')(X_2 - X_2') > 0) - 1 \\
&= \tau(X_1, X_2).
\end{aligned}
$$

It follows immediately from (9.7) that $\rho = \sin(\pi \tau(Y_1, Y_2)/2)$. Therefore, the estimate $\widehat{\tau}$ of $\tau(Y_1, Y_2)$ from the sample from the distribution of (Y_1, Y_2) gives an estimate $\widehat{\rho} = \sin(\pi \widehat{\tau}/2)$ of ρ.

Example 9.14 (Investments in foreign stocks). Consider a Swedish investor about to invest Swedish kronor (SEK) in foreign telecom stocks. The current share prices of British Telecom (BT) and Deutsche Telekom (DT) are 185.5 British pounds (GBP)

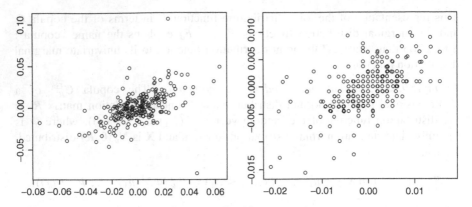

Fig. 9.7 *Scatter plot* to *left* shows log-return pairs, British Telecom in pounds on the *x*-axis and Deutsche Telekom in euros on the *y*-axis. The *scatter plot* to the *right shows* log-return pairs, SEK/GBP on the *x*-axis and SEK/EUR on the *y*-axis

and 9.26 euros (EUR), respectively. The current SEK/GBP exchange rate is 0.0942 (*x* kronor can be exchanged for 0.0942*x* pounds). The current SEK/EUR exchange rate is 0.1098.

The investor has obtained a sample of four-dimensional vectors of share prices, in the local currencies, and exchange rates from the 249 most recent (trading) days. We assume that the investor believes that the information in the data is relevant for assessing future portfolio values, and that no additional information on which to base model selection is available. The scatter plots for the stock log-return pairs and for the exchange-rate log-return pairs are shown in Fig. 9.7.

The investor is about to invest the amounts w_1 and w_2 kronor in the two foreign telecom stocks and wants to model the portfolio value V_1 in kronor tomorrow. Let A_t, B_t, C_t, D_t denote the time t share prices (BT and DT) and exchange rates (SEK/GBP and SEK/EUR). Let $X_A = \log(A_1/A_0)$ be the log return from today until tomorrow for BT in GBP and similarly for X_B, X_C, X_D. If h_1 and h_2 are the number of shares of BT and DT bought, then

$$\frac{A_0}{C_0}h_1 = w_1 \quad \text{and} \quad \frac{B_0}{D_0}h_2 = w_2.$$

The portfolio value in kronor tomorrow is therefore

$$V_1 = h_1 \frac{A_1}{C_1} + h_2 \frac{B_1}{D_1}$$

$$= w_1 \frac{A_1}{A_0} \left(\frac{C_1}{C_0}\right)^{-1} + w_2 \frac{B_1}{B_0} \left(\frac{D_1}{D_0}\right)^{-1}$$

$$= w_1 \exp\{X_A - X_C\} + w_2 \exp\{X_B - X_D\}.$$

If the investor has already decided on a particular portfolio, i.e., has chosen the portfolio weights w_1 and w_2, then the log-return data may be used to generate a sample from the distribution of V_1 by viewing V_1 as a function of (X_A, X_B, X_C, X_D). This sample can be transformed into a sample from the distribution of the portfolio log return $\log(V_1/V_0)$, where $V_0 = w_1 + w_2$, and a parametric model can be chosen for the portfolio log return.

Here we want to allow the investor to vary the portfolio weights in order to choose an optimal (according to some criterion left unspecified) portfolio. Therefore, instead of setting up a model for V_1 directly, we set a model for the joint log-return distribution of (X_A, X_B, X_C, X_D) from which the model for V_1 is easily inferred.

The Student's t location-scale family of distributions is a natural choice of parametric family for log returns. Maximum-likelihood estimation of the parameter triple (μ, σ, ν) of the Student's t location-scale family on the samples of daily log returns gives the following estimates:

$$
\begin{array}{ll}
(-6 \cdot 10^{-4}, 0.013, 3.7) & \text{(British Telecom in pounds)}, \\
(2 \cdot 10^{-4}, 0.015, 7.7) & \text{(Deutsche Telekom in euros)}, \\
(2 \cdot 10^{-4}, 0.006, 9.6) & \text{(SEK/GBP)}, \\
(8 \cdot 10^{-5}, 0.004, 8.6) & \text{(SEK/EUR)}.
\end{array}
$$

There is no a priori reason for the log-return distributions to be symmetric; the polynomial normal model in Example 8.10 is also a natural model for the log returns. The estimated parameters $(\theta_0, \theta_1, \theta_2, \theta_3)$ based on the samples of daily log returns are

$$
\begin{array}{ll}
(3.1, 142.6, -1.4, 15.5) \cdot 10^{-4} & \text{(British Telecom in pounds)}, \\
(-8.8, 120.1, \quad 9.2, 22.1) \cdot 10^{-4} & \text{(Deutsche Telekom in euros)}, \\
(4.0, \quad 53.5, -3.7, \quad 3.6) \cdot 10^{-4} & \text{(SEK/GBP)}, \\
(1.9, \quad 31.1, -2.0, \quad 3.5) \cdot 10^{-4} & \text{(SEK/EUR)}.
\end{array}
$$

The conditions $\theta_3 > 0$ and $3\theta_1\theta_3 - \theta_2^2 > 0$ ensuring that the third-degree polynomial is strictly increasing is satisfied for estimated parameter vectors. Figure 9.8 shows the empirical quantiles of the log returns of BT and DT against those of the fitted parametric distributions. By comparing the two upper q–q plots we find that the polynomial normal model captures the asymmetry between the left and right tails in BT log-return data, whereas the Student's t model does not.

We now proceed to the modeling of the dependence between the log returns. The sample correlations between log returns of the stocks and log returns of the exchange rates is approximately zero, and there are no obvious economic reasons not to assume independence between the log-return pairs (X_A, X_B) and (X_C, X_D) of stocks and exchange rates, respectively. We therefore assume that the log-return pairs (X_A, X_B) and (X_C, X_D) are independent and that the distribution functions of the two log-return pairs are of the form, with subscripts s for stocks and e for exchange rates,

$$
P(X_A \leq x_A, X_B \leq x_B) = C^t_{\nu_s, \rho_s}(F_A(x_A), F_B(x_B)),
$$

$$
P(X_C \leq x_C, X_D \leq x_D) = C^t_{\nu_e, \rho_e}(F_C(x_C), F_D(x_D)),
$$

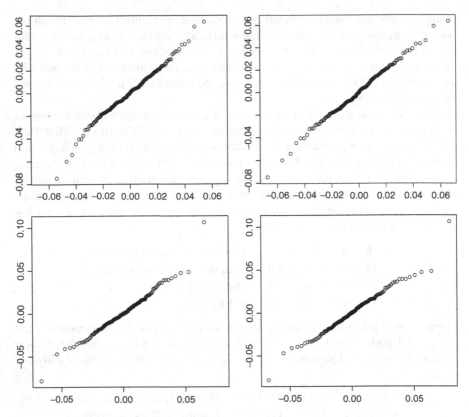

Fig. 9.8 *Upper plots*: empirical quantiles of British Telecom log-return data (y-axes) against quantiles of fitted distributions (x-axes): Student's t model to the *left* and polynomial normal model to the *right*. *Lower plots*: empirical quantiles of Deutsche Telekom log-return data (y-axes) against quantiles of fitted distributions (x-axes): Student's t model to the *left* and polynomial normal model to the *right*

where F_A, F_B, F_C, F_D denote the distribution functions of X_A, X_B, X_C, X_D. Student's t copula is a flexible parametric family for the dependence structure of the log-return pairs. Set $U_A = F_A(X_A)$ and similarly for U_B, U_C, U_D. The assumption of Student's t copulas as models for the dependence structure for the log-return pairs requires that $(U_A, U_B) \stackrel{\mathrm{d}}{=} (1 - U_A, 1 - U_B)$ and $(U_C, U_D) \stackrel{\mathrm{d}}{=} (1 - U_C, 1 - U_D)$. Whatever choice of models for the individual log returns X_A, X_B, X_C, X_D among the sets of models given above, the log-return data give no reasons to reject the hypothesis that $(U_A, U_B) \stackrel{\mathrm{d}}{=} (1 - U_A, 1 - U_B)$ and $(U_C, U_D) \stackrel{\mathrm{d}}{=} (1 - U_C, 1 - U_D)$ (Fig. 9.9).

We may now estimate ρ_s and ρ_e by $\widehat{\rho}_s = \sin(\pi \widehat{\tau}_s / 2)$ and $\widehat{\rho}_e = \sin(\pi \widehat{\tau}_e / 2)$, and the estimate of (ρ_s, ρ_e) is approximately $(0.62, 0.61)$. Under the assumption that the marginal distribution functions F_A, F_B, F_C, F_D of the joint log-return

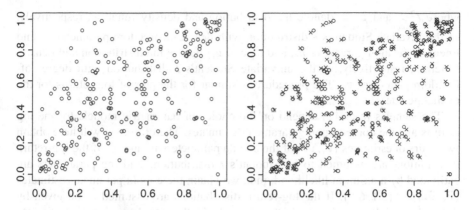

Fig. 9.9 *Left scatter plot*: sample points in (9.12) obtained by componentwise transformation of original log-return pairs for stocks by fitted Student's *t* location-scale distribution functions. *Right scatter plot*: with corresponding sample points for componentwise transformation by distribution functions of fitted polynomial normal models added, marked by *times symbol*, to illustrate the effect of the componentwise transformations

distribution equal the estimated marginal distribution functions $\widehat{F}_A, \widehat{F}_B, \widehat{F}_C, \widehat{F}_D$, we may transform the samples

$$\{(X_A^1, X_B^1), \ldots, (X_A^{248}, X_B^{248})\} \quad \text{and} \quad \{(X_C^1, X_D^1), \ldots, (X_C^{248}, X_D^{248})\}$$

into the samples

$$\{(U_A^1, U_B^1), \ldots, (U_A^{248}, U_B^{248})\} \quad \text{and} \quad \{(U_C^1, U_D^1), \ldots, (U_C^{248}, U_D^{248})\} \qquad (9.12)$$

from Student's *t* copulas, where $U_A^k = \widehat{F}_A(X_A^k)$, and similarly for U_B^k, U_C^k, U_D^k. In the case of a polynomial normal model choice, dropping subscripts for notational convenience, $\widehat{F}(x) = \Phi(\widehat{g}^{-1}(x))$, and $\widehat{g}^{-1}(x)$ is obtained as the (here unique real) solution y to the polynomial equation $\widehat{\theta}_0 + \widehat{\theta}_1 y + \widehat{\theta}_2 y^2 + \widehat{\theta}_3 y^3 = x$. Under the further assumption that the linear correlation parameters ρ_s, ρ_e equal the estimates $\widehat{\rho}_s, \widehat{\rho}_e$, the two samples in (9.12) are samples from two Student's *t* copulas whose parameters are known except for the degree-of-freedom parameters ν_s and ν_e. The unknown parameters can be estimated by maximum likelihood, and the bivariate density function of Student's *t* copula corresponding to the pair of log returns for stocks is given by

$$c^t_{\nu_s, \widehat{\rho}_s}(u_1, u_2) = \frac{\partial^2}{\partial u_1 \partial u_2} t^2_{\nu_s, \widehat{\rho}_s}(t_{\nu_s}^{-1}(u_1), t_{\nu_s}^{-1}(u_2)) = \frac{g^2_{\nu_s, \widehat{\rho}_s}(t_{\nu_s}^{-1}(u_1), t_{\nu_s}^{-1}(u_2))}{g_{\nu_s}(t_{\nu_s}^{-1}(u_1)) g_{\nu_s}(t_{\nu_s}^{-1}(u_2))},$$

where $t^2_{\nu_s,\rho_s}$ and $g^2_{\nu_s,\rho_s}$ denote the distribution and density function, respectively, of the bivariate Student's t distribution with degree-of-freedom parameter ν_s and linear correlation parameter $\widehat{\rho}_s$, and t_{ν_s} and g_{ν_s} denote the distribution and density function, respectively, of the univariate Student's t distribution with degree-of-freedom parameter ν_s. The procedure is similar for the pair of log returns for the exchange rates.

The samples in (9.12) depend on the choice of parametric models for the log returns and the corresponding parameter estimates. Therefore, we will here obtain two pairs of estimates $(\widehat{\nu}_s, \widehat{\nu}_e)$ of the copula parameters ν_s and ν_e. If the log-return distributions are assumed to be Student's t distributions and the parameters are estimated by maximum likelihood, then we obtain the copula parameter estimates $(\widehat{\nu}_s, \widehat{\nu}_e) \approx (5.1, 6.8)$. If the log-return distributions are assumed to be given by the polynomial normal model, then we obtain the copula parameter estimates $(\widehat{\nu}_s, \widehat{\nu}_e) \approx (3.6, 5.5)$.

Now that the two models for the joint log-return distribution of the vector (X_A, X_B, X_C, X_D) are set up and their parameters estimated, we evaluate the models in terms of how close the resulting distribution of the portfolio log return

$$\log(V_1/V_0), \quad V_1 = \frac{V_0}{2}\exp\{X_A - X_C\} + \frac{V_0}{2}\exp\{X_B - X_D\}$$

is to the empirical distribution of the portfolio log return. The joint log-return models do not give closed-form expressions for the distributions of the portfolio log return. However, the portfolio log-return distributions are straightforward to simulate from. We simulate 10^5 outcomes of $\log(V_1/V_0)$, according to the chosen model, by simulating outcomes (Z_A, Z_B) and (Z_C, Z_D) of independent Student's t-distributed random vectors and using the formula

$$\log\left(\frac{1}{2}\exp\{\widehat{F}_A^{-1}(t_{\nu_s}(Z_A)) - \widehat{F}_C^{-1}(t_{\nu_e}(Z_C))\} + \frac{1}{2}\exp\{\widehat{F}_B^{-1}(t_{\nu_s}(Z_B)) - \widehat{F}_D^{-1}(t_{\nu_e}(Z_D))\}\right),$$

where (Z_A, Z_B) has a bivariate standard Student's t distribution with degree-of-freedom parameter $\widehat{\nu}_s$ and linear correlation parameter $\widehat{\rho}_s$, and (Z_C, Z_D) has a bivariate standard Student's t distribution with degree-of-freedom parameter $\widehat{\nu}_e$ and linear correlation parameter $\widehat{\rho}_e$. Finally, we compare the empirical distributions of the simulated samples of size 10^5 to the empirical distribution based on the original log-return sample. The result is shown in Fig. 9.10. Both models give a good fit to the log-return data.

If $\mathbf{X} = (X_1, \ldots, X_d)$ is a random vector with continuous marginal distribution functions $F_1, \ldots, F_d$, and if $G_1, \ldots, G_d$ are any given distribution functions, then the random vector $\mathbf{Y} = (G_1^{-1}(F_1(X_1)), \ldots, G_d^{-1}(F_d(X_d)))$ has marginal distribution functions $G_1, \ldots, G_d$ and has inherited the dependence structure or copula from vector $\mathbf{X}$. However, it may happen that the distribution functions $F_1, \ldots, F_d$ cannot be determined explicitly. Another option is to consider a family of models for vectors $(U_1, \ldots, U_d)$ with components that are uniformly distributed on $(0, 1)$ and consider models of the form $\mathbf{Y} = (G_1^{-1}(U_1), \ldots, G_d^{-1}(U_d))$.

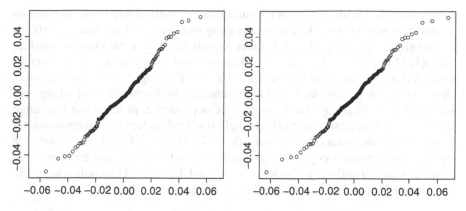

Fig. 9.10 These two q–q plots show the empirical quantiles of the portfolio log returns (y-axes) for $w_1 = w_2 = V_0/2$ against the quantiles of two models for $\log(V_1/V_0)$ (x-axes). The plot to the *left* corresponds to the model for (X_A, X_B, X_C, X_D) with Student's t marginal distributions, and the plot to the *right* corresponds to the model for (X_A, X_B, X_C, X_D) with polynomial normal marginal distributions

Example 9.15 (Archimedean copulas). Consider a strictly positive random variable X with a density f and Laplace transform $\Psi(t) = E[e^{-tX}]$. A useful family of copulas called Archimedean copulas is based on the fact that $\Psi(-\log(V)/X)$ is uniformly distributed on $(0,1)$ if V is uniformly distributed on $(0,1)$ and independent of X. To verify this claim we first note that

$$\Psi(t) = \int_0^\infty e^{-tx} f(x)dx \quad \text{and} \quad \Psi'(t) = -\int_0^\infty xe^{-tx} f(x)dx < 0,$$

so Ψ is nonnegative, continuous, and strictly decreasing on $[0, \infty)$. For any $u \in (0,1)$ we can now verify that

$$P\left(\Psi\left(\frac{-\log V}{X}\right) \leq u\right) = E\left[P\left(\Psi\left(\frac{-\log V}{X}\right) \leq u \mid X\right)\right]$$

$$= E\left[P\left(V \leq e^{-\Psi^{-1}(u)X} \mid X\right)\right]$$

$$= E\left[e^{-\Psi^{-1}(u)X}\right]$$

$$= \Psi(\Psi^{-1}(u)) = u.$$

It follows that if $V_1, \ldots, V_d$ are uniformly distributed on $(0,1)$ and independent of X, then the distribution function C of

$$\mathbf{U} = \left(\Psi\left(\frac{-\log V_1}{X}\right), \ldots, \Psi\left(\frac{-\log V_d}{X}\right)\right) \tag{9.13}$$

is a copula. We should always aim to understand a multivariate model through its stochastic representation. Here Ψ is decreasing with $\Psi(0) = 1$ and $\lim_{t\to\infty} \Psi(t) = 0$. Therefore, we observe that if X takes a small value, then the random variables $-\log(V_k)/X$, for $k = 1, \ldots, d$, are all likely to take large values, which implies small values for the random variables $U_k = \Psi(-\log(V_k)/X)$. In particular, choosing a random variable X that has a relatively high probability of taking very small values is likely to lead to asymptotic dependence, in the sense that small values for one component are likely to imply small values for other components, for a model with the stochastic representation $(G_1^{-1}(U_1), \ldots, G_d^{-1}(U_d))$. Simulation from an Archimedean copula C as above is straightforward: just independently simulate standard uniform variates $V_1, \ldots, V_d$ and X, and set $\mathbf{U}$ according to (9.13). Note that the copula can be expressed explicitly as

$$
\begin{aligned}
C(u_1, \ldots, u_d) &= \mathrm{P}(U_1 \le u_1, \ldots, U_d \le u_d) \\
&= \mathrm{E}\left[\mathrm{P}\left(V_1 \le e^{-\Psi^{-1}(u_1)X}, \ldots, V_d \le e^{-\Psi^{-1}(u_d)X} \mid X \right) \right] \\
&= \mathrm{E}\left[e^{-(\Psi^{-1}(u_1) + \cdots + \Psi^{-1}(u_d))X} \right] \\
&= \Psi(\Psi^{-1}(u_1) + \cdots + \Psi^{-1}(u_d)).
\end{aligned}
\tag{9.14}
$$

Example 9.16 (Clayton copula). If X has a Gamma$(1/\theta, 1)$ distribution, then X has density function $f(x) = x^{1/\theta - 1} e^{-x}/\Gamma(1/\theta)$ and Laplace transform

$$
\Psi(t) = \mathrm{E}[e^{-tX}] = \int_0^\infty e^{-tx} \frac{1}{\Gamma(1/\theta)} x^{1/\theta - 1} e^{-x} \mathrm{d}x = (t+1)^{-1/\theta}.
$$

This choice of Ψ gives the Clayton copula. Solving $\Psi(\Psi^{-1}(u)) = u$ for $\Psi^{-1}(u)$ gives $\Psi^{-1}(u) = u^{-\theta} - 1$. Therefore, the copula expression (9.14) takes the form

$$
C_\theta^{\mathrm{Cl}}(\mathbf{u}) = (u_1^{-\theta} + \cdots + u_d^{-\theta} - d + 1)^{-1/\theta}.
$$

Applying l'Hôpital's rule shows that the Clayton copula has lower tail dependence in the sense that

$$
\begin{aligned}
\lim_{u\to 0} \mathrm{P}(U_k \le u \mid U_j \le u) &= \lim_{u\to 0} \frac{(2u^{-\theta} - 1)^{-1/\theta}}{u} \\
&= \lim_{u\to 0} \frac{\frac{d}{du}(2u^{-\theta} - 1)^{-1/\theta}}{\frac{d}{du} u} \\
&= \lim_{u\to 0} 2u^{-\theta-1}(2u^{-\theta} - 1)^{-1/\theta - 1} \\
&= 2^{-1/\theta}.
\end{aligned}
$$

If $\theta = 1$, then both X, and the random variables $-\log V_k$ are standard exponentially distributed. In particular, we may write

$$(U_1, \ldots, U_d) \stackrel{d}{=} \left(\frac{E_0}{E_0 + E_1}, \ldots, \frac{E_0}{E_0 + E_d} \right),$$

where $E_0, E_1, \ldots, E_d$ are independent and standard exponentially distributed. We see that for all the U_k to take small values, we need E_0 to take a small value. However, for all the U_k to take large values, we need E_0 to take a large value and for all $E_1, \ldots, E_d$ to take small values. The latter is less likely, and therefore a reasonable guess is that the Clayton copula does not have upper tail dependence: $\lim_{u \to 0} P(U_k > u \mid U_j > u) = 0$. An application of l'Hôpital's rule verifies this claim. Samples from the Clayton copula are illustrated graphically in Fig. 9.11.

9.4.1 Misconceptions of Correlation and Dependence

Now we turn to common misconceptions of linear correlation. We have seen that given any two univariate distribution functions F_1 and F_2 and copula function C, $F(x_1, x_2) = C(F_1(x_1), F_2(x_2))$ is a bivariate distribution function with marginal distribution functions F_1 and F_2. It is typically hard to know which copula C to choose, and it is therefore tempting to ask for a bivariate distribution with given marginal distribution functions F_1 and F_2 and a given linear correlation coefficient ρ. However, we will see that this question is ill-posed in the sense that the set of bivariate distributions fulfilling the requirement may be empty.

To this end we first consider an integral representation of the covariance between two random variables in terms of their joint distribution function and their marginal distribution functions.

Proposition 9.8. *If (X_1, X_2) has distribution function F and marginal distribution functions F_1 and F_2 and the covariance $\text{Cov}(X_1, X_2)$ exists finitely, then*

$$\text{Cov}(X_1, X_2) = \int_{-\infty}^{\infty} \int_{-\infty}^{\infty} (F(x_1, x_2) - F_1(x_1) F_2(x_2)) dx_1 dx_2.$$

Proof. Let (Y_1, Y_2) be an independent copy of (X_1, X_2), and note that

$$E[(X_1 - Y_1)(X_2 - Y_2)] = E[X_1 X_2] - E[X_1 Y_2] + E[Y_1 Y_2] - E[Y_1 X_2] = 2 \text{Cov}(X_1, X_2).$$

Writing

$$(X_1 - Y_1) = \int_{-\infty}^{\infty} (I\{Y_1 \le x_1\} - I\{X_1 \le x_1\}) dx_1,$$

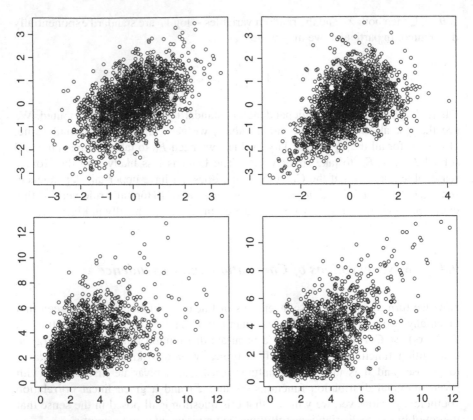

Fig. 9.11 The *upper two scatter plots* show samples of size 2,000 from two bivariate distributions with standard normal marginal distributions. The *left plot* shows a sample from the bivariate standard normal distribution with linear correlation coefficient 0.5 and the *right plot* shows a sample from a bivariate Clayton copula with parameter $\theta = 1$, componentwise transformed to standard normal marginal distributions. The *two lower scatter plots* show samples of size 2,000 from two bivariate distributions with Gamma(3, 1) marginal distributions. The *left plot* corresponds to the copula of a bivariate standard normal distribution with linear correlation 0.5, and the *right plot* corresponds to the copula of the vector (U_1, U_2) such that $(1 - U_1, 1 - U_2)$ has a bivariate Clayton copula with parameter $\theta = 1$

and similarly for $(X_2 - Y_2)$, we find that

$$\mathrm{E}[(X_1 - Y_1)(X_2 - Y_2)]$$

$$= \mathrm{E}\left[\int_{-\infty}^{\infty} (I\{Y_1 \leq x_1\} - I\{X_1 \leq x_1\}) dx_1 \int_{-\infty}^{\infty} (I\{Y_2 \leq x_2\} - I\{X_2 \leq x_2\}) dx_2\right]$$

$$= \mathrm{E}\left[\int_{-\infty}^{\infty} \int_{-\infty}^{\infty} (I\{Y_1 \leq x_1\} - I\{X_1 \leq x_1\})(I\{Y_2 \leq x_2\} - I\{X_2 \leq x_2\}) dx_1 dx_2\right]$$

$$= \int_{-\infty}^{\infty} \int_{-\infty}^{\infty} \mathrm{E}[I\{Y_1 \leq x_1\} - I\{X_1 \leq x_1\}] \, \mathrm{E}[I\{Y_2 \leq x_2\} - I\{X_2 \leq x_2\}] dx_1 dx_2$$

$$= 2 \int_{-\infty}^{\infty} \int_{-\infty}^{\infty} (F(x_1, x_2) - F_1(x_1)F_2(x_2)) dx_1 dx_2,$$

from which the conclusion follows. □

To determine which joint distribution function gives the minimal and maximal covariance (and therefore also linear correlation), we need to determine sharp upper and lower bounds on F in terms of F_1 and F_2. Note that

$$\min(\mathrm{P}(X_1 \leq x_1), \mathrm{P}(X_2 \leq x_2)) \geq \mathrm{P}(X_1 \leq x_1, X_2 \leq x_2)$$

$$= 1 - \mathrm{P}(X_1 > x_1 \text{ or } X_2 > x_2)$$

$$\geq 1 - (\mathrm{P}(X_1 > x_1) + \mathrm{P}(X_2 > x_2))$$

$$= \mathrm{P}(X_1 \leq x_1) + \mathrm{P}(X_2 \leq x_2) - 1,$$

so

$$\max(F_1(x_1) + F_2(x_2) - 1, 0) \leq F(x_1, x_2) \leq \min(F_1(x_1), F_2(x_2)). \qquad (9.15)$$

If $(X_1, X_2) = (F_1^{-1}(U), F_2^{-1}(U))$, then statement (i) of Proposition 6.1 implies that

$$F(x_1, x_2) = \mathrm{P}(F_1^{-1}(U) \leq x_1, F_2^{-1}(U) \leq x_2)$$

$$= \mathrm{P}(U \leq F_1(x_1), U \leq F_2(x_2))$$

$$= \min(F_1(x_1), F_2(x_2)),$$

so the upper bound is attained. In this case, X_1 and X_2 are said to be comonotonic. If $(X_1, X_2) = (F_1^{-1}(U), F_2^{-1}(1-U))$, then statement (i) of Proposition 6.1 implies that

$$F(x_1, x_2) = \mathrm{P}(F_1^{-1}(U) \leq x_1, F_2^{-1}(1-U) \leq x_2)$$

$$= \mathrm{P}(U \leq F_1(x_1), 1 - U \leq F_2(x_2))$$

$$= \max(F_1(x_1) + F_2(x_2) - 1, 0),$$

so the lower bound is also attained. In this case, X_1 and X_2 are said to be countermonotonic.

Proposition 9.9. *Let F_1 and F_2 be distribution functions for random variables with nonzero finite variances. The set of linear correlation coefficients $\rho(F)$ for the set of bivariate distribution functions F with marginal distribution functions F_1 and F_2 form a closed interval $[\rho_{\min}, \rho_{\max}]$ with $0 \in (\rho_{\min}, \rho_{\max})$ such that $\rho(F) = \rho_{\min}$ if and only if $F(x_1, x_2) = \max(F_1(x_1) + F_2(x_2) - 1, 0)$ and $\rho = \rho_{\max}$ if and only if $F(x_1, x_2) = \min(F_1(x_1), F_2(x_2))$.*

Proof. The existence of attainable minimum and maximum linear correlation values $\rho_{\min}, \rho_{\max}$ follows immediately from Proposition 9.8 and the bounds in (9.15). Taking $F(x_1, x_2) = F_1(x_1)F_2(x_2)$ shows that $0 \in [\rho_{\min}, \rho_{\max}]$. By Proposition 9.8, $\rho_{\max} = 0$ would imply that $\min(F_1(x_1), F_2(x_2)) = F_1(x_1)F_2(x_2)$ for all x_1, x_2, which in turn implies that either F_1 or F_2 takes only the values 0 and 1. Such distribution functions correspond to constant random variables for which the variance is zero. We conclude that $\rho_{\max} > 0$. A similar argument shows that $\rho_{\min} < 0$. It remains to show that any value in $[\rho_{\min}, \rho_{\max}]$ is attainable. For $\lambda \in [0, 1]$ the function

$$F_\lambda(x_1, x_2) = \lambda \max(F_1(x_1) + F_2(x_2) - 1, 0) + (1 - \lambda) \min(F_1(x_1), F_2(x_2))$$

is a distribution function since it is the distribution function of the random vector

$$I(F_1^{-1}(U), F_2^{-1}(1 - U)) + (1 - I)(F_1^{-1}(U), F_2^{-1}(U)),$$

where I and U are independent, I takes the value 1 with probability λ and the value 0 otherwise, and U is uniformly distributed on $(0, 1)$. Moreover, $F_\lambda(x_1, 1) = F_1(x_1)$ and $F_\lambda(1, x_2) = F_2(x_2)$. Varying $\lambda \in [0, 1]$ shows that all values in the interval $[\rho_{\min}, \rho_{\max}]$ are attainable correlation values. $\square$

Example 9.17 (A bad stress test). Consider potential aggregate losses X and Y in two lines of business for an insurance company. Suppose that X is Exp(α)-distributed and that Y is Pa(α)-distributed with an unspecified dependence structure. To perform a stress test, the chief risk officer asks an actuary to assign a high linear correlation to the pair (X, Y) and study the effect on the quantile values for the sum $X + Y$. This problem is ill-posed. The correlation coefficient does not exist for $\alpha \leq 2$ since

$$E[Y^2] = \lim_{x \to \infty} \int_1^x y^2 \alpha y^{-\alpha - 1} dy = \lim_{x \to \infty} \frac{\alpha}{2 - \alpha}(x^{2 - \alpha} - 1)$$

does not exist finitely for $\alpha \leq 2$. Moreover, for $\alpha > 2$ not all correlation values are possible for the pair (X, Y), and for each attainable correlation value there are infinitely many possible joint distributions for (X, Y) that may produce very different distributions for $X + Y$.

To compute the upper bound $\rho_{\max}$ of the attainable correlation values, we note that $Y \stackrel{d}{=} e^X$ and that X and Y are comonotonic if $Y = e^X$. In particular, $\rho_{\max} = \text{Cor}(X, e^X)$. The means and variances of the Exp(α) and Pa(α) distributions are given by

$$E[X] = \frac{1}{\alpha}, \quad \text{Var}(X) = \frac{1}{\alpha^2}, \quad E[e^X] = \frac{\alpha}{\alpha - 1}, \quad \text{Var}(e^X) = \frac{\alpha}{(\alpha - 1)^2(\alpha - 2)},$$

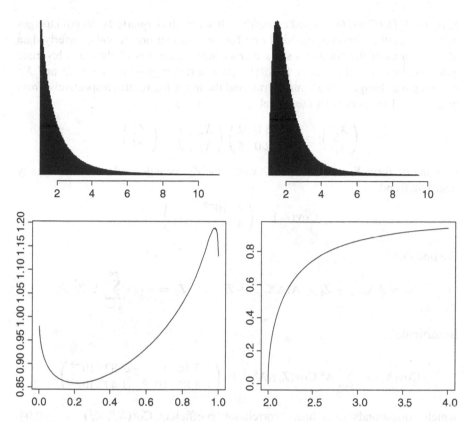

Fig. 9.12 The *upper two plots* show histograms of the distribution of $X + Y$, based on samples of size 10^6 and with the values corresponding to very high quantiles omitted, where X is $\mathrm{Exp}(\alpha)$-distributed and Y is $\mathrm{Pa}(\alpha)$-distributed with $\alpha = 2.1$. The *left histogram* corresponds to $Y = e^X$, and the *right histogram* corresponds to X and Y independent. *Lower right plot*: empirical quantiles of $X + e^X$ divided by empirical quantiles of $X + Y$ with X and Y independent. *Lower right plot*: $\rho_{\max}$ as a function of α

and integration by parts can be used to compute the covariance

$$\mathrm{Cov}(X, e^X) = \mathrm{E}[Xe^X] - \mathrm{E}[X]\,\mathrm{E}[e^X] = \int_0^\infty x\alpha e^{(1-\alpha)x}\,dx - \frac{1}{\alpha - 1} = \frac{1}{(\alpha - 1)^2}.$$

We find that

$$\rho_{\max} = \frac{\mathrm{Cov}(X, e^X)}{\mathrm{Var}(X)^{1/2}\,\mathrm{Var}(e^X)^{1/2}} = \frac{(\alpha^2 - 2\alpha)^{1/2}}{\alpha - 1}.$$

The lower right plot in Fig. 9.12 shows $\rho_{\max}$ as a function of α. For instance, $\alpha = 2.1$ gives $\rho_{\max} \approx 0.4$, which may indicate weak dependence, although it corresponds to comonotonicity. The histograms in Fig. 9.12 show the distribution of $X + Y$ in the case of comonotonicity (left plot) and independence (right plot) for X and Y for $\alpha = 2.1$.

Example 9.18 (Correlation and causality). If we analyze quarterly data of changes in the 3-month, zero-coupon bond rate for government bonds and quarterly data of log returns of the country's stock market index, then it is likely that a bivariate autoregressive model of order 1, AR(1), gives a rather good fit. With X_t^1 and X_t^2 denoting the change in the 3-month rate and the index log return, respectively, from quarter $t - 1$ to t, consider the model

$$\begin{pmatrix} X_t^1 \\ X_t^2 \end{pmatrix} = \begin{pmatrix} 0.45 & 0.02 \\ -9.2 & 0.35 \end{pmatrix} \begin{pmatrix} X_{t-1}^1 \\ X_{t-1}^2 \end{pmatrix} + \begin{pmatrix} Z_t^1 \\ Z_t^2 \end{pmatrix},$$

or in matrix form $\mathbf{X}_t = \mathbf{A}\mathbf{X}_{t-1} + \mathbf{Z}_t$, where the $\mathbf{Z}_k$ are independent and identically distributed and

$$\mathrm{Cov}(\mathbf{Z}_t) = \begin{pmatrix} 2 \cdot 10^{-5} & 0 \\ 0 & 10^{-2} \end{pmatrix}.$$

We find that

$$\mathbf{X}_t = \mathbf{A}\mathbf{X}_{t-1} + \mathbf{Z}_t = \mathbf{A}(\mathbf{A}\mathbf{X}_{t-2} + \mathbf{Z}_{t-1}) + \mathbf{Z}_t = \cdots = \sum_{k=0}^{\infty} \mathbf{A}^k \mathbf{Z}_{t-k}.$$

In particular,

$$\mathrm{Cov}(\mathbf{X}_t) = \sum_{k=0}^{\infty} \mathbf{A}^k \, \mathrm{Cov}(\mathbf{Z}_t)(\mathbf{A}^k)^{\mathrm{T}} \approx \begin{pmatrix} 3.18 \cdot 10^{-5} & -2.82 \cdot 10^{-5} \\ -2.82 \cdot 10^{-5} & 1.47 \cdot 10^{-2} \end{pmatrix},$$

which corresponds to a linear correlation coefficient $\mathrm{Cor}(X_t^1, X_t^2) \approx -0.04$. However,

$$\mathrm{Cov}(X_t^2, X_{t-1}^1) = \mathrm{Cov}(-9.2X_{t-1}^1 + 0.35X_{t-1}^2 + Z_t^2, X_{t-1}^1)$$
$$= -9.2\,\mathrm{Var}(X_t^1) + 0.35\,\mathrm{Cov}(X_t^1, X_t^2),$$

which gives

$$\mathrm{Cor}(X_t^2, X_{t-1}^1) = -9.2 \left(\frac{\mathrm{Var}(X_t^1)}{\mathrm{Var}(X_t^2)} \right)^{1/2} + 0.35\,\mathrm{Cor}(X_t^1, X_t^2) \approx -0.44$$

reflecting the fact that the stock market typically reacts negatively to increasing interest rates (the present value of future dividends decreases) and positively to decreasing interest rates. Similarly, $\mathrm{Cor}(X_t^1, X_{t-1}^2) \approx 0.41$, which may reflect the fact that central banks raise interest rates to cool down an overheated economy and lower interest rates to boost a struggling economy. The main point is that the linear correlation coefficient $\mathrm{Cor}(X_t^1, X_t^2) \approx -0.04$ that could be estimated on the pairs of interest rate changes and index log returns says very little about the dependencies between interest rate changes and index log returns. Here we have two rather strong causal dependencies that essentially net out when only considering the dependence among the components of the random vector (X_t^1, X_t^2).

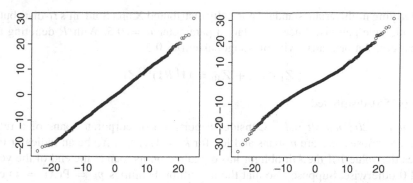

Fig. 9.13 q–q plots of simulated samples of size 2,000 against a normal distribution with zero mean and variance 55. The first distribution (*left*) is the sum of the components of a ten-dimensional standard normal distributed vector with pairwise linear correlation 0.5. The second distribution (*right*) is the sum of the components on a ten-dimensional random vector with standard normal univariate marginal distributions and the dependence structure of a ten-dimensional Student's t distribution with one degree of freedom and pairwise linear correlation 0.5

Example 9.19 (Asymptotic dependence). We know from Proposition 9.5 that the components of a bivariate standard normally distributed vector (X_1, X_2) with linear correlation $\rho < 1$ are asymptotically independent in the sense that $\lim_{x \to -\infty} P(X_2 \leq x \mid X_1 \leq x) = 0$. In this case, an extreme value for one component is not likely to make the other component take an extreme value. Combining Proposition 9.6 and Example 8.2 implies that the components of a bivariate standard Student's t_ν-distributed vector (Y_1, Y_2) with linear correlation $\rho \in (0, 1)$ are asymptotically dependent in the sense that $\lim_{x \to -\infty} P(Y_2 \leq x \mid Y_1 \leq x) = \lambda > 0$. In this case, an extreme value for one component makes it likely that the other component will take an extreme value.

Consider the random vector $(U_1, U_2) = (\Phi(X_1), \Phi(X_2))$, whose distribution function is called a Gaussian copula, and the random vector $(V_1, V_2) = (t_\nu(Y_1), t_\nu(Y_2))$, whose distribution function is called a t_ν copula. If G is a distribution function and $p \in (0, 1)$ is small, then the probability that both components of the vector $(Z_1, Z_2) = (G^{-1}(V_1), G^{-1}(V_2))$ take values smaller than $G^{-1}(p)$ is approximately

$$P(Z_1 \leq G^{-1}(p), Z_2 \leq G^{-1}(p)) = P(V_1 \leq p) P(V_2 \leq p \mid V_1 \leq p) \approx \lambda p,$$

whereas the corresponding probability of joint extremes for the vector $(W_1, W_2) = (G^{-1}(U_1), G^{-1}(U_2))$ is of the order p^2. As a consequence, the left tail of $Z_1 + Z_2$ will be heavier than that of $W_1 + W_2$. The influence of the (lack of) asymptotic dependence of the (Gaussian) t_ν copula of a random vector on the tail behavior of the sum of its components is valid for vectors of arbitrary dimension. Figure 9.13 illustrates this effect graphically in terms of q–q plots for 10-dimensional random vectors $\mathbf{Z}$ and $\mathbf{W}$, where $G = \Phi$ is the standard normal distribution function and the underlying multivariate standard normally distributed $\mathbf{X}$ and Student's t_1-distributed $\mathbf{Y}$ both have pairwise linear correlation parameter $\rho = 0.5$. With R denoting the linear correlation matrix with off-diagonal entries 0.5,

$$Z_1 + \cdots + Z_{10} \stackrel{\mathrm{d}}{=} (\mathbf{1}^{\mathsf{T}} R \mathbf{1})^{1/2} Z_1$$

is $N(0, 55)$-distributed.

Example 9.20 (Default risk). Consider a portfolio of corporate loans of a retail bank. Suppose there are n loans and let, for $k = 1, \ldots, n$, X_k be an indicator that takes the value 1 if the kth obligor has defaulted on its loan at the end of the year, and 0 otherwise. Suppose also that the default probabilities $p_k = \mathrm{P}(X_k = 1)$ can be accurately estimated and may be considered as known. A common estimation approach is to divide the obligors into m homogeneous groups so that all obligors belonging to the same group have the same default probability. The estimates of default probabilities can then be based on the relative frequencies of defaults over the years for the different groups.

The random variable $N = X_1 + \cdots + X_n$, representing the total number of defaults within the current year, is likely to be of interest to the bank. However, the default probabilities only determine the marginal distributions and not the full multivariate distribution of the random vector $(X_1, \ldots, X_n)$. To specify a multivariate model for the default indicators, it is common to consider a vector $(Y_1, \ldots, Y_n)$ of so-called latent variables. The latent variable Y_k may represent the difference between the values of the assets and liabilities of the kth obligor at the end of the year, and a threshold d_k is determined so that $Y_k \leq d_k$ corresponds to default for obligor k. We may now express the probability that the first k among the n loans default as, assuming that the Y_k have continuous distribution functions,

$$
\begin{aligned}
p_{1\ldots k} &= \mathrm{P}(Y_1 \leq d_1, \ldots, Y_k \leq d_k) \\
&= C(\mathrm{P}(Y_1 \leq d_1), \ldots, \mathrm{P}(Y_k \leq d_k), 1, \ldots, 1) \\
&= C(p_1, \ldots, p_k, 1, \ldots, 1),
\end{aligned}
$$

where C denotes the copula of $(Y_1, \ldots, Y_n)$. Joint default probabilities of this type will depend heavily on the choice of copula C. To illustrate this point, we consider a numerical example.

Consider a loan portfolio with $n = 1{,}000$ obligors, and suppose that the default probability of each obligor is equal to $p = 0.05$, i.e., $p_k = 0.05$ for each k. We consider four different copula models for the latent variable vector: (a) C is a Gaussian copula with pairwise correlation parameter $\rho = 0$, (b) C is a Gaussian copula with pairwise correlation parameter $\rho = 0.1$, (c) C is a Student's t_3 copula with pairwise correlation parameter $\rho = 0$, and (d) C is a Student's t_3 copula with pairwise correlation parameter $\rho = 0.1$.

For each model we generate a sample of size 10^5 from the resulting model for N, the total number of defaults, and illustrate the distribution of N in terms of the histograms shown in Fig. 9.14. The histograms show clearly that zero correlation for

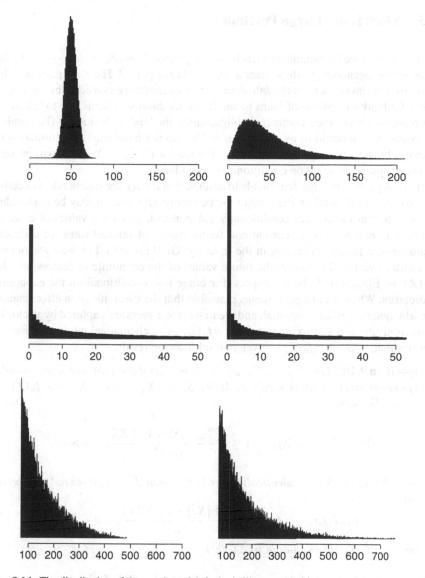

Fig. 9.14 The distribution of the number of defaults is illustrated in histograms based on samples of size 10^5 for the sum of 10^3 default indicators. The *histograms* correspond to the following latent variable models: Gaussian with $\rho = 0$ (*upper left*), Gaussian with $\rho = 0.1$ (*upper right*), Student's t_3 with $\rho = 0$ (*middle* and *lower left*), Student's t_3 with $\rho = 0.1$ (*middle* and *lower right*)

the underlying Student's t distribution is far from independence. For the Gaussian copula, zero correlation is equivalent to independence. The histograms also show the impact on the distribution of N of the small change in the correlation parameter ρ from 0 to 0.1.

9.5   Models for Large Portfolios

In this section we investigate models for the aggregated loss $S_n = X_1 + \cdots + X_n$ for a large homogeneous portfolio over a specified time period. Here X_k represent the loss from an investment in the kth asset. As an example we consider the aggregate loss of a bank's portfolio of loans to small and medium size firms due to failure of borrowers to honor their contracted obligations to the lender (the bank). The number of assets, n, is thought of as very large, and we do not have enough information to accurately specify an n-dimensional distribution for $(X_1, \ldots, X_n)$. We will present a cruder approach based on conditional independence.

In many cases, it is not reasonable to assume that the X_k are independent because the losses may depend on the state of the economy. However, it may be reasonable to assume that the X_k are conditionally independent, given the values of a set of economic indicators (e.g., current and future values of interest rates for different maturities, capacity utilization in the industry, GDP growth). Let the components of random vector $\mathbf{Z}$ represent the future values of the economic indicators, and let $f_n(\mathbf{Z}) = \mathrm{E}[S_n/n \mid \mathbf{Z}]$ be the expected average loss conditional on the economic indicators. When n is large, it seems plausible that the diversification effect causes the idiosyncratic risks to be small and the main risk drivers are captured by vector $\mathbf{Z}$. This motivates the approximation $S_n \approx n f_n(\mathbf{Z})$. A mathematical motivation for the approximation $S_n \approx n f_n(\mathbf{Z})$ is given in the following result.

Proposition 9.10. *Let $X_1, \ldots, X_n$ be random variables that are conditionally independent given random vector $\mathbf{Z}$. Write $S_n = X_1 + \cdots + X_n$ and $f_n(\mathbf{Z}) = \mathrm{E}[S_n/n \mid \mathbf{Z}]$. Then*

$$\mathrm{P}(|S_n/n - f_n(\mathbf{Z})| > \varepsilon) \leq \frac{\sum_{k=1}^n \mathrm{E}[\mathrm{Var}(X_k \mid \mathbf{Z})]}{(n\varepsilon)^2}, \quad \varepsilon > 0.$$

If, in addition, the X_k are identically distributed, then $f = f_n$ does not depend on n and

$$\mathrm{P}(|S_n/n - f(\mathbf{Z})| > \varepsilon) \leq \frac{\mathrm{E}[X_1^2] - \mathrm{E}[f(\mathbf{Z})^2]}{n\varepsilon^2}, \quad \varepsilon > 0.$$

If, further, the X_k take values in $\{0, 1\}$, then

$$\mathrm{P}(|S_n/n - f_n(\mathbf{Z})| > \varepsilon) \leq \frac{\mathrm{E}[f(\mathbf{Z})] - \mathrm{E}[f(\mathbf{Z})^2]}{n\varepsilon^2}.$$

Proof. An application of Chebyshev's inequality gives

$$\mathrm{P}(|S_n/n - f_n(\mathbf{Z})| > \varepsilon) = \mathrm{E}[\mathrm{P}(|S_n - n f_n(\mathbf{Z})| > n\varepsilon \mid \mathbf{Z})]$$
$$\leq \frac{\mathrm{E}[\mathrm{Var}(S_n \mid \mathbf{Z})]}{n^2\varepsilon^2}.$$

Because the X_k are conditionally independent given $\mathbf{Z}$, it follows that

$$E[\mathrm{Var}(S_n \mid \mathbf{Z})] = \sum_{k=1}^{n} E[\mathrm{Var}(X_k \mid \mathbf{Z})],$$

which proves the first claim. The second claim follows from the first claim because

$$E[\mathrm{Var}(X_k \mid \mathbf{Z})] = E[\mathrm{Var}(X_1 \mid \mathbf{Z})] = E[E[X_1^2 \mid \mathbf{Z}] - (E[X_1 \mid \mathbf{Z}])^2] = E[X_1^2] - E[f(\mathbf{Z})^2].$$

Moreover, if X_1 takes a value in $\{0, 1\}$, then $E[X_1^2 \mid \mathbf{Z}] = E[X_1 \mid \mathbf{Z}] = f(\mathbf{Z})$. This completes the proof. □

Proposition 9.10 not only motivates the approximation $S_n \approx n f_n(\mathbf{Z})$; it also provides an upper bound for tail probabilities for the aggregated loss S_n. For instance, combining Proposition 9.10 and the inequality

$$P(S_n > s) = P(S_n > s, |S_n - n f_n(\mathbf{Z})| \leq \varepsilon n) + P(S_n > s, |S_n - n f_n(\mathbf{Z})| > \varepsilon n)$$
$$\leq P(n f_n(\mathbf{Z}) > s - \varepsilon n) + P(|S_n - n f_n(\mathbf{Z})| > \varepsilon n), \quad \varepsilon > 0,$$

gives an upper bound for $P(S_n > s)$. The upper bound for the tail probability gives an upper bound for the quantile. If the X_k are identically distributed and conditionally independent given $\mathbf{Z}$, then, with $C = E[X_1^2] - E[f(\mathbf{Z})^2]$,

$$F_{S_n}^{-1}(q) = \min\{s : F_{S_n}(s) \geq q\}$$
$$= \min\{s : P(S_n > s) \leq 1 - q\}$$
$$\leq \min\{s : P(n f(\mathbf{Z}) > s - \varepsilon n) + C/(\varepsilon^2 n) \leq 1 - q\}$$
$$= n(\varepsilon + F_{f(\mathbf{Z})}^{-1}(q + C/(\varepsilon^2 n))), \quad \varepsilon > 0.$$

In particular,

$$F_{S_n}^{-1}(q) \leq n \min_{\varepsilon > 0} \left(\varepsilon + F_{f(\mathbf{Z})}^{-1}(q + C/(\varepsilon^2 n)) \right), \quad C = E[X_1^2] - E[f(\mathbf{Z})^2]. \quad (9.16)$$

The upper bound for quantile (9.16) can be used to derive upper bounds for risk measures such as VaR and ES.

Example 9.21 (A large homogeneous loan portfolio). Consider a large portfolio of loans to small and medium size firms and suppose that we want to analyze the distribution of aggregate losses from now until 1 year from now due to defaults. Write X_k for the loss on the kth loan and $S_n = X_1 + \cdots + X_n$ for the aggregated loss. In this case, X_k can be written as $X_k = L_k I_k$, where I_k is the default indicator that takes the value 1 if the kth obligor defaults and 0 otherwise, and L_k is the amount of money lost if the kth obligor defaults. If the default probabilities $P(I_k = 1)$ are of similar size and the loss given default variables L_k are statistically similar, then the loan portfolio can be considered homogeneous.

A particularly nice situation is where the L_k are identical and deterministic, $L_k = l$, for each $k = 1, \ldots, n$. In this case

$$f_n(\mathbf{Z}) = \mathrm{E}\left[\frac{S_n}{n} \mid \mathbf{Z}\right] = \mathrm{E}\left[\frac{1}{n}\sum_{k=1}^{n} L_k I_k \mid \mathbf{Z}\right] = l\,\mathrm{E}\left[\frac{N_n}{n} \mid \mathbf{Z}\right],$$

where $N_n = I_1 + \cdots + I_n$ is the number of defaults. The fraction of defaults, given the economic indicators, is written as $p_n(\mathbf{Z}) = \mathrm{E}[N_n/n \mid \mathbf{Z}]$. That is, $f_n(\mathbf{Z}) = l p_n(\mathbf{Z})$, and the aggregated loss can be approximated by $S_n \approx n l p_n(\mathbf{Z})$. If, in addition, the default indicators are identically distributed, then $p_n(\mathbf{Z}) = p(\mathbf{Z})$ does not depend on n, and the last statement of Proposition 9.10 leads to

$$\mathrm{P}(|N_n/n - p(\mathbf{Z})| > \varepsilon) \leq \frac{\mathrm{E}[p(\mathbf{Z})(1 - p(\mathbf{Z}))]}{n\varepsilon^2}.$$

9.5.1 Beta Mixture Model

In this section, we will illustrate the modeling approach presented in the previous example for a specific choice of model for $N = N_n$ defaults and $p(\mathbf{Z})$ fraction of defaults. Write $N = I_1 + \cdots + I_n$, where the I_k are identically distributed, independent, and Bernoulli distributed with parameter Z conditional on the random variable $Z = f(\mathbf{Z})$, which we take to be Beta(a, b)-distributed. We do not give any economic interpretation of the Beta(a, b)-distributed Z and choose this model only because it is a particularly simple model to work with in terms of both analytical and numerical computations.

The assumption that Z is Beta(a, b)-distributed implies that Z has the density function

$$g(z) = \frac{1}{\beta(a, b)} z^{a-1}(1 - z)^{b-1}, \quad a, b > 0, z \in (0, 1),$$

where $\beta(a, b)$ can be expressed in terms of the Gamma function as

$$\beta(a, b) = \int_0^1 z^{a-1}(1 - z)^{b-1}\,\mathrm{d}z = \frac{\Gamma(a)\Gamma(b)}{\Gamma(a + b)}.$$

Using the property $\Gamma(z + 1) = z\Gamma(z)$ of the Gamma function we find that

$$\mathrm{E}[Z] = \frac{1}{\beta(a, b)} \int_0^1 z^a(1 - z)^{b-1}\,\mathrm{d}z = \frac{\beta(a + 1, b)}{\beta(a, b)} = \frac{a}{a + b},$$

$$\mathrm{E}[Z^2] = \frac{\beta(a + 2, b)}{\beta(a, b)} = \frac{a(a + 1)}{(a + b)(a + b + 1)}.$$

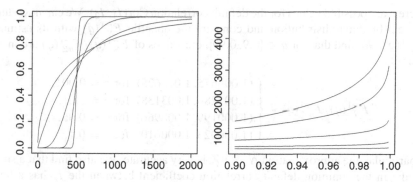

Fig. 9.15 Distribution functions (*left*) and quantile functions (*right*) for beta-binomial distributions with $n = 10^4$, $p = 0.05$, and $(a, b) = ((1-c)/c)(p, 1-p)$ for $c = 0, 0.001, 0.01, 0.05, 0.1$ ($c = 0$ gives the Bin(n, p) distribution)

Conditional on Z, the number of defaults N has a Bin(n, Z) distribution, and therefore the distribution of N is given by

$$P(N = k) = \binom{n}{k} \int_0^1 z^k (1 - z)^{n-k} g(z) \mathrm{d}z$$

$$= \binom{n}{k} \frac{1}{\beta(a, b)} \int_0^1 z^{a+k-1} (1 - z)^{n-k+b-1} \mathrm{d}z$$

$$= \binom{n}{k} \frac{\beta(a + k, b + n - k)}{\beta(a, b)},$$

which is called the beta-binomial distribution. The distribution function of the beta-binomial distribution is illustrated in Fig. 9.15. The expected number of defaults is easily computed:

$$E[N] = E[E[N \mid Z]] = E[nZ] = n \frac{a}{a + b}.$$

In addition, the individual default probability is $P(I_1 = 1) = E[E[I_1 \mid Z]] = E[Z]$, the pairwise default probability is $P(I_1 = I_2 = 1) = E[Z^2]$, and the default correlation is

$$\mathrm{Cor}(I_1, I_2) = \frac{E[I_1 I_2] - E[I_1]^2}{E[I_1^2] - E[I_1]^2} = \frac{E[Z^2] - E[Z]^2}{E[Z] - E[Z]^2} = \frac{1}{a + b + 1}.$$

To analyze the model, we fix the common individual default probability at $p = P(I_1 = 1)$. This implies that we allow only parameter pairs (a, b) for which $p = a/(a + b)$, i.e., pairs (a, b) satisfying

$$(a, b) = \frac{1 - c}{c}(p, 1 - p), \quad c \in (0, 1),$$

where c are possible values for the default correlation $\mathrm{Cor}(I_1, I_2)$. We can now study the beta-binomial distribution and compare the quantile $F_N^{-1}(q)$ with its estimate $nF_Z^{-1}(q)$. We find that for $q \in [0.9, 0.99]$, the values of $F_N^{-1}(q)/F_{nZ}^{-1}(q)$ are in the intervals

$$F_N^{-1}(q)/F_{nZ}^{-1}(q) \in \begin{cases} (1.006675, 1.015625) & \text{for } c = 0.001, \\ (1.001286, 1.003138) & \text{for } c = 0.01, \\ (1.000199, 1.000966) & \text{for } c = 0.05, \\ (1.000023, 1.000610) & \text{for } c = 0.1. \end{cases}$$

In particular, the approximation $N \approx nZ$ is very accurate. We also find that a small change in the common default correlation coefficient between the I_k has a huge effect on the distribution of $N = I_1 + \cdots + I_n$. This is seen in Fig. 9.15, which shows distribution functions and quantile functions for beta-binomial models with $p = 0.05$, $n = 10^4$, and different correlation coefficients. Figure 9.15 illustrates clearly that only specifying the individual default probability p says very little about the distribution of N. Every choice of $(a, b) = ((1 - c)/c)(p, 1 - p)$, $c > 0$, gives default probability p. Let $Z_{c,p}$ be Beta-distributed with the parameters a, b above. Then, for every $\varepsilon > 0$,

$$\mathrm{P}(|Z_{c,p} - p| > \varepsilon) \le \frac{\mathrm{Var}(Z_{c,p})}{\varepsilon^2} = \frac{p(1 - p)c}{\varepsilon^2}.$$

In particular, if $N_{c,p}$ is beta-binomially distributed with mixture variable $Z_{c,p}$, then

$$\mathrm{P}(N_{c,p} = k) = \mathrm{E}\left[\binom{n}{k} Z_{c,p}^k (1 - Z_{c,p})^{n-k}\right]$$

$$= \mathrm{E}\left[\binom{n}{k} Z_{c,p}^k (1 - Z_{c,p})^{n-k}; |Z_{c,p} - p| \le c^{1/3}\right]$$

$$+ \mathrm{E}\left[\binom{n}{k} Z_{c,p}^k (1 - Z_{c,p})^{n-k}; |Z_{c,p} - p| > c^{1/3}\right]$$

$$\le \max_{|t| \le c^{1/3}} \binom{n}{k}(p + t)^k (1 - (p + t))^{n-k} + p(1 - p)c^{1/3}$$

$$\to \binom{n}{k} p^k (1 - p)^{n-k} \quad \text{as } c \to 0.$$

The lower bound is constructed similarly. We conclude that $N_{c,p}$ converges in distribution to $\mathrm{Bin}(n, p)$ as $c \to 0$. This is also seen in Fig. 9.15.

9.6 Notes and Comments

Much more material on elliptical distributions can be found in the book [16] by Kai-Tai Fang, Samuel Kotz, and Kai Wang Ng.

For further material on multivariate elliptical and copula-based models, dependence concepts, and applications in financial risk management we refer the reader to the book [31] by Alexander McNeil, Rüdiger Frey, and Paul Embrechts. Much material on models and methods for portfolio credit risk, which we have only touched upon here, can be found in [31]. Moreover, techniques for parameter estimation for copula models, a topic we have not considered at all, are presented and illustrated in [31].

A statement equivalent to Proposition 9.6 appears in the book Chap. [12] by Paul Embrechts, Alexander McNeil and Daniel Straumann It can be proved by considering the conditional density of one component of a bivariate Student's t-distributed vector given a value of its other component. However, the asymptotic dependence (or tail dependence) property of the Student's t distribution is a consequence of a more general fact that says that pairs of components of an elliptically distributed random vector are asymptotically dependent if the distribution functions of its components are regularly varying. A proof of this more general fact, which also applies to Proposition 9.6, can be found in the article [22] by Henrik Hult and Filip Lindskog. The statement in Proposition 9.7 appears in the book Chap. [25] by Filip Lindskog, Alexander McNeil, and Uwe Schmock. and in the article [15] by Hong-Bin Fang, Kai-Tai Fang, and Samuel Kotz.

The reader seeking more information about copulas in general is encouraged to consult the books [23] by Harry Joe and [35] by Roger Nelson.

9.7 Exercises

In the exercises below, it is assumed, whenever applicable, that you can take positions corresponding to fractions of assets.

Exercise 9.1 (Risk minimization). Consider the value L of a liability and values $X_1, \ldots, X_d$ of assets at time $T > 0$ that may be used to hedge the liability. Suppose that L and the X_k have finite variances, and let ρ be a translation-invariant and positively homogeneous risk measure.

(a) Show that if $(X_1, \ldots, X_d, L)$ has an elliptical distribution, then the portfolio weights $h_0, h_1, \ldots, h_d$ minimizing

$$\mathrm{E}[(h_0 + h_1 X_1 + \cdots + h_d X_d - L)^2],$$

   i.e., the optimal quadratic hedge, minimize $\rho(h_0 + h_1 X_1 + \cdots + h_d X_d - L)$.
(b) Show, by an explicit example, that the conclusion in (a) does not hold in general when $(X_1, \ldots, X_d, L)$ does not have an elliptical distribution.

Exercise 9.2 (Allocation invariance). Let $\mathbf{X} = (X_1,\ldots,X_d)^{\mathsf{T}}$ and $\mathbf{Y} = (Y_1,\ldots,Y_d)^{\mathsf{T}}$ be random vectors having normal variance mixture distributions with identical dispersion matrices and identical location vectors $R_0\mathbf{1}$, where R_0 is the return on a risk-free asset. Vectors $\mathbf{X}$ and $\mathbf{Y}$ represent returns on $2d$ risky assets. Let $V_{\mathbf{X}}(\mathbf{w})$ and $V_{\mathbf{Y}}(\mathbf{w})$ denote the values at the end of the investment horizon for an investment of the capital V_0 in positions in the risk-free asset and in the assets with return vectors $\mathbf{X}$ and $\mathbf{Y}$, respectively, where $\mathbf{w}$ is a vector of monetary portfolio weights corresponding to the positions in the risky assets.

(a) Show that if ρ is a translation-invariant and positively homogeneous risk measure, then
$$\frac{\rho(V_{\mathbf{X}}(\mathbf{w}) - V_0 R_0)}{\rho(V_{\mathbf{Y}}(\mathbf{w}) - V_0 R_0)} \tag{9.17}$$
does not depend on the allocation of the initial capital or on the common dispersion matrix of the return vectors.

(b) Suppose that $\mathbf{X}$ has a Student's t distribution with four degrees of freedom, that $\mathbf{Y}$ has a normal distribution, and that $\rho = \mathrm{VaR}_p$, and compute the expression in (9.17) as a function of p for $p \le 0.05$.

Exercise 9.3 (Asymptotic dependence). Consider a random vector (X_1, X_2) whose components are equally distributed and use Propositions 9.5 and 9.6 to compute $\lim_{x\to\infty} P(X_2 > x \mid X_1 > x)$ in the following two cases:

(a) X_1 and X_2 are Student's t-distributed with four degrees of freedom, and (X_1, X_2) has a Gaussian copula with linear correlation parameter 0.5.
(b) X_1 and X_2 are Student's t-distributed with four degrees of freedom, and (X_1, X_2) has a Student's t copula with linear correlation parameter 0.5 and degrees of freedom parameter 6.

Exercise 9.4 (Comonotonic additive risk). Show that if X_1 and X_2 are comonotone random variables, then $\mathrm{VaR}_p(X_1 + X_2) = \mathrm{VaR}_p(X_1) + \mathrm{VaR}_p(X_2)$ and $\rho_\phi(X_1 + X_2) = \rho_\phi(X_1) + \rho_\phi(X_2)$ for any spectral risk measure ρ_ϕ defined in (6.18).

Exercise 9.5 (Kendall's tau). Let Ψ be the Laplace transform of a strictly positive random variable, and consider the random pair (U_1, U_2) whose distribution function is the copula $C(u_1, u_2) = \Psi(\Psi^{-1}(u_1) + \Psi^{-1}(u_2))$.

(a) Show that $\tau(U_1, U_2) = 4\,\mathrm{E}[C(U_1, U_2)] - 1$.
(b) It can be shown that $P(C(U_1, U_2) \le v) = v - \Psi^{-1}(v)/(\Psi^{-1})'(v)$ for v in $(0, 1)$. Use this relation to show that
$$\tau(U_1, U_2) = 1 + 4 \int_0^1 \frac{\Psi^{-1}(v)}{(\Psi^{-1})'(v)}\,dv.$$

(c) Compute $\tau(U_1, U_2)$ when $C = C_\theta^{\mathrm{Cl}}$ is a Clayton copula.

Exercise 9.6 (Credit rating migration). Consider the two corporate bonds in Exercise 4.6. Let the credit ratings be numbered from 1 to 4 and correspond to the ratings Excellent, Good, Poor, and Default in Exercise 4.6. Let (X_1, X_2) denote the pair of credit ratings of the two issuers after 1 year with the distribution given in Table 4.1.

(a) Find a copula C such that $P(X_1 \leq x_1, X_2 \leq x_2) = C(P(X_1 \leq x_1), P(X_2 \leq x_2))$ for all (x_1, x_2).
(b) The copula C of (X_1, X_2) in (a) can be well approximated by a Gaussian copula. Investigate numerically what value of the correlation parameter in the Gaussian copula gives a good approximation of the copula of (X_1, X_2) in (a).

Exercise 9.7 (Portfolio default risk). Consider a latent variable model for a homogeneous portfolio of n risky loans. Let p be the default probability for each loan, let $Y, Y_1, \ldots, Y_n$ be independent and standard normally distributed, and let $\rho \in (0, 1)$ be a parameter. The default indicators are modeled as

$$X_k = \begin{cases} 1 \text{ if } \sqrt{\rho}Y + \sqrt{1 - \rho}Y_k \leq \Phi^{-1}(p), \\ 0 \text{ otherwise,} \end{cases} \qquad (9.18)$$

where Φ denotes the standard normal distribution function.

(a) Determine the random variable $\Theta = g(Y)$ such that the default indicators are conditionally independent and $\mathrm{Be}(\theta)$-distributed given $\Theta = \theta$.
(b) Show that the following formula holds for the q-quantile of Θ:

$$F_\Theta^{-1}(q) = \Phi\left(\Phi^{-1}(q)\frac{\sqrt{\rho}}{\sqrt{1-\rho}} + \Phi^{-1}(p)\frac{1}{\sqrt{1-\rho}}\right).$$

(c) Consider a loan portfolio of a bank consisting of one thousand loans, each of size one million dollars. Suppose that, for each of the loans, the probability of default within 1 year is 3%, and in case of default the bank makes a loss equal to 25% of the size of the loan. Suppose further that the bank makes a profit of $10,000 per year from interest payments on each loan that does not default and nothing on those that do. The bank decides to set aside an amount of buffer capital that equals its estimate of $\mathrm{ES}_{0.01}(S)$, where S is the profit from interest income minus the loss from defaults over a 1-year period. Estimate the size of the buffer capital under the assumption that the default indicators are given by (9.18) with $\rho = 0.2$ and that the bank may invest in a risk-free, 1-year, zero-coupon bond with a zero rate of 3%.

Exercise 9.8 (Potential death spiral). Consider a life insurance company with a liability cash flow with long duration. The value of the liability 1 year from now is denoted by L and increases in value when interest rates decline. The premium received for insuring the liability is $V_0 = 1.1\,E[L]$. The insurer invests its capital in a fixed-income portfolio with 1-year return R_1 and in a stock market portfolio with 1-year return R_2. The vector (R_1, R_2, L) is, for simplicity, assumed to have a multivariate Student's t distribution with four degrees of freedom. Its mean vector and correlation matrix are given by

$$
E\begin{pmatrix} R_1 \\ R_2 \\ L \end{pmatrix} = \begin{pmatrix} 1.02 \\ 1.10 \\ 1.2 \cdot 10^7 \end{pmatrix} \quad \text{and} \quad \text{Cor}\begin{pmatrix} R_1 \\ R_2 \\ L \end{pmatrix} = \begin{pmatrix} 1 & 0.3 & 0.9 \\ 0.3 & 1 & 0.2 \\ 0.9 & 0.2 & 1 \end{pmatrix}.
$$

The standard deviations of R_1, R_2, and L are given by 0.005, 0.05, and $1.2 \cdot 10^5$, respectively.

Let w_1, w_2 be the amount invested in the fixed-income portfolio and the stock market portfolio, respectively. The insurer invests the initial capital V_0 in the two portfolios so that the expected value of its asset portfolio has an expected return of 1.06.

(a) Determine w_1 and w_2.
(b) Is the insurer solvent in the sense that $\text{VaR}_{0.005}(A - L) \leq 0$?
(c) Suppose there is an instantaneous decline of 15% in the value of the stock market portfolio. Does the insurer remain solvent? If not, determine how the insurer must adjust the asset portfolio weights w_1 and w_2 simply to become solvent in the sense that $\text{VaR}_{0.005}(A - L) = 0$.
(d) Compute the expected return of the insurer's adjusted asset portfolio determined in (c).

Comment: A simultaneous decline in the value of stocks and in interest rates is particularly dangerous to an insurer with a liability having a long duration. The reduction in the value of the insurer's capital forces the insurer to adjust its asset allocation away from stocks to less risky fixed-income instruments to remain solvent. The adjusted allocation has a lower expected return, which makes it difficult for the insurer to make up for the suffered losses. Moreover, insurance companies often have large amounts of capital invested in the stock market, and a forced sale of large positions in stocks and an increase in the demand for safe bonds could reduce both the prices of stocks and the interest rates even more. This phenomenon, sometimes referred to as a death spiral, makes the insurer stuck in a near-insolvent state with an asset portfolio that is unlikely to generate good returns.

Project 10 (Scenario-based risk analysis). Consider a stylized model of a life insurer. The insurer faces a liability cash flow of 100 each year for the next 30 years. The current zero rates are given in Table 9.1, from which the current value of the liability can be computed. In the market there is a short supply of bonds with maturities longer than 10 years. Therefore, the insurer has purchased a bond portfolio with payments only within the next 10 years. The bond portfolio has the cash flow given in Table 9.1. The insurer has also invested in a stock portfolio. The initial capital of the insurer is 30% more than the current value of the liability. The insurer invests 70% of the initial capital in the bond portfolio and 30% of the initial capital in the stock portfolio. The objective in this project is to identify the most dangerous extreme scenario.

Suppose that there are two risk factors in the model, the log return Y_1 of the stock portfolio and the size Y_2 of a parallel shift of the zero-rate curve. The risk factors are assumed to have a bivariate normal distribution, means μ_1, μ_2, standard deviations

Table 9.1 Annual cash flow of bond portfolio and current zero rates

Time	1	2	3	4	5	6	7	8	9	10
Bond payment	4	2	3	1	4	2	3	1	5	5
Zero rate (%)	2.86	3.24	3.55	3.93	4.27	4.62	4.96	5.30	5.55	5.80
Time	11	12	13	14	15	16	17	18	19	20
Bond payment	0	0	0	0	0	0	0	0	0	0
Zero rate (%)	6.05	6.30	6.45	6.60	6.74	6.90	7.00	7.21	7.32	7.32
Time	21	22	23	24	25	26	27	28	29	30
Bond payment	0	0	0	0	0	0	0	0	0	0
Zero rate (%)	7.40	7.48	7.56	7.64	7.70	7.77	7.83	7.90	7.95	8.00

σ_1, σ_2, and linear correlation coefficient ρ given by

$$\mu_1 = 0.08, \quad \mu_2 = 0, \quad \sigma_1 = 0.2, \quad \sigma_2 = 0.01, \quad \rho = 0.1.$$

Consider equally likely extreme scenarios in the following sense. The risk factors can be represented via two independent standard normally distributed random variables Z_1 and Z_2 as

$$Y_1 = \mu_1 + \sigma_1 Z_1,$$
$$Y_2 = \sigma_2 \left(\rho Z_1 + \sqrt{1 - \rho^2} Z_2 \right).$$

All scenarios with $\sqrt{Z_1^2 + Z_2^2} = 3$ can be viewed as equally likely extreme scenarios corresponding to three-standard-deviation movements. The extreme scenarios for Z_1, Z_2 translate into extreme scenarios for the risk factors Y_1, Y_2 by the relation above.

(a) Plot the value of the insurer's portfolio, assets minus liabilities, in 1 year for all the equally likely extreme scenarios.
(b) Identify which scenario for Y_1, Y_2 leads to the worst outcome for the value of the insurer's assets minus that of the liabilities in 1 year.
(c) Repeat the analysis outlined above and find the most dangerous scenario when (Y_1, Y_2) has another bivariate elliptical distribution.

Project 11 (Tail dependence in large portfolios). Let $Z_1, \ldots, Z_{50}$ represent log returns from today until tomorrow for 50 hypothetical financial assets. Suppose that Z_k has a Student's t distribution with three degrees of freedom and standard deviation 0.01 for each k and that $\tau(Z_j, Z_k) = 0.4$ for $j \neq k$.

Consider an investment of \$20,000 in long positions in each of the assets. Let V_0 and V_1 be the portfolio value today and tomorrow, respectively. Investigate the effect of tail dependence on the distribution of the portfolio value V_1 tomorrow and the distribution of the portfolio log return $\log(V_1/V_0)$ by simulating from the distribution of V_1. Simulate from the distribution of V_1 under the assumption that

(a) $(Z_1, \ldots, Z_{50})$ has a Gaussian copula.
(b) $(Z_1, \ldots, Z_{50})$ has a t_4-copula.
(c) $(Z_1, \ldots, Z_{50})$ has a Clayton copula.
(d) How large a sample size is needed to get stable estimates of $\mathrm{VaR}_{0.01}(V_1 - V_0)$ and $\mathrm{ES}_{0.01}(V_1 - V_0)$? Explain the differences in the estimates of $\mathrm{VaR}_{0.01}(V_1 - V_0)$ and $\mathrm{ES}_{0.01}(V_1 - V_0)$ in the three cases (a)–(c).
(e) Compare the results in (a)–(d) to the results when \$1 million is invested in only one of the assets.
(f) Suppose that the Z_k are equally distributed and have a left-skewed polynomial normal distribution with zero mean and standard deviation 0.01. Study and explain the effect of the log-return distribution of the Z_k on the distribution of V_1 and the portfolio risk by simulating from the distribution of V_1 under assumptions (a)–(c).

References

1. Acerbi, C.: Spectral measures of risk: a coherent representation of subjective risk aversion. J. Bank. Finan. **26**, 1505–1518 (2002)
2. Acerbi, C., Tasche, D.: On the coherence of Expected Shortfall. J. Bank. Finan. **26**, 1487–1503 (2002)
3. Arrow, K.J.: Essays in the Theory of Risk Bearing. Markham, Chicago (1971)
4. Artzner, P., Delbaen, F., Eber, J.-M., Heath, D.: Coherent measures of risk. Math. Finan. **9**, 203–228 (1999)
5. Black, F.: The pricing of commodity contracts. J. Finan. Econ. **3**, 167–179 (1976)
6. Black, F., Scholes, M.: The pricing of options and corporate liabilities. J. Polit. Econ. **81**, 637–659 (1973)
7. Boyd, S., Vandenberghe, L.: Convex Optimization. Cambridge University Press, Cambridge (2004)
8. Carr, P., Madan, D.: Optimal positioning in derivative securities. Quant. Finan. **1**, 19–37 (2001)
9. Casella, G., Berger, R.L.: Statistical Inference, 2nd edn. Duxbury Press, Belmont (2002)
10. Dalang, R.C., Morton, A., Willinger, W.: Equivalent martingale measures and no-arbitrage in stochastic securities market models. Stochastics Stochastics Rep. **29**, 185–201 (1990)
11. Efron, B., Tibshirani, R.: An Introduction to the Bootstrap. Chapman & Hall, New York (1993)
12. Embrechts, P., McNeil, A., Straumann, D.: Correlation and dependence in risk management: properties and pitfalls. In: Dempster, M. (ed.) Risk Management: Value at Risk and Beyond. Cambridge University Press, Cambridge (2002)
13. Embrechts, P., Klüppelberg, C., Mikosch, T.: Modelling Extremal Events for Insurance and Finance. Springer, New York (1997)
14. England, P., Verrall, R.: Analytic and bootstrap estimates of prediction errors in claims reserving. Insurance Math. Econ. **25**, 281–293 (1999)
15. Fang, H.B., Fang, K.T., Kotz, S.: The meta-elliptical distributions with given marginals. J. Multivar. Anal. **82**, 1–16 (2002).
16. Fang, K.-T., Kotz, S., Ng, K.-W.: Symmetric Multivariate and Related Distributions. Chapman & Hall, London (1987)
17. Föllmer, H., Schied, A.: Stochastic Finance: An Introduction in Discrete Time, 3rd edn. de Gruyter, Berlin (2011)
18. Gerber, H.U.: Life Insurance Mathematics, 3rd edn. Springer, New York (2010)
19. Gilboa, I.: Theory of Decision Under Uncertainty. Cambridge University Press, Cambridge, UK (2009)
20. de Haan, L., Ferreira, A.: Extreme Value Theory: An Introduction. Springer, New York (2006)
21. Hull, J.C.: Options, Futures, and Other Derivatives, 8th edn. Prentice-Hall, Englewood Cliffs (2012)

H. Hult et al., *Risk and Portfolio Analysis: Principles and Methods*, Springer Series in Operations Research and Financial Engineering, DOI 10.1007/978-1-4614-4103-8, © Springer Science+Business Media New York 2012

22. Hult, H., Lindskog, F.: Multivariate extremes, aggregation and dependence in elliptical distributions. Adv. Appl. Prob. **34**, 587–608 (2002)
23. Joe, H.: Multivariate Models and Dependence Concepts. Chapman & Hall, New York (1997)
24. Jorion, P.: Value-at-Risk: The New Benchmark for Managing Financial Risk, 3rd edn. McGraw-Hill, New York (2007)
25. Lindskog, F., McNeil, A.J., Schmock, U.: Kendall's tau for elliptical distributions. In: Bol, G., Nakhaeizadeh, G., Rachev, S.T., Ridder, T., Vollmer, K.-H. (eds.) Credit Risk: Measurement, Evaluation and Management, pp. 149–156. Physica, Heidelberg (2003)
26. Luenberger, D.G.: Linear and Nonlinear Programming, 2nd edn. Addison-Wesley, Reading (1989)
27. Macaulay, F.R., Frank, M.: Redington and the emergence of modern fixed income analysis. In: Poitras, G. (ed.) Pioneers of Financial Economics, vol. 2: Twentieth-Century Contributions. Edward Elgar Publishing, Cheltenham (2006)
28. Mack, T.: Distribution free calculation of the standard error of chain ladder reserve estimates. ASTIN Bull. **23**, 213–225 (1993)
29. Markowitz, H.M.: Portfolio selection. J. Finan. **7**, 77–91 (1952)
30. Markowitz, H.M.: Portfolio Selection: Efficient Diversification of Investments. Wiley, New York (1959)
31. McNeil, A., Frey, R., Embrechts, P.: Quantitative Risk Management: Concepts, Techniques, and Tools. Princeton University Press, Princeton (2005)
32. Merton, R.C.: Theory of rational option pricing. Bell J. Econ. Manag. Sci. **4**, 141–183 (1973)
33. Meucci, A.: Risk and Asset Allocation. Springer, New York (2005)
34. Mikosch, T.: Non-Life Insurance Mathematics, 2nd edn. Springer, New York (2009)
35. Nelsen, R.B.: An Introduction to Copulas. Springer, New York (1999)
36. Pratt, J.W.: Risk aversion in the small and in the large. Econometrica **32**, 122–136 (1964)
37. Resnick, S.I.: Heavy-Tail Phenomena: Probabilistic and Statistical Modeling. Springer, New York (2007)
38. Rockafellar, R.T., Uryasev, S.: Optimization of conditional value-at-risk. J. Risk **2**, 493–517 (2000)
39. Rockafellar, R.T., Uryasev, S., Zabarankin, M.: Master funds in portfolio analysis with general deviation measures. J. Bank. Finan. **30** , 743–778 (2005)
40. Rockafellar, R.T., Uryasev, S., Zabarankin, M.: Generalized deviations in risk analysis. Finan. Stochastics **10**, 51–74 (2006)
41. Rockafellar, R.T., Uryasev, S., Zabarankin, M.: Optimality conditions in portfolio analysis with general deviation measures. Math. Program. Ser. B **108**, 515–540 (2006)
42. Samuelson, P.A.: Proof that properly anticipated prices fluctuate randomly. Ind. Manag. Rev. **6**, 41–49 (1965)
43. Savage, L.J.: The Foundations of Statistics. Wiley, New York (1954)
44. Sharpe, W.F.: Capital asset prices: a theory of market equilibrium under conditions of risk. J. Finan. **19**, 425–442 (1964)
45. Taylor, G.C., Ashe, F.R.: Second moments of estimates of outstanding claims. J. Econometr. **23**, 37–61 (1983)
46. Wütrich, M., Merz, M.: Stochastic Claims Reserving Methods in Insurance. Wiley, New York (2008)

Index

H. Hult et al., *Risk and Portfolio Analysis: Principles and Methods*, Springer Series
in Operations Research and Financial Engineering, DOI 10.1007/978-1-4614-4103-8,
© Springer Science+Business Media New York 2012